# BRASSINOSTEROIDS

# BRASSINOSTEROIDS

## Bioactivity and Crop Productivity

Edited by

S. HAYAT

*Department of Botany,*
*Aligarth Muslim University, India*

and

A. AHMAD

*Department of Botany,*
*Aligarth Muslim University, India*

KLUWER ACADEMIC PUBLISHERS

DORDRECHT / BOSTON / LONDON

A C.I.P. Catalogue record for this book is available from the Library of Congress.

ISBN 1-4020-1710-3

---

Published by Kluwer Academic Publishers,
P.O. Box 17, 3300 AA Dordrecht, The Netherlands.

Sold and distributed in North, Central and South America
by Kluwer Academic Publishers,
101 Philip Drive, Norwell, MA 02061, U.S.A.

In all other countries, sold and distributed
by Kluwer Academic Publishers,
P.O. Box 322, 3300 AH Dordrecht, The Netherlands.

*Printed on acid-free paper*





Printed in the Netherlands.

Dedicated
to the memory of Sir Syed Ahmad Khan,
founder of the Aligarh Muslim University, Aligarh, India

# Contents

# Contents

# PREFACE

The entire range of the developmental processes in plants is regulated by the shift in the hormonal concentration, tissue sensitivity and their interaction with the factors operating around the plants. Out of the recognized hormones, attention has largely been focused on five (Auxins, Gibberellins, Cytokinin, Abscisic acid and Ethylene). However, in this book, the information about the most recent group of phytohormones (Brassinosteroids) has been compiled by us. It is a class of over 40 polyhydroxylated sterol derivatives, ubiquitously distributed throughout the plant kingdom. A large portion of these steroids is restricted to the reproductive organs (pollens and immature seeds). Moreover, their strong growth-inducing capacity, recognized as early as prior to their identification in 1979, tempted the scientists to visualize the practical importance of this group of phytohormones. The brassin solution, from rape pollen, was used in a collaborative project by the scientists of Brazil and U.S.A. in a pre-sowing seed treatment to augment the yield. This was followed by large-scale scientific programmes in U.S., Japan, China, Germany and erstwhile U.S.S.R., after the isolation of the brassinosteroids. This approach suits best in today's context where plants are targeted only as producers and hormones are employed to get desired results.

Chapter 1 of this book (which embodies a total of 10 chapters), gives a comprehensive survey of the hitherto known brassinosteroids, isolated from lower and higher plants. Chapter 2 deals with the history of brassinosteroids with a physiological approach. The recent progress in brassinosteroid research in relation to hormone perception and signal transduction is discussed in Chapter 3. A summarized version of the synthesis of several brassinosteroid analogs with structural variations, compared with available steroids such as diosgenin, hecogenis, solasodine, solanidine and bile acids comprises Chapter 4. Besides this, the main results relating to the application of these novel phytohormones to plants under field conditions have also been discussed. The use of brassinosteroids in seed germination is documented in Chapter 5. The successful utilization of desired brassinosteroids in the *in vitro* propagation of plants and their parts to manipulate the desired exposures has been explained in Chapter 6. The possible mechanism of action and practical applicability of 24-epibrassinolide, in agriculture, is covered in Chapter 7. An approach for enhancing the biological activity of a brassinosteroid by involving its administration, as a guest, in an inclusion complex of plant growth inactive compound, has been described in Chapter 8. Chapter 9 includes a summary of the results of field-grown crops supplemented with brassinosteroids, obtained during ten years by agricultural scientists of Russia and Belarus. Lastly, Chapter 10 covers some important aspects of plant metabolism that determine crop productivity under the influence of brassinosteroids.

This book is not an encyclopedic review. However the various chapters incorporate both theoretical and practical aspects and may serve as baseline information for future researches through which significant developments are possible. It is intended that this book will be useful to students, teachers and researchers, both in universities and research institutes, especially in relation to biological and agricultural sciences.

With great pleasure, we extend our sincere thanks to all the contributors for their timely response, their excellent and up-to-date contributions and consistent support and cooperation. We express our deep sense of gratitude to Professor M.M.R.K. Afridi who introduced us with this discipline and has been a great source of inspiration. Special thanks are extended to Dr. B.N. Vyas, General Manager, Godrej Agrovet Ltd., Mumbai, India, who was instrumental in leading us into this field of research by gifting generous samples of 28-homobrassinolide. We are also thankful to our departmental colleagues who helped us in the preparation of the manuscript. Thanks are also due to Mr. M. Shakir who spent considerable time at the computer for proper formatting of various chapters. We gratefully acknowledge the encouragement and support of Mr. Naseem Ahmad, I.A.S., Vice-Chancellor of the Aligarh Muslim University, during the preparation of the manuscript.

We are extremely thankful to Kluwer Academic Publishers, The Netherlands for expeditious acceptance of our proposal and completion of the review process. Subsequent cooperation and understanding of their staff, especially Dr. J.A.C. Flipsen and Ms. Noeline Gibson is also gratefully acknowledged.

We express our sincere thanks to the members of our family for all the support they provided and the neglect and loss they suffered during the preparation of this book.

Finally, we are thankful to the Almighty God who provided and guided all the channels to work in cohesion and coordination right from the conception of the idea to the development of the final version of this treatise Brassinosteroids: Bioactivity and Crop Productivity, until the successful completion of the job.

**S. Hayat**
**A. Ahmad**

## Contributors

A.Ahmad

Plant Physiology Section, Department of Botany, Aligarh Muslim University, Aligarh-202002, INDIA

Mariangela de Burgos Martins de Azevedo

STQ - Scientia Tecnologia Química, CIETEC - Centro Incubador de Empresas Tecnológicas, Av. Prof. Lineu Prestes 2242, Cidade Universitária USP, CEP 05508-000 São Paulo SP. BRAZIL.

Andrzej Bajguz

University of Bialystok, Institute of Biology, Swierkowa 20 B, 15-950 Bialystok, POLAND

Gerard J.Bishop

Institute of Biological Sciences, University of Wales Aberystwyth, Aberystwyth, Ceredigion SY 23 3DA, Wales, U.K.

Julie Castle

Institute of Biological Sciences, University of Wales Aberystwyth, Aberystwyth, Ceredigion SY 23 3DA, Wales, U.K.

Chen Ji-chu

Shanghai Institute of Plant Physiology,Academia,Sinica,Shanghai,CHINA

Q.Fariduddin

Plant Physiology Section, Department of Botany, Aligarh Muslim University, Aligarh-202002, INDIA

Martin Fellner*

University of Washington, Department of Botany, 407 Hitchcock Hall, Seattle, WA 98195, USA

Lydia R. Galagovsky

Department of Botany-SCB, Centro Politecnico-UFPR, CP. 19031 Curitibia, PR-BRAZIL

S.Hayat

Plant Physiology Section, Department of Botany, Aligarh Muslim University, Aligarh-202002, INDIA

Vladimir A. Khripach

Institute of Bio-organic Chemistry, Academy of Sciences of Belarus, Minsk, BELARUS

Nataliya B. Khripach
Institute of Bio-organic Chemistry, Academy of Sciences of Belarus, Minsk, BELARUS

Gerhard Leubner-Metzger
Institut für Biologie II, Botanik, Albert-Ludwigs-Universität, Schänzlestr. 1, D-79104 Freiburg i. Br., GERMANY

Francisco Coll Manchado
Centro de Estudios de Productos Naturales. Facultad de Química. Universidad de La Habana. Zapata y G. Vedado. C.P. 10 400. Ciudad de la Habana. CUBA

Teresa Montoya
Institute of Biological Sciences, University of Wales Aberystwyth, Aberystwyth, Ceredigion SY 23 3DA, Wales, U.K.

Adaucto B. Pereira-Netto
Department of Botany-SCB, Centro Politecnico-UFPR, CP. 19031 Curitibia, PR-BRAZIL

Javier A. Ramirez
Department of Botany-SCB, Centro Politecnico-UFPR, CP. 19031 Curitibia, PR-BRAZIL

Caridad Robaina Rodríguez
Centro de Estudios de Productos Naturales. Facultad de Química. Universidad de La Habana. Zapata y G. Vedado. C.P. 10 400. Ciudad de la Habana. CUBA

Silvia Schaefer
Department of Botany-SCB, Centro Politecnico-UFPR, CP. 19031 Curitibia, PR-BRAZIL

Andrzej Tretyn
Nicholas Copernicus University, Institute of General and Molecular Biology, Gagarina 9, 87-100 Torun, POLAND

Miriam Núñez Vázquez
Departamento de Fisiología y Bioquímica Vegetal. Instituto Nacional de Ciencias Agrícolas. Gaveta Postal No. 1. San José de las Lajas. C.P. 32 700. La Habana. CUBA.

Vladimir N. Zhabinskii
Institute of Bio-organic Chemistry, Academy of Sciences of Belarus, Minsk, BELARUS

Zhao Yu Ju
Shanghai Institute of Plant Physiology,Academia,Sinica,Shanghai,CHINA

Marco António Teixeira Zullo
Instituto Agronômico, Laboratório de Fitoquímica, Caixa Postal 28, 13001-970 Campinas, SP, BRAZIL

*Present address: Institute of Experimental Botany,Academy of Sciences of the Czech Republic, Slechtitelu 11, 783 71 Olomouc-Holice, CZECH REPUBLIC
E-mail: emfee@prfholnt.upol.cz

CHAPTER 1

ANDRZEJ BAJGUZ AND ANDRZEJ TRETYN

# THE CHEMICAL STRUCTURES AND OCCURRENCE OF BRASSINOSTEROIDS IN PLANTS

Brassinosteroids are plant hormones with high-promoting activity. Brassinosteroids are hydroxylated derivatives of cholestane and their structural variations comprise the substitution pattern at rings A and B as well as the C-17 side chain. They can be classified as $C_{27}$, $C_{28}$, and $C_{29}$ compounds, depending on the alkyl-substitution pattern of side chain. Up to now 65 free brassinosteroids and 5 brassinosteroid conjugates have been characterized. This chapter gives a comprehensive survey on the hitherto known brassinosteroids isolated from lower and higher plants. The occurrence of brassinosteroids has been demonstrated in almost every part of plants such as pollen, flower, shoot, vascular cambium, leaf, fruit, seed, and root.

## INTRODUCTION

Brassinosteroids (BRs) represent a new sixth class of plant hormones with wide occurrence in the plant kingdom in addition to auxins, gibberellins, cytokinins, abscisic acid and ethylene. They are also growth-promoting plant hormones with structures similar to animal steroidal hormones – ecdysteroids. BRs have unique biological effects on plant growth and development (Sasse, 1997, 1999). BRs are phytohormones, controlling important developmental functions, such as promotion of cell elongation and division, photomorphogenesis, fertility, seed germination, senescence, retardation of abscission, promotion of ethylene biosynthesis. However, their physiological functions in plants are not fully understood to date. In addition to their role in plant development, BRs have the ability to protect plants from various environmental stresses, including drought, extreme temperatures, heavy metals, herbicidal injury and salinity (Sasse, 1999). The biosynthetic and metabolic pathways with enzymatic studies and the molecular mode of action of BRs have been investigated (Clouse and Feldmann, 1999; Bishop and Yokota, 2001; Friedrichsen and Chory, 2001; Müssig and Altmann, 2001; Schneider, 2002). Recently, the first BR-biosynthesis inhibitor, brassinazole (Brz), was discovered. Brz, a triazole derivative, inhibits plant growth, but this effect can be reversed by the application of a mixture of brassinolide and Brz (Asami and Yoshida, 1999).

This chapter describes the structural characteristics of BRs and their distribution in the plant kingdom including BRs isolated for the first time in plants.

*S.Hayat and A.Ahmad (eds.), Brassinosteroids,* 1-44.

## CHEMICAL STRUCTURE OF BRASSINOSTEROIDS

The history of BRs started in 1979 when Grove *et al.* (1979) isolated from pollen of rape (*Brassica napus*), brassinolide (BL). Its structure was determinated by spectroscopic analysis (EI-MS, FAB-MS, NMR) and X-ray diffraction to be (22*R*,23*R*,24*S*)-2α,3α,22,23-tetrahydroxy-24-methyl-B-homo-7-oxa-5α-cholestan-6-one. The second BR, termed castasterone (CS), was isolated in 1982 by Yokota *et al.* (1982a) from the insect galls of chestnut (*Castanea crenata*). The structure of CS was established as (22*R*,23*R*,24*S*)-2α,3α,22,23-tetrahydroxy-24-methyl-5α-cholestan-6-one (Yokota, 1999a, b). Since the discovery of BL, the natural occurrence of 70 BRs (65 unconjugated and 5 conjugated compounds) of this group has been detected.

BRs are derived from the 5α-cholestane skeleton and their structural variations come from the type and position of functionality in the A/B rings and the side chain (Fig. 1) (Yokota, 1997).

With respect to the A-ring, BRs having vicinal hydroxyl groups at C-2α and C-3α. BRs with an α-hydroxyl, β-hydroxyl or ketone at position C-3 are precursors of BRs having 2α,3α-vicinal hydroxyls. On the other hand, BR with 2α,3β-, 2β,3α- or 2β,3β-vicinal hydroxyls probably may be metabolites of 2α,3α-vicinal hydroxyls. The two 2α,3α-vicinal hydroxyl groups at the A-ring represent a general structural feature of most active BRs, such as BL and CS. Decreasing order of activity 2α,3α > 2α,3β > 2β,3α > 2β,3β shown by structure-activity relationship suggests that the α-oriented hydroxyl group at C-2 is essential for greater biological activity of BRs in plants. Biogenic precursors, like typhasterol (TY) and teasterone (TE), have only one hydroxyl group in the A-ring. Also BRs with an 2,3-epoxide group in the A-ring – secasterone (SE) and its epimers (2,3-diepiSE and 24-epiSE) have been found. There are two BRs having a 3-oxo group, such as 3-dehydroteasterone (3-DT) and 3-dehydro-6-deoxoteasterone (6-deoxo-3-DT) but also BRs having additional hydroxyl in the A-ring at position C-1α or C-1β, such as 3-epi-1α-hydroxycastasterone (3-epi-1α-OH-CS) and 1β-hydroxycastasterone (1β-OH-CS). Furthermore, the structures of BRs with double bond in the A- ($\Delta^{2,3}$ or $\Delta^{4,5}$) or B-ring ($\Delta^{5,6}$) have been discovered (Table 1) (Mandava, 1988; Kim, 1991; Adam and Petzold, 1994; Bishop *et al.*, 1999; Fujioka, 1999; Fujioka *et al.*, 2002; Antonchick *et al.*, 2003).

With respect to the B-ring oxidation stage, BRs are divided into 7-oxalactone (12 compounds), 6-oxo (6-ketone) (34 compounds) and 6-deoxo (non-oxidized) (21 compounds) types. As a fourth type, there is only one BR with hydroxyl group at C-6, such as 6α-hydroxycastasterone (6α-OH-CS). On the other hand, two compounds, such as (22*S*)-22-hydroxycampestrol (22-OHCR) and 28-nor-(22*S*)-22-hydroxycampestrol (28-nor-22-OHCR) represented a fifth type of BRs (Table 1). In general, 7-oxalactone BRs have stronger biological activity than 6-oxo type, and 6-deoxo type. Sometimes 6-oxo BRs have an activity similar to 7-oxalactone compounds, but non-oxidized BRs reveal almost no activity in the bean internode test or very weak in the rice lamina inclination test (Kim, 1991; Bishop *et al.*, 1999; Fujioka, 1999).

Furthermore, with respect to the A/B ring fuctionalities the hitherto clarified members can be divided into following groups:



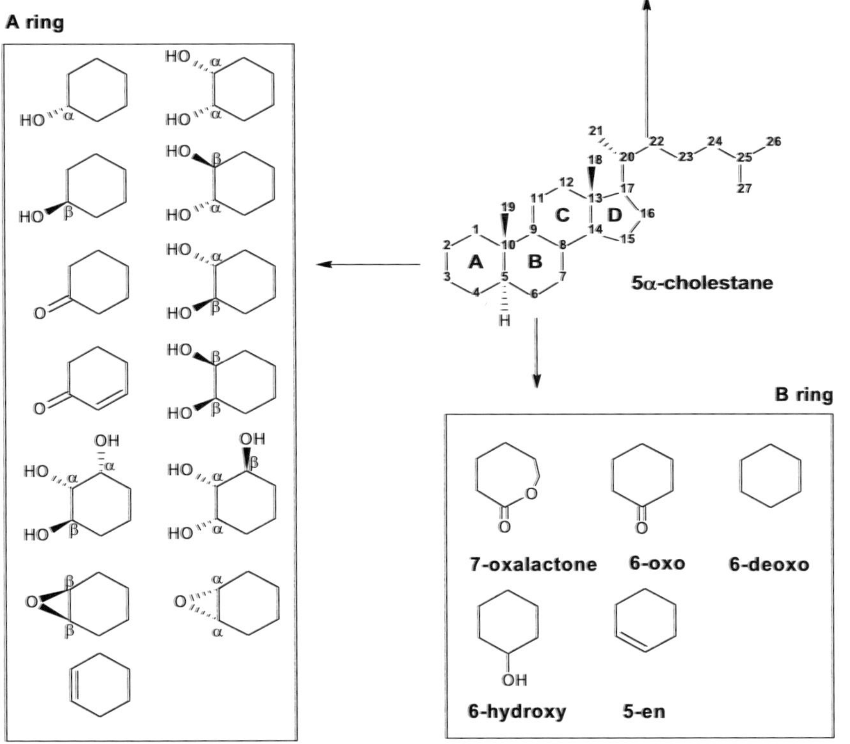


*Figure 1. Different substituents in the a- and b-rings and side chain of naturally occurring brassinosteroids*

- BRs with 7-membered 7-oxalactone-B-ring and vicinal 2α,3α-hydroxyl groups;
- 6-oxocompounds with a 6-membered B-ring having two hydroxyl groups at position C-2 and C-3;
- 6-oxocompounds with 2β,3β-oriented epoxide group;
- 6-oxocompounds with an additional hydroxyl group at position C-1 (α or β);
- 6-oxocompounds with 3-oxo group;
- BRs without oxygen functions in the B-ring;
- BRs having hydroxyl group at position C-6;
- BRs having double bond ($\Delta^{2,3}$ or $\Delta^{4,5}$) in the A-ring (Table 1) (Fujioka, 1999; Fujioka *et al.*, 2002; Antonchick *et al.*, 2003).

According to the cholestane side chain, BRs are divided into eleven types with different substituents at C-23, C-24 and C-25: 23-oxo, 24*S*-methyl, 24*R*-methyl, 24-methylene, 24*S*-ethyl, 24-ethylidene, 24-methylene-25-methyl, 24-methyl-25-methyl, without substituent at C-23, without substituent at C-24 and without substituents at C-23, C-24 (Table 2) (Sakurai and Fujioka, 1993; Fujioka, 1999; Watanabe *et al.*, 2000; Fujioka *et al.*, 2002; Antonchick *et al.*, 2003).

Unconjugated BRs are grouped into $C_{27}$, $C_{28}$ and $C_{28}$ steroids whose chemical structures have been presented in figures 2-4. These classifications result basically from different alkyl substitutions in the side chain. The presence of a saturated alkyl (a methyl or an ethyl group) at position C-24 and a methyl at C-25 makes BRs biologically more active. Most of BRs carry an *S*-oriented alkyl group at C-24. Nevertheless, there are five exceptions among BRs which have *R*-oriented alkyl, for example 24-epiBL or 24-epiCS. Also BRs without substituent at C-23 and/or C-24 have been found (Table 2) (Fujioka, 1999). All of these alkyl substituents are also common structural features of plant sterols. It is suggest that BRs are derived from sterols carrying the same side chain. The $C_{27}$ BRs (13 compounds) having no substituent at C-24 may come from cholesterol (Fig. 2). The $C_{28}$ BRs (39 compounds) carrying either an α-methyl, β-methyl or methylene group may be derived from campesterol, 24-epicampesterol or 24-methylenecholesterol, respectively (Fig. 3). The $C_{29}$ BRs (13 compounds) with an α-ethyl group may came from sitosterol (Fig. 4). Furthermore, the $C_{29}$ BRs carrying a methylene at C-24 and an additional methyl group at C-25 may be derived from 24-methylene-25-methylcholesterol (Yokota, 1999b).

In addition to free 65 BRs also 5 sugar and fatty acid conjugates have been identified in plants (Fig. 5). 25-Methyldolichosterone-23-β-*D*-glucoside (25-MeDS-Glu) and its 2β isomer from *Phaseolus vulgaris* seeds and teasterone-3β-*D*-glucoside (TE-3-Glu), teasterone-3-laurate (TE-3-La) and teasterone-3-myristate (TE-3-My) from *Lilium longiflorum* pollen were isolated as endogenous BRs (Abe *et al.*, 2001).

## OCCURRENCE OF BRASSINOSTEROIDS

Since the discovery of BL, 70 BRs, among them 65 unconjugated and 5 conjugated BRs, have been isolated from 60 plant species including 51 angiosperms (12 monocotyledons and 39 dicotyledons) (Table 4-7), 6 gymnosperms (Table 8), 1 pteridophyte (*Equisetum arvense*), 1 bryophyte (*Marchantia polymorpha*) and 1

***Table 1.*** *Division of brassinosteroids according to the.B-ring and orientation of hydroxyl, ketone and epoxide groups at position C-1, C-2, C-3 and C-6, and double bond in the A-ring.*

| *Carbon position* | *Type of brassinosteroids* | | | | |
|---|---|---|---|---|---|
| | *7-oxalactone* | *6-oxo* | *6-deoxo* | *6-hydroxy* | *5-en* |
| C(2α,3α)[a] | brassinolide (BL)<br>24-epiBL<br>28-norBL<br>28-homoBL<br>dolicholide (DL)<br>28-homoDL<br>23-dehydroBL | castasterone (CS)<br>24-epiCS<br>28-norCS<br>28-homoCS<br>25-methylCS (25-MeCS)<br>dolichosterone (DS)<br>28-homoDS<br>25-MeDS<br>23-*O*-β-D-glucopyranosyl-25-MeDS (25-MeDS-Glu) | 6-deoxoCS<br>6-deoxo-24-epiCS<br>6-deoxo-28-norCS<br>6-deoxoDS<br>6-deoxo-28-homoDS<br>6-deoxo-25-MeDS | | |
| C(2α,3β)[a] | 3-epiBL<br>3-epi-23-dehydroBL | 3-epiCS<br>3,24-diepiCS | 3-epi-6-deoxoCS | | |

***Table 1.*** *Division of brassinosteroids according to the B-ring and orientation of hydroxyl, ketone and epoxide groups at position C-1, C-2, C-3 and C-6, and double bond in the A-ring - continued.*

| *Carbon position* | *Type of brassinosteroids* | | | | |
|---|---|---|---|---|---|
| | *7-oxalactone* | *6-oxo* | *6-deoxo* | *6-hydroxy* | *5-en* |
| C(2β,3α)[a] | 2-epi-23-dehydroBL | 2-epiCS<br>2-epi-25-MeDS<br>23-*O*-β-D-glucopyranosyl-2-epi-25-MeDS (2-epi-25-MeDS-Glu) | | | |
| C(2β,3β)[a] | 2,3-diepi-23-dehydroBL | 2,3-diepiCS<br>2,3-diepi-25-MeDS | | | |
| C(3α)[a] | 2-deoxyBL | typhasterol (TY)<br>28-homoTY<br>28-norTY<br>2-deoxy-25-MeDS | 6-deoxoTY<br>6-deoxoTY-28-norTY<br>6-deoxo-28-norTE<br>3-epi-6-deoxoCT<br>3-epi-6-deoxo-28-norCT | | |

***Table 1.*** *Division of brassinosteroids according to the B-ring and orientation of hydroxyl, ketone and epoxide groups at position C-1, C-2, C-3 and C-6, and double bond in the A-ring - continued.*

| *Carbon position* | *Type of brassinosteroids* | | | | |
|---|---|---|---|---|---|
| | *7-oxalactone* | *6-oxo* | *6-deoxo* | *6-hydroxy* | *5-en* |
| C(3β)[a] | | teasterone (TE)<br>28-homoTE<br>TE-3-myristate (TE-3-My)<br>TE-3-laurate (TE-3-La)<br>TE-3-*O*-β-D-glucoside (TE-3-Glu)<br>3-epi-2-deoxy-25-MeDS<br>cathasterone (CT) | 6-deoxoTE<br>6-deoxoCT<br>6-deoxoCT-28-norCT | | 22-OHCR<br>28-nor-22-OHCR |
| C(1α,2α,3β)[a] | | 3-epi-1α-OH-CS | | | |
| C(1β,2α,3α)[a] | | 1β-OH-CS | | | |
| C(2α,3α,6α)[a] | | | | 6α-OH-CS | |

***Table 1.*** *Division of brassinosteroids according to the B-ring and orientation of hydroxyl, ketone and epoxide groups at position C-1, C-2, C-3 and C-6, and double bond in the A-ring - continued.*

| *Carbon position* | *Type of brassinosteroids* | | | | |
|---|---|---|---|---|---|
| | *7-oxalactone* | *6-oxo* | *6-deoxo* | *6-hydroxy* | *5-en* |
| C3[b] | | 3-dehydroTE (3-DT) | 3-dehydro-6-deoxoTE<br>3-dehydro-6-deoxo-28-norTE<br>22-OH-3-one<br>28-nor-22-OH-3-one | | |
| C(2β,3β)[c] | | secasterone (SE)<br>2,3-diepiSE<br>24-epiSE | | | |
| C3, $\Delta^{4,5}$ | | | 22-OH-4-en-3-one<br>28-nor-22-OH-4-en-3-one | | |
| $\Delta^{2,3}$ | | secasterol | | | |

[a] – hydroxyl group, [b] – ketone group, [c] – epoxide group

*Table 2. Division of brassinosteroids according to different substituents in the side chain.*

| *Type* | *Representatives* | *Total number* |
|---|---|---|
| 23-oxo | 23-dehydroBL, 2-epi-23-dehydroBL, 3-epi-23-dehydroBL, 2,3-diepi-23-dehydroBL | 4 |
| 24*S*-methyl | BL, 3-epiBL, CS, 2-epiCS, 3-epiCS, 2,3-diepiCS, TY, TE, TE-3-La, TE-3-My, 6-deoxoCS, 3-epi-6-deoxoCS, 3-DT, SE, 2,3-diepiSE, 6-deoxoTY, 2-deoxyBL, 3-epi-1α-OH-CS, 1β-OH-CS, 6α-OH-CS, TE-3-Glu, 6-deoxoTE, 3-dehydro-6-deoxoTE, secasterol | 24 |
| 24*R*-methyl | 24-epiBL, 24-epiCS, 3,24-diepiCS, 6-deoxo-24-epiCS, 24-epiSE | 5 |
| 24-methylene | DL, DS, 6-deoxoDS | 3 |
| 24*S*-ethyl | 28-homoBL, 28-homoCS, 28-homoTE, 28-homoTY | 4 |
| 24-ethylidene | 28-homoDL, 28-homoDS, 6-deoxo-28-homoDS | 3 |
| 24-methylene-25-methyl | 25-MeDS, 2-epi-25-MeDS, 2,3-diepi-25-MeDS, 2-deoxy-25-MeDS, 3-epi-2-deoxy-25-MeDS, 6-deoxo-25-MeDS, 25-MeDS-Glu, 2-epi-25-MeDS-Glu | 8 |
| 24*S*-methyl-25-methyl | 25-MeCS | 1 |
| without substituent at C-23 | CT, 6-deoxoCT, 3-epi-6-deoxoCT, 22-OH-4-en-3-one, 22-OH-3-one, 22-OHCR | 6 |
| without substituent at C-24 | 28-norBL, 28-norCS, 28-norTY, 6-deoxo-28-norCS, 6-deoxo-28-norTY, 6-deoxo-28-norTE, 3-dehydro-6-deoxo-28-norTE | 7 |
| without substituents at C-23, C-24 | 6-deoxo-28-norCT, 28-nor-22-OHCR, 28-nor-22-OH-4-en-3-one, 28-nor-22-OH-3-one, 3-epi-6-deoxo-28-norCT | 5 |

chlorophyte, the alga (*Hydrodictyon reticulatum*) (Table 9). Thus the BRs are widely distributed in the plant kingdom, including higher and lower plants. Table 3 summarizes from 1979 to the present day the history of isolation for the first time naturally occurring BRs in plants.

BRs were detected in all plant organs such as pollen, anthers, seeds, leaves, stems, roots, flowers, and grain. Another interesting tissues are insect and crown galls. The galls of *Castanea crenata* and *Distylium racemosum* have higher levels of BRs (several μg/kg) than the normal, healthy tissues. Another tissue with BRs content is the crown gall cells of *Catharanthus roseus* which have higher contents of BL and CS (ca. 30-40 μg/kg) than the normal cells. Also, young growing tissues contain higher levels of BRs than mature tissues. Generally, pollen and immature seeds are especially rich source of BRs, while the concentrations in vegetative tissues are very low compared to those of other plant hormones. In the pollen of *Cupressus arizonica* the concentration of 6-deoxoTY can be about 6400-fold greater than BL. Pollen and immature seeds are the richest sources with ranges of 1-100 ng $g^{-1}$ fresh weight, while shoots and leaves usually have lower amounts of 0.01-0.1 ng $g^{-1}$ fresh weight. BRs occur endogenously at quite low levels. Compared to the pollen and immature seeds, the other plant parts contain BRs in the nanogram or subnanogram levels of BRs per gram fresh weight. The highest concentration of BR, 6.4 mg 6-deoxoTY per 1 kg pollen, was detected in *Cupressus arizonica* (Griffiths *et al*., 1995; Clouse and Sasse, 1998; Fujioka, 1999).

Among the BRs, CS is the most widely distributed (50 plant species), followed by BL (34), TY (25), 6-deoxoCS (19), TE (19), and 28-norCS (12). Furthermore from 2 to 10 BRs are distributed in a limited number of plant species, it means that 24-epiCS was isolated in 8 plant species, DS – 7, 3-DT – 7, 6-deoxoTY – 5, 28-homoCS – 4, 24-epiBL – 4, DL – 3, 6-deoxoTE – 3, 6-deoxoDS – 3, 28-norBL – 2, 28-homoTE – 2, 2-deoxyBL – 2. To the present day 34 other BRs and 5 BR conjugates have been found in only one plant species. Among all naturally occurring BRs, CS and BL are the most important BRs because of their wide distribution as well as their potent biological activity (Kim, 1991; Fujioka, 1999).

Among the plant sources investigated, immature seeds of *Phaseolus vulgaris* contain a wide array of BRs, this is 25 free BRs and 2 conjugates. The wide occurrences of BRs were also found in the dwarf mutant of *Catharanthus roseus* (19 compounds), *Arabidopsis thaliana* (18 compounds), *Cryptomeria japonica* and *Cupressus arizonica* (9 compounds), *Dolichos lablab*, *Oryza sativa*, *Thea sinensis* and *Secale cereale* (8 compounds), *Lilium longiflorum* (7 compounds), *Distylium racemosum* (6 compounds).

## DISTRIBUTION IN MONOCOTYLEDONS

The occurrence of BRs in monocotyledons has been demonstrated from four families including twelve plant species (Table 4). BRs are represented by 18 various compounds: 7-oxalactone (1, BL), 6-oxo (16, including two conjugates) and 6-deoxo (1 – 6-deoxoCS) types. Seven BRs, such as SE, 2,3-diepiSE, TY, 3-DT, TE-3-La, TE-3-My and secasterol were isolated for the first time in plants (Table 3).

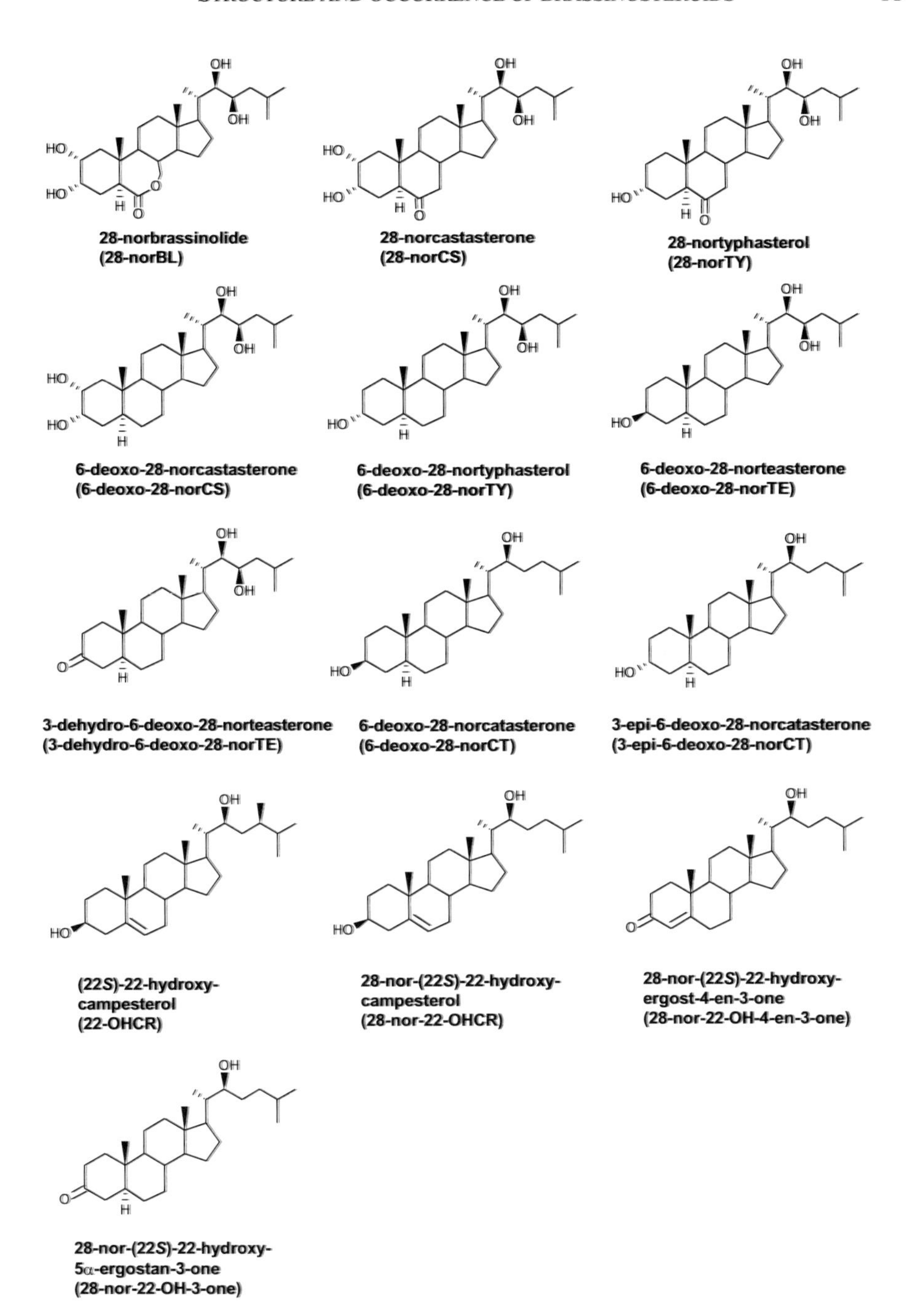


*Figure 2. Chemical structures of $C_{27}$ brassinosteroids.*

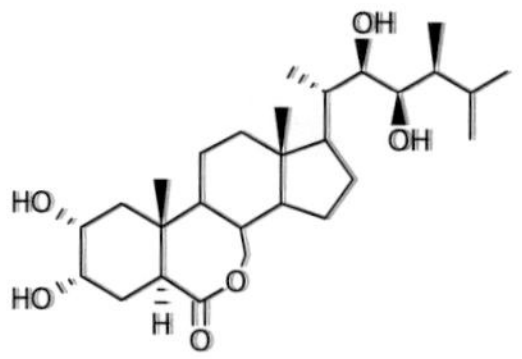

brassinolide
(BL)

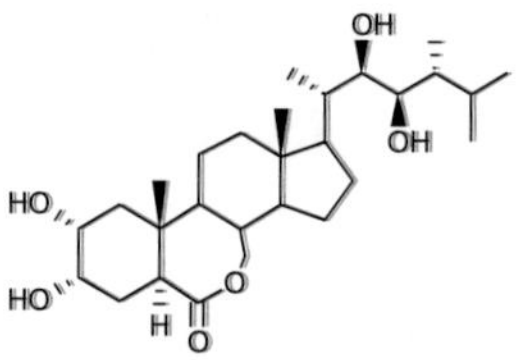

24-epibrassinolide
(24-epiBL)

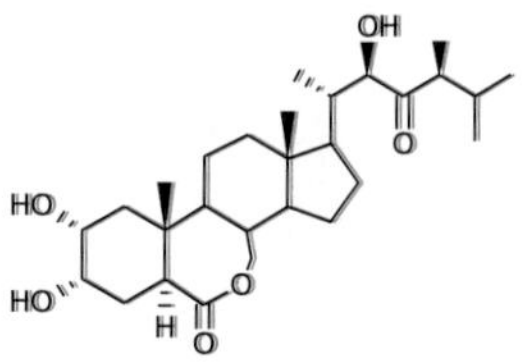

23-dehydrobrassinolide
(23-dehydroBL) (cryptolide)

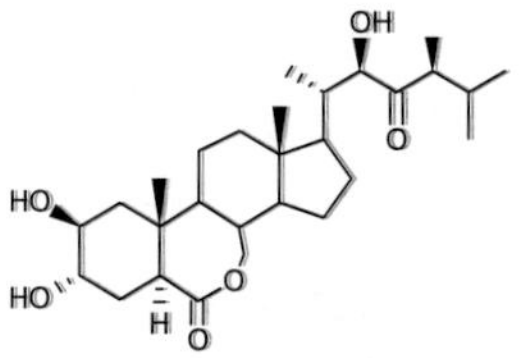

2-epi-23-dehydrobrassinolide
(2-epi-23-dehydroBL)

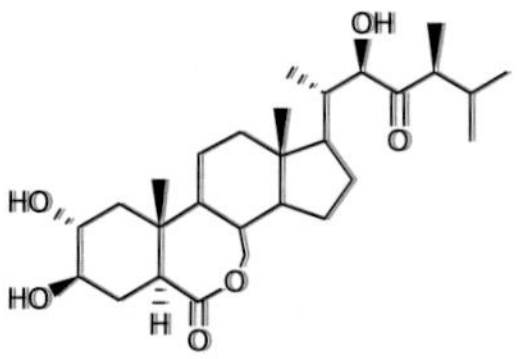

3-epi-23-dehydrobrassinolide
(3-epi-23-dehydroBL)

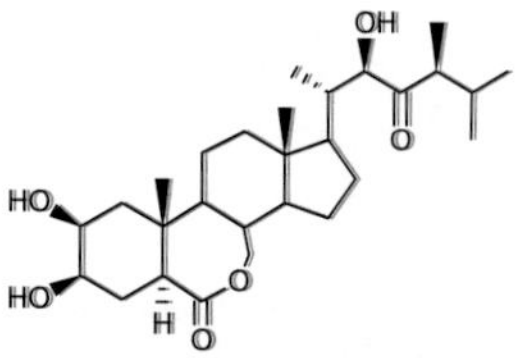

2,3-diepi-23-dehydrobrassinolide
(2,3-diepi-23-dehydroBL)

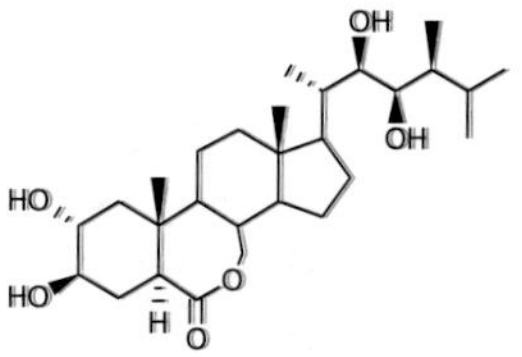

3-epibrassinolide
(3-epiBL)

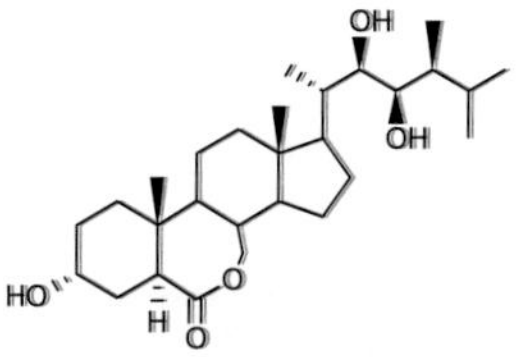

2-deoxybrassinolide
(2-deoxyBL)

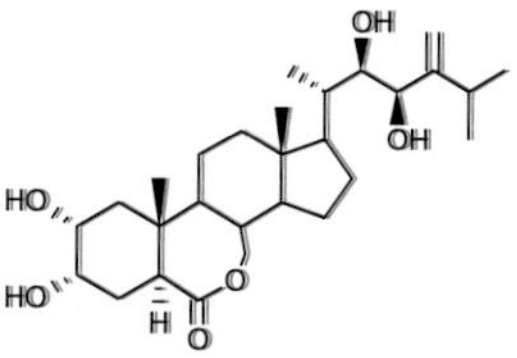

dolicholide
(DL)

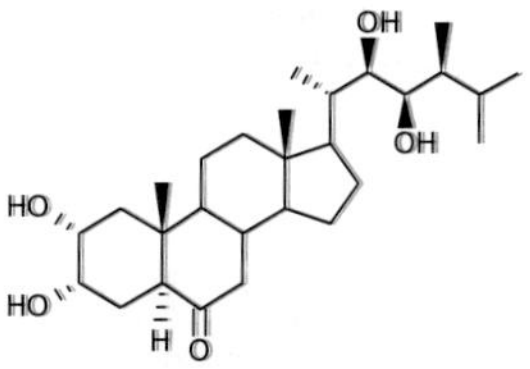

castasterone
(CS)

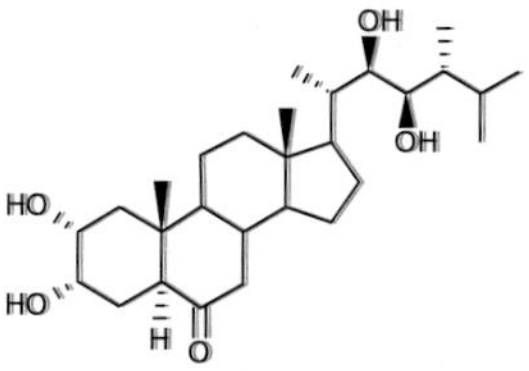

24-epicastasterone
(24-epiCS)

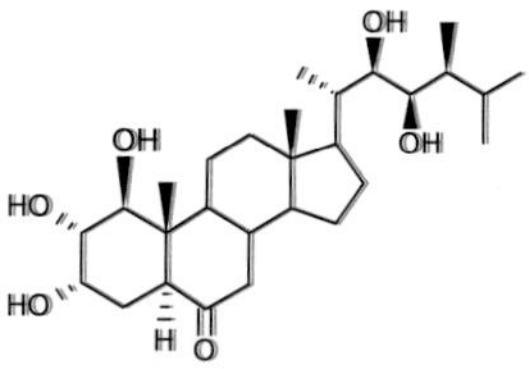

1β-hydroxycastasterone
(1β-OH-CS)

*Figure 3. Chemical structures of $C_{28}$ brassinosteroids – continued.*

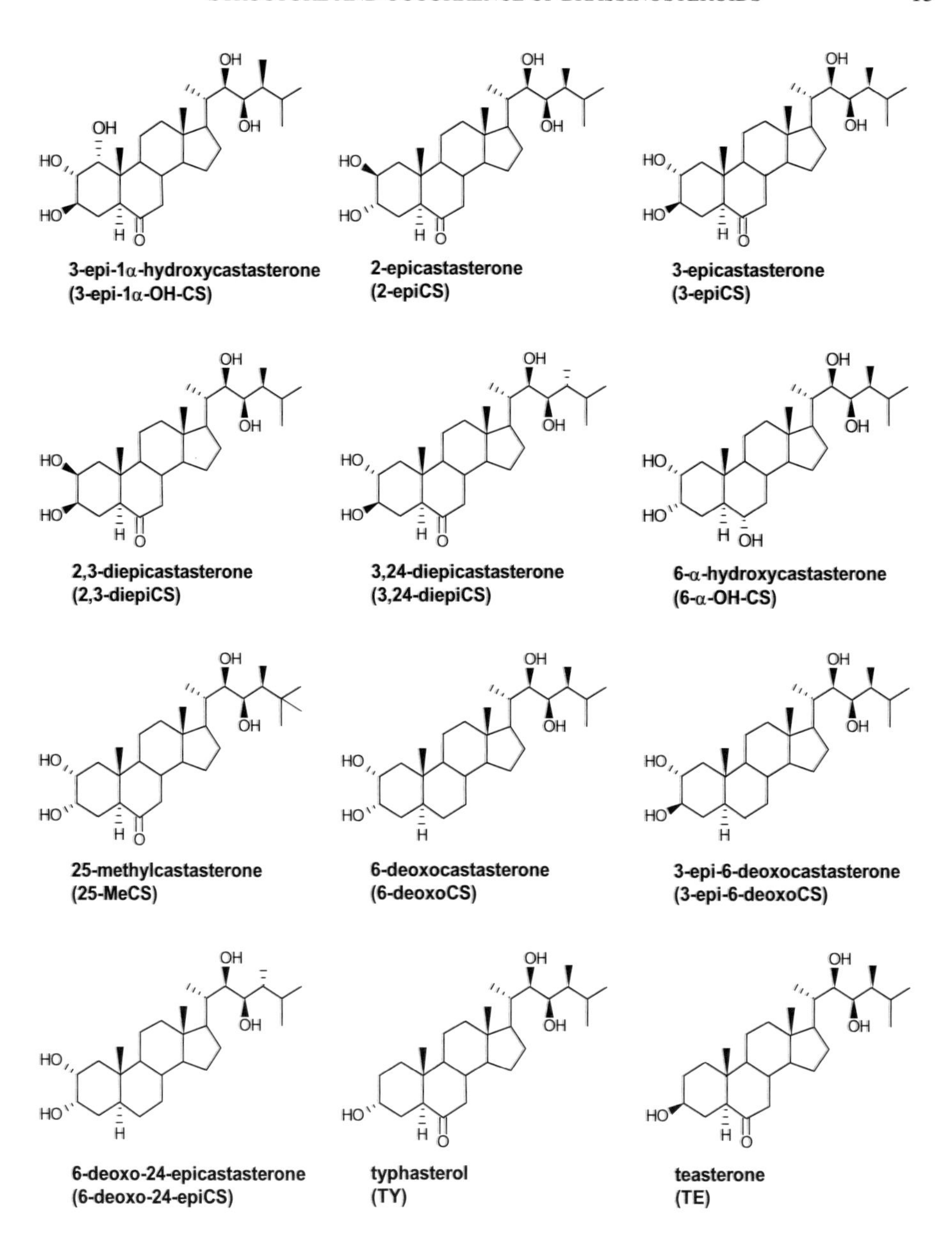


*Figure 3. Chemical structures of $C_{28}$ brassinosteroids – continued.*

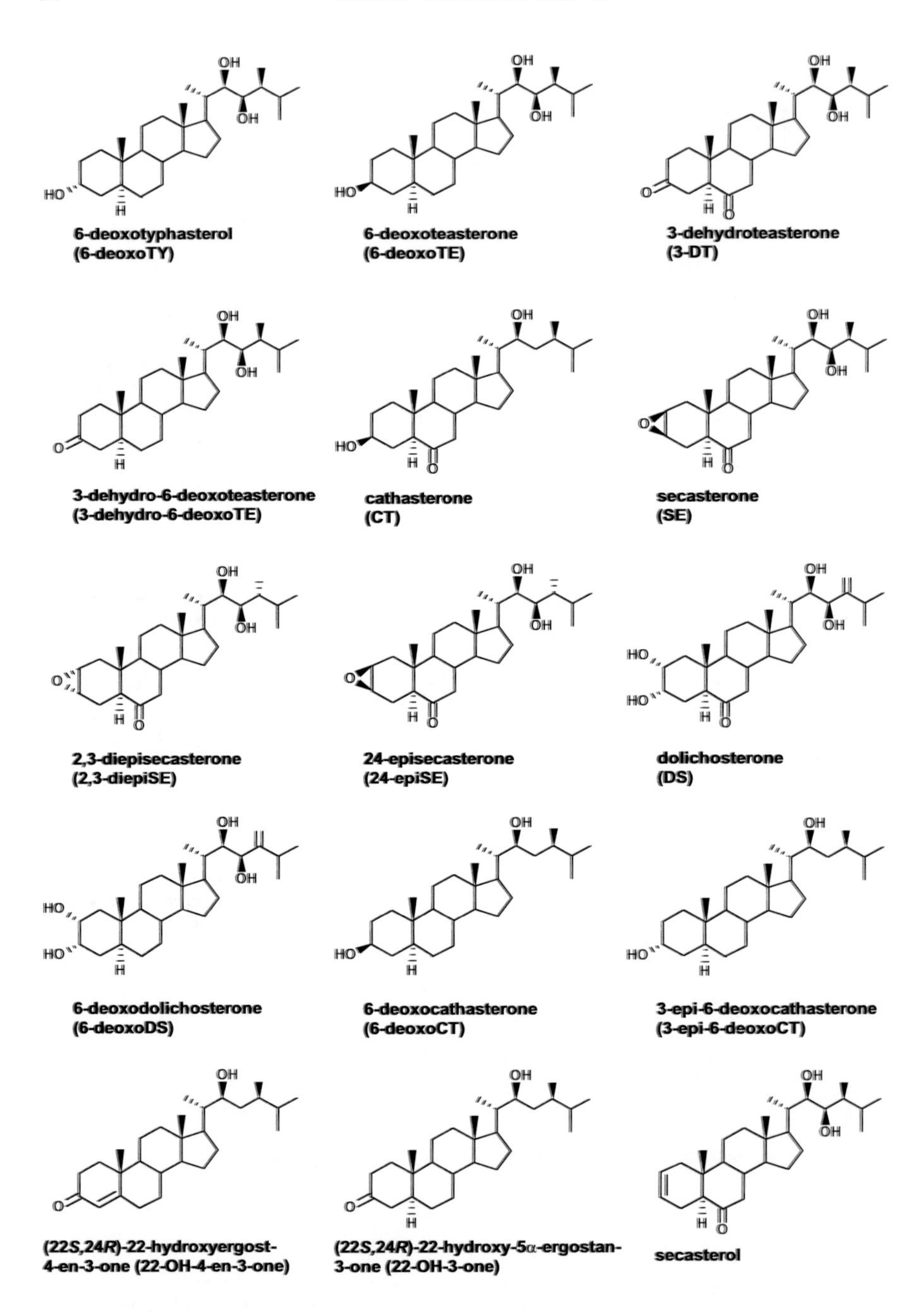


*Figure 3. Chemical structures of $C_{28}$ brassinosteroids – continued.*

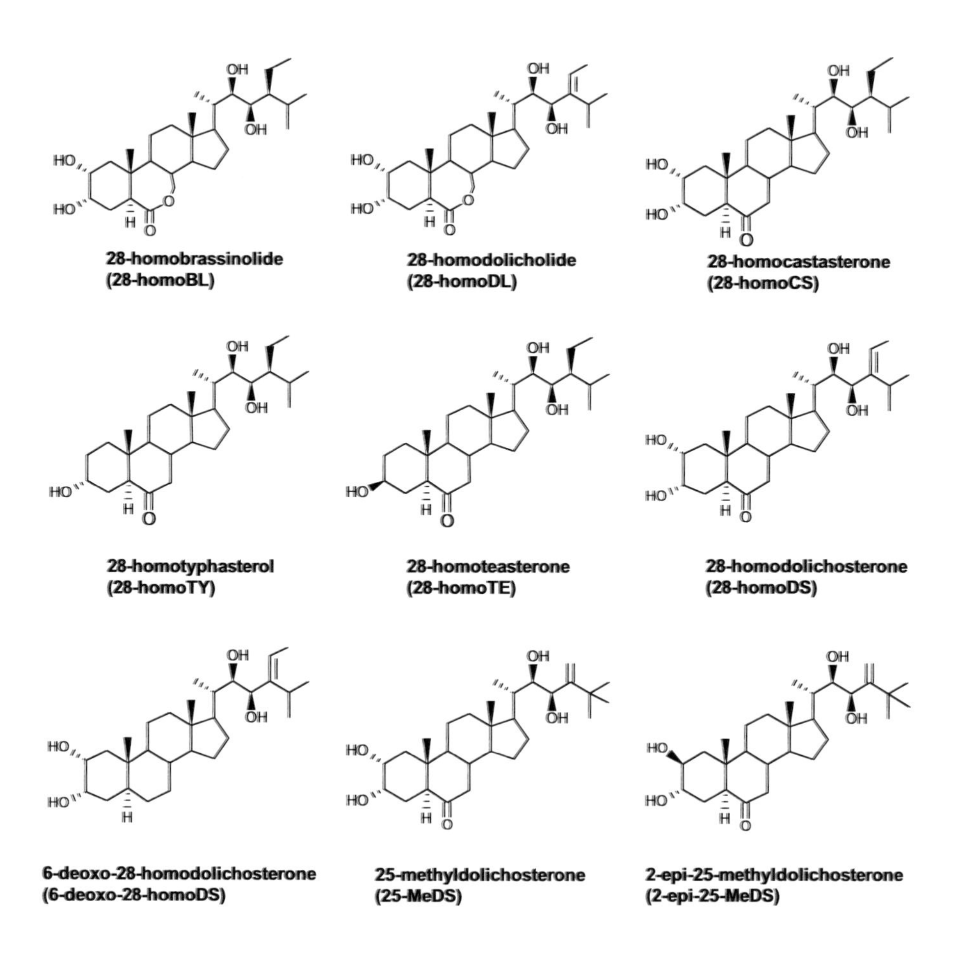


*Figure 4. Chemical structures of $C_{29}$ brassinosteroids.*

## DISTRIBUTION IN DICOTYLEDONS

The presence of BRs in dicotyledons has been reported from three subclasses. The first, the Apetalae is represented by 6 families including 8 plant species (Table 5).

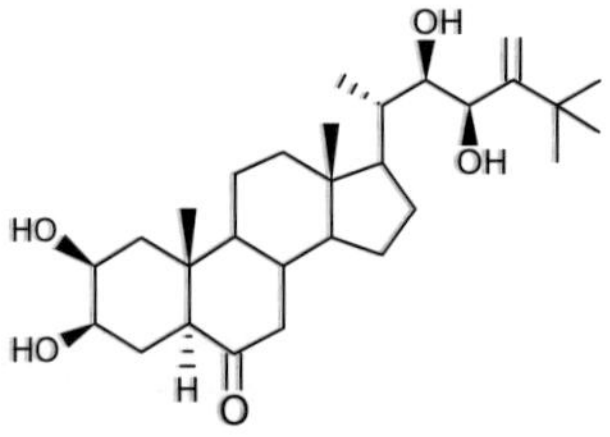

**2,3-diepi-25-methyldolichosterone (2,3-diepi-25-MeDS)**

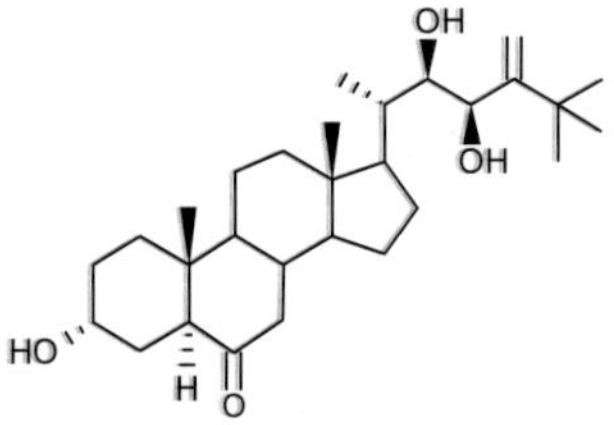

**2-deoxy-25-methyldolichosterone (2-deoxy-25-MeDS)**

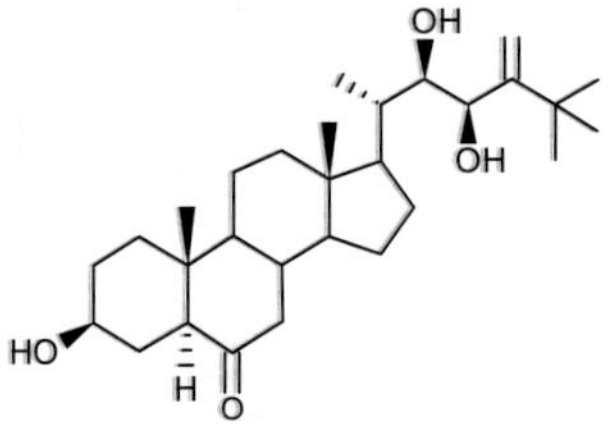

**3-epi-2-deoxy--25-methyldolichosterone (3-epi-2-deoxy-25-MeDS)**

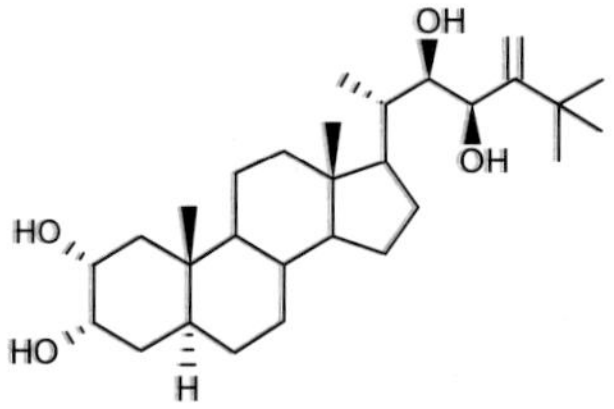

**6-deoxo-25-methyldolichosterone (6-deoxo-25-MeDS)**

*Figure 4. Chemical structures of $C_{29}$ brassinosteroids – continued.*

Total quantity of BRs amount to 7 various compounds. The second, the Chloripetalae is represented by 8 families including 21 plant species (Table 6). There are 49 BRs, among them 25 compounds belong to 6-oxo type, 17 belong to 6-deoxo type, 5 belong to 7-oxalactone type and 2 belong to 5-en type. Furthermore, from immature seeds of *Phaseolus vulgaris* a large quantity of 23 unconjugated and 2 conjugated BRs have been isolated so far. Among plants of this subclass, 44 BRs were detected for the first time. The third, the Sympetalae is represented by 7 families including 10 plant species (Table 7). Total quantity of BRs amount to 23 compounds of which 12 were isolated for the first time (from *Catharanthus roseus* and tomato) (Table 3). Among the BRs, compounds belong to 6-deoxo type are widely distributed (13), 6 belong to 6-oxo type, 2 belong to 5-en type, and one compound belong to 7-oxalactone (BL) and 6-hydroxy (6α-OH-CS) types.

**23-O-β-glucopyranosyl-
25-methyldolichosterone
(25-MeDS-Glu)**

**23-O-β-glucopyranosyl-2-epi-
25-methyldolichosterone
(2-epi-25-MeDS-Glu)**

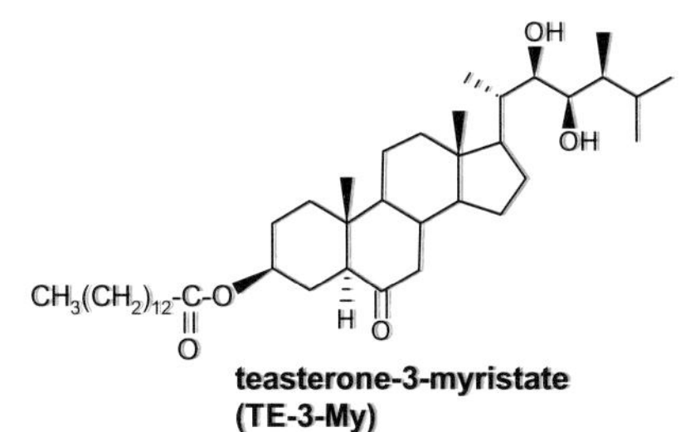


**teasterone-3-myristate
(TE-3-My)**

**teasterone-3-laurate
(TE-3-La)**

**teasterone-3-O-β-D-glucoside
(TE-3-Glu)**

*Figure 5. Chemical structures of brassinosteroid conjugates.*

*Table 3. Brassinosteroids isolated for the first time in plants.*

| *No.* | *Common name* | *Chemical name* | *Plant* | *References* |
|---|---|---|---|---|
| 1. | brassinolide | (22*R*,23*R*,24*S*)-2α,3α,22,23-tetrahydroxy-24-methyl-B-homo-7-oxa-5α-cholestan-6-one | *Brassica napus* L. | Grove et al., 1979 |
| 2. | castasterone | (22*R*,23*R*,24*S*)-2α,3α,22,23-tetrahydroxy-24-methyl-5α-cholestan-6-one | *Castanea crenata* Sieb. et Zucc. | Yokota et al., 1982a |
| 3. | dolicholide | (22*R*,23*R*)-2α,3α,22,23-tetrahydroxy-B-homo-7-oxa-5α-ergost-24(28)-en-6-one | *Dolichos lablab* L. | Yokota et al., 1982b |
| 4. | 28-norcastasterone | (22*R*,23*R*)-2α,3α,22,23-tetrahydroxy-5α-cholestan-6-one | *Brassica campestris var. pekinensis* L. *Thea sinensis* L. | Abe et al., 1983 |
| 5. | 28-homocastasterone | (22*R*,23*R*,24*S*)-2α,3α,22,23-tetrahydroxy-24-ethyl-5α-cholestan-6-one | *Brassica campestris var. pekinensis* L. *Thea sinensis* L. | Abe et al., 1983 |
| 6. | 28-norbrassinolide | (22*R*,23*R*)-2α,3α,22,23-tetrahydroxy-B-homo-7-oxa-5α-cholestan-6-one | *Brassica campestris var. pekinensis* L. | Abe et al., 1983 |
| 7. | dolichosterone | (22*R*,23*R*)-2α,3α,22,23-tetrahydroxy-5α-ergost-24(28)-en-6-one | *Dolichos lablab* L. | Baba et al., 1983 |
| 8. | 28-homodolichosterone | (22*R*,23*R*,24(28)*E*)-24(28)-ethylidene-2α,3α,22,23-tetrahydroxy-5α-cholestan-6-one | *Dolichos lablab* L. | Baba et al., 1983 |
| 9. | typhasterol (2-deoxycastasterone) | (22*R*,23*R*,24*S*)-3α,22,23-trihydroxy-24-methyl-5α-cholestan-6-one | *Typha latifolia* G.F.W. Mey | Schneider et al., 1983 |
| 10. | 28-homodolicholide | (22*R*,23*R*,24(28)*E*)-24(28)-ethylidene-2α,3α,22,23-tetrahydroxy-B-homo-7-oxa-5α-cholestan-6-one | *Dolichos lablab* L. | Yokota et al., 1983b |

*Table 3. Brassinosteroids isolated for the first time in plants - continued.*

| *No.* | *Common name* | *Chemical name* | *Plant* | *References* |
|---|---|---|---|---|
| 11. | 6-deoxocastasterone | (22*R*,23*R*,24*S*)-2α,3α,22,23-tetrahydroxy-24-methyl-5α-cholestane | *Phaseolus vulgaris* L. | Yokota et al., 1983c |
| 12. | 6-deoxodolichosterone | (22*R*,23*R*)-2α,3α,22,23-tetrahydroxy-5α-ergost-24(28)-ene | *Phaseolus vulgaris* L. | Yokota et al., 1983c |
| 13. | 28-homobrassinolide | (22*R*,23*R*,24*S*)-2α,3α,22,23-tetrahydroxy-24-etylo-B-homo-7-oxa-5α-cholestan-6-one | *Brassica campestris var. pekinensis* L. | Ikekawa et al., 1984 |
| 14. | teasterone | (22*R*,23*R*,24*S*)-3β,22,23-trihydroxy-24-methyl-5α-cholestan-6-onee | *Thea sinensis* L. | Abe et al., 1984a |
| 15. | 23-*O*-β-D-glucopyranosyl-25-methyldolichosterone | (22*R*,23*R*)-2α,3α,22-trihydroxy-23-*O*-β-D-glucopyranosyl-25-methyl-5α-ergost-24(28)-en-6-one | *Phaseolus vulgaris* L. | Yokota et al., 1987a |
| 16. | 23-*O*-β-D-glucopyranosyl-2-epi-25-methyldolichosterone | (22*R*,23*R*)-2β,3α,22-trihydroxy-23-*O*-β-D-glucopyranosyl-25-methyl-5α-ergost-24(28)-en-6-one | *Phaseolus vulgaris* L. | Yokota et al., 1987a |
| 17. | 24-epicastasterone | (22*R*,23*R*,24*R*)-2α,3α,22,23-tetrahydroxy-24-methyl-5α-cholestan-6-one | *Hydrodictyon reticulatum* (L.) Lager. | Yokota et al., 1987b |
| 18. | 6-deoxo-28-homodolichosterone | (22*R*,23*R*,24(28)*E*)-24(28)-ethylidene-2α,3α,22,23-tetrahydroxy-5α-cholestane | *Phaseolus vulgaris* L. | Yokota et al., 1987c |
| 19. | 25-methyldolichosterone | (22*R*,23*R*)-2α,3α,22,23-tetrahydroxy-25-methyl-5α-ergost-24(28)-en-6-one | *Phaseolus vulgaris* L. | Kim et al., 1987 |
| 20. | 24-epibrassinolide | (22*R*,23*R*,24*R*)-2α,3α,22,23-tetrahydroxy-24-methyl-B-homo-7-oxa-5α-cholestan-6-one | *Vicia faba* L. | Ikekawa et al., 1988 |

*Table 3. Brassinosteroids isolated for the first time in plants - continued.*

| *No.* | *Common name* | *Chemical name* | *Plant* | *References* |
|---|---|---|---|---|
| 21. | 2-epicastasterone | (22*R*,23*R*,24*S*)-2β,3α,22,23-tetrahydroxy-24-methyl-5α-cholestan-6-one | *Phaseolus vulgaris* L. | Kim, 1991 |
| 22. | 3-epicastasterone | (22*R*,23*R*,24*S*)-2α,3β,22,23-tetrahydroxy-24-methyl-5α-cholestan-6-one | *Phaseolus vulgaris* L. | Kim, 1991 |
| 23. | 2,3-diepicastasterone | (22*R*,23*R*,24*S*)-2β,3β,22,23-tetrahydroxy-24-methyl-5α-cholestan-6-one | *Phaseolus vulgaris* L. | Kim, 1991 |
| 24. | 3,24-diepicastasterone | (22*R*,23*R*,24*R*)-2α,3β,22,23-tetrahydroxy-24-methyl-5α-cholestan-6-one | *Phaseolus vulgaris* L. | Kim, 1991 |
| 25. | 2,3-diepi-25-methyldolichosterone | (22*R*,23*R*)-2β,3β,22,23-tetrahydroxy-25-methyl-5α-ergost-24(28)-en-6-one | *Phaseolus vulgaris* L. | Kim, 1991 |
| 26. | 3-epi-2-deoxy-25-methyldolichosterone | (22*R*,23*R*)-3β,22,23-trihydroxy-25-methyl-5α-ergost-24(28)-en-6-one | *Phaseolus vulgaris* L. | Kim, 1991 |
| 27. | 2-deoxy-25-methyldolichosterone | (22*R*,23*R*)-3α,22,23-trihydroxy-25-methyl-5α-ergost-24(28)-en-6-one | *Phaseolus vulgaris* L. | Kim, 1991 |
| 28. | 2-epi-25-methyldolichosterone | (22*R*,23*R*)-2β,3α,22,23-tetrahydroxy-25-methyl-5α-ergost-24(28)-en-6-one | *Phaseolus vulgaris* L. | Kim, 1991 |
| 29. | 6-deoxo-25-methyldolichosterone | (22*R*,23*R*)-2α,3α,22,23-tetrahydroxy-25-methyl-5α-ergost-24(28)-ene | *Phaseolus vulgaris* L. | Kim, 1991 |
| 30. | 3-epi-6-deoxocastasterone | (22*R*,23*R*,24*S*)-2α,3β,22,23-tetrahydroxy-24-methyl-5α-cholestane | *Phaseolus vulgaris* L. | Kim, 1991 |
| 31. | 3-epi-1α-hydroxy-castasterone | (22*R*,23*R*,24*S*)-1α,2α,3β,22,23-pentahydroxy-24-methyl-5α-cholestan-6-one | *Phaseolus vulgaris* L. | Kim, 1991 |

*Table 3. Brassinosteroids isolated for the first time in plants - continued.*

| | | | | |
|---|---|---|---|---|
| 32. | 1β-hydroxycastasterone | (22*R*,23*R*,24*S*)-1β,2α,3α,22,23-pentahydroxy-24-methyl-5α-cholestan-6-one | *Phaseolus vulgaris* L. | Kim, 1991 |
| 33. | 28-homoteasterone | (22*R*,23*R*,24*S*)-3α,22,23-trihydroxy-24-ethyl-5α-cholestan-6-one | *Raphanus sativus* L. | Schmidt et al., 1993b |
| 34. | 25-methylcastasterone | (22*R*,23*R*,24*S*)-2α,3α,22,23-tetrahydroxy-24,25-dimethyl-5α-cholestan-6-one | *Lolium perenne* L. | Taylor et al., 1993 |
| 35. | 3-dehydroteasterone (3-oxoteasterone) | (22*R*,23*R*,24*S*)-22,23-dihydroxy-24-methyl-5α-cholestan-3,6-dione | *Lilium longiflorum* Thunb. | Abe et al., 1994 |
| 36. | teasterone-3-myristate | (22*R*,23*R*,24*S*)-22,23-dihydroxy-3β-myristate-24-methyl-5α-cholestan-6-one | *Lilium longiflorum* Thunb. | Asakawa et al., 1994 |
| 37. | cathasterone | (22*S*,24*R*)-3β,22-dihydroxy-24-methyl-5α-cholestan-6-one | *Catharanthus roseus* G. Don. | Fujioka et al., 1995 |
| 38. | 6-deoxoteasterone | (22*R*,23*R*,24*S*)-2β,22,23-trihydroxy-24-methyl-5α-cholestane | *Catharanthus roseus* G. Don. | Fujioka et al., 1995 |
| 39. | 3-dehydro-6-deoxoteasterone | (22*R*,23*R*,24*S*)-22,23-dihydroxy-24-methyl-5α-cholestan-3-one | *Cupressus arizonica* Greene | Griffiths et al., 1995 |
| 40. | 6-deoxotyphasterol | (22*R*,23*R*,24*S*)-3α,22,23-trihydroxy-24-methyl-5α-cholestane | *Cupressus arizonica* Greene | Griffiths et al., 1995 |
| 41. | 6-deoxo-24-epicastasterone | (22*R*,23*R*,24*R*)-2α,3α,22,23-tetrahydroxy-24-methyl-5α-cholestane | *Ornithopus sativus* Brot. | Spengler et al., 1995 |
| 42. | 6-deoxo-28-norcastasterone | (22*R*,23*R*)-2α,3α,22,23-tetrahydroxy-5α-cholestane | *Ornithopus sativus* Brot. | Spengler et al., 1995 |

*Table 3. Brassinosteroids isolated for the first time in plants - continued.*

| *No.* | *Common name* | *Chemical name* | *Plant* | *References* |
|---|---|---|---|---|
| 43. | secasterone | (22*R*,23*R*,24*S*)-22,23-dihydroxy-2β,3β-epoxy-24-methyl-5α-cholestan-6-one | *Secale cereale* L. | Schmidt et al., 1995b |
| 44. | 2-deoxybrassinolide | (22*R*,23*R*,24*S*)-3α,22,23-trihydroxy-24-methyl-B-homo-7-oxa-5α-cholestan-6-one | *Apium graveolens* L. | Schmidt et al., 1995c |
| 45. | 28-homotyphasterol | (22*R*,23*R*,24*S*)-3α,22,23-trihydroxy-24-ethyl-5α-cholestan-6-one | *Oryza sativa* L. | Abe et al., 1995a |
| 46. | teasterone-3-laurate | (22*R*,23*R*,24*S*)-22,23-dihydroxy-3β-laurate-24-methyl-5α-cholestan-6-one | *Lilium longiflorum* Thunb. | Asakawa et al., 1996 |
| 47. | 23-dehydrobrassinolide | (22*R*,24*S*)-2α,3α,22-trihydroxy-24-methyl-B-homo-7-oxa-5α-cholestan-6,23-dione | *Cryptomeria japonica* D. Don. | Yokota et al., 1998 |
| 48. | 24-episecasterone | (22*R*,23*R*,24*R*)-22,23-dihydroxy-2β,3β-epoxy-24-methyl-5α-cholestan-6-one | *Lychnis viscaria* L. | Friebe et al., 1999 |
| 49. | 6α-hydroxycastasterone | (22*R*,23*R*,24*S*)-2α,3α,6α,22,23-pentahydroxy-24-methyl-5α-cholestane | *Lycopersicon esculentum* Mill. | Bishop et al., 1999 |
| 50. | 2-epi-23-dehydrobrassinolide | (22*R*,24*S*)-2β,3α,22-trihydroxy-24-methyl-B-homo-7-oxa-5α-cholestan-6,23-dione | *Cryptomeria japonica* D. Don. | Watanabe et al., 2000 |
| 51. | 3-epi-23-dehydrobrassinolide | (22*R*,24*S*)-2α,3β,22-trihydroxy-24-methyl-B-homo-7-oxa-5α-cholestan-6,23-dione | *Cryptomeria japonica* D. Don. | Watanabe et al., 2000 |
| 52. | 2,3-diepi-23-dehydrobrassinolide | (22*R*,24*S*)-2β,3β,22-trihydroxy-24-methyl-B-homo-7-oxa-5α-cholestan-6,23-dione | *Cryptomeria japonica* D. Don. | Watanabe et al., 2000 |

*Table 3. Brassinosteroids isolated for the first time in plants - continued.*

| | | | | |
|---|---|---|---|---|
| 53. | teasterone-3-*O*-β-D-glucoside | (22*R*,23*R*,24*S*)-22,23-dihydroxy-3-*O*-β-D-glucopyranosyl-24-methyl-5α-cholestan-6-one | *Lilium longiflorum* Thunb. | Soeno et al., 2000 |
| 54. | 28-nortyphasterol | (22*R*,23*R*)-3α,22,23-trihydroxy-5α-cholestan-6-one | *Arabidopsis thaliana* (L.) Heynh. | Fujioka et al., 2000a |
| 55. | 6-deoxocathasterone | (22*S*,24*R*)-3β,22-dihydroxy-24-methyl-5α-cholestane | *Catharanthus roseus* G. Don. | Fujioka et al., 2000b |
| 56. | 3-epi-6-deoxocathasterone | (22*S*,24*R*)-3α,22-dihydroxy-24-methyl-5α-cholestane | *Catharanthus roseus* G. Don. | Fujioka et al., 2000b |
| 57. | 6-deoxo-28-norcathasterone | (22*S*,24*R*)-3β,22-dihydroxy-24-methyl-5α-cholestane | *Lycopersicon esculentum* Mill. | Yokota et al., 2001 |
| 58. | 6-deoxo-28-nortyphasterol | (22*R*,23*R*)-3α,22,23-trihydroxy-5α-cholestane | *Lycopersicon esculentum* Mill. | Yokota et al., 2001 |
| 59. | 3-epibrassinolide | (22*R*,23*R*,24*S*)-2α,3β,22,23-tetrahydroxy-24-methyl-B-homo-7-oxa-5α-cholestan-6-one | *Arabidopsis thaliana* (L.) Heynh. | Konstantinova et al., 2001 |
| 60. | (22*S*)-22-hydroxycampestrol | (22*S*)-22-hydroksyergost-5-en-3β-ol | *Arabidopsis thaliana* (L.) Heynh. | Choe et al., 2001 |
| 61. | 6-deoxo-28-norteasterone | (22*R*,23*R*)-3β,22,23-trihydroxy-5α-cholestane | *Thea sinensis* L. | Kaur et al., 2002 |
| 62. | 3-dehydro-6-deoxo-28-norteasterone | (22*R*,23*R*)-22,23-dihydroxy-5α-cholestan-3-one | *Thea sinensis* L. | Kaur et al., 2002 |
| 63. | 28-nor-22-hydroxycampesterol | 28-nor-(22S)-22-hydroxyergost-5-en-3β-ol | *Catharanthus roseus* G. Don.<br>*Arabidopsis thaliana* (L.) Heynh. | Fujioka et al., 2002 |

*Table 3. Brassinosteroids isolated for the first time in plants - continued.*

| *No.* | *Common name* | *Chemical name* | *Plant* | *References* |
|---|---|---|---|---|
| 64. | 22-OH-4-en-3-one | (22*S*,24*R*)-22-hydroxyergost-4-en-3-one | *Catharanthus roseus* G. Don. *Arabidopsis thaliana* (L.) Heynh. | Fujioka et al., 2002 |
| 65. | 28-nor-22-OH-4-en-3-one | (22*S*)-22-hydroxyergost-4-en-3-one | *Catharanthus roseus* G. Don. *Arabidopsis thaliana* (L.) Heynh. | Fujioka et al., 2002 |
| 66. | 22-OH-3-one | (22*S*,24*R*)-22-hydroxy-5α-ergostan-3-one | *Catharanthus roseus* G. Don. *Arabidopsis thaliana* (L.) Heynh. | Fujioka et al., 2002 |
| 67. | 28-nor-22-OH-3-one | (22*S*)-22-hydroxy-5α-ergostan-3-one | *Catharanthus roseus* G. Don. *Arabidopsis thaliana* (L.) Heynh. | Fujioka et al., 2002 |
| 68. | 3-epi-6-deoxo-28-norcathasterone | (22*S*)-3α,22-dihydroxy-5α-cholestane | *Catharanthus roseus* G. Don. *Arabidopsis thaliana* (L.) Heynh. | Fujioka et al., 2002 |
| 69. | secasterol | (22*R*,23*R*,24*S*)-22,23-dihydroxy-24-mthyl-5α-cholest-2-en-6-one | *Secale cereale* L. | Antonchick et al., 2003 |
| 70. | 2,3-diepisecasterone | (22*R*,23*R*,24*S*)-22,23-dihydroxy-2α,3α-epoxy-24-methyl-5α-cholestan-6-one | *Secale cereale* L. | Antonchick et al., 2003 |

*Table 4. The occurrence of brassinosteroids in the monocotyledons.*

| *Family / Species* | *Plant parts* | *Brassinosteroid* | *Isolated quantity (μg/kg fresh wt.)* | *References* |
|---|---|---|---|---|
| **Arecaceae** | | | | |
| *Phoenix dactylifera* L. | Pollen | 24-epiCS | | Zaki et al., 1993 |
| **Gramineae** | | | | |
| *Lolium perenne* L. | Pollen | 25-MeCS | 0.001 | Taylor et al., 1993 |
| *Oryza sativa* L. | Shoot | CS<br>DS<br>BL | 0.014<br>0.008 | Abe et al., 1984b; Abe, 1991 |
| | Bran | 6-deoxoCS, 28-homoTE, 28-homoTY | | Abe et al., 1995a |
| | Seeds | CS, TE, 6-deoxoCS | | Park et al., 1994b |
| *Phalaris canariensis* L. | Seeds | CS<br>TE | 5<br>0.7 | Shimada et al., 1996 |
| *Triticum aestivum* L. | Grain | CS, TY, TE, 6-deoxoCS, 3-DT | | Yokota et al., 1994 |
| *Secale cereale* L. | Seeds | CS, TY, TE, 6-deoxoCS, 28-norCS, SE | | Schmidt et al., 1995b |
| | Leaves | SE<br>2,3-diepiSE<br>secasterol | 0.052<br>0.02 | Antonchick et al., 2003 |
| | roots | SE<br>2,3-diepiSE | 0.107<br>0.032 | Antonchick et al., 2003 |

*Table 4. The occurrence of brassinosteroids in the monocotyledons - continued.*

| *Family / Species* | *Plant parts* | *Brassinosteroid* | *Isolated quantity (μg/kg fresh wt.)* | *References* |
|---|---|---|---|---|
| **Gramineae** | | | | |
| *Zea mays* L.<br>- dent corn | pollen | CS<br>TY<br>TE | 120<br>6.6<br>4.1 | Suzuki et al., 1986 |
| | roots | CS | 0.3 | Kim et al., 2000a |
| - sweet corn | pollen | CS<br>28-norCS<br>DS | 27.2<br>18.3<br>16.9 | Gamoh et al., 1990 |
| **Liliaceae** | | | | |
| *Erythronium japonicum* Decne | pollen<br>anther | TY | 5 | Yasuta et al., 1995 |
| *Lilium elegans* Thunb. | pollen | CS<br>TY<br>BL<br>TE | 10-50<br>10-50<br>1-5<br>1-5 | Suzuki et al., 1994b; Yasuta et al., 1995 |
| *Lilium longiflorum* Thunb. | Pollen<br>Anther | 3-DT<br>TY<br>3-Glu-TE<br>TE<br>BL, CS, 3-La-TE, 3-My-TE | 3180<br>2440<br>720<br>20 | Abe, 1991; Abe et al., 1994; Asakawa et al., 1994, 1996; Soeno et al., 2000 |
| *Tulipa gesneriana* L. | Pollen | TY | | Abe, 1991 |
| **Typhaceae** | | | | |
| *Typha latifolia* G.F.W. Mey | Pollen | TY, TE | 68 | Schneider et al., 1983; Abe, 1991 |

*Table 5. The occurrence of brassinosteroids in the dicotyledons – the Apetalae.*

| *Family / Species* | *Plant parts* | *Brassinosteroid* | *Isolated quantity (µg/kg fresh wt.)* | *References* |
|---|---|---|---|---|
| **Betulaceae** | | | | |
| *Alnus glutinosa* (L.) Gaertn. | Pollen | BL, CS | | Plattner et al., 1986 |
| **Cannabaceae** | | | | |
| *Cannabis sativa* L. | Seeds | TE<br>CS | 1800<br>600 | Takatsuto et al., 1996b |
| **Caryophyllaceae** | | | | |
| *Gypsophilla perfoliata* L. | Seeds | 24-epiBL | | Schmidt et al., 1996 |
| *Lychnis viscaria* L. | Seeds | 24-epiCS, 24-epiSE | | Friebe et al., 1999 |
| **Chenophyllaceae** | | | | |
| *Beta vulgaris* L. | Seeds | CS, 24-epiCS | | Schmidt et al., 1994 |
| **Fagaceae** | | | | |
| *Castanea crenata* Sieb. et Zucc. | Galls | 6-deoxoCS<br>BL<br>CS | 9-26<br>4-12<br>1 | Yokota et al., 1982a; Ikeda et al., 1983 |
| | Shoot leaves | 6-deoxoCS<br>CS | 15-30<br>2-6 | Arima et al., 1984 |
| **Polygonaceae** | | | | |
| *Fagopyrum esculentum* Moench | pollen | CS<br>BL | 7.1<br>5 | Takatsuto et al., 1990b |
| *Rheum rhabarbarum* L. | panicles | BL, CS, 24-epiCS | | Schmidt et al., 1995a |

*Table 6. The occurrence of brassinosteroids in the dicotyledons – the Chloripetalae.*

| *Family / Species* | *Plant parts* | *Brassinosteroid* | *Isolated quantity (μg/kg fresh wt.)* | *References* |
|---|---|---|---|---|
| **Apiaceae** | | | | |
| *Apium graveolens* L. | seeds | 2-deoxyBL | | Schmidt et al., 1995c |
| *Daucus carota* ssp. *sativus* L. | seeds | BL, CS, 24-epiCS | | Schmidt et al., 1998 |
| **Brassicaceae** | | | | |
| *Arabidopsis thaliana* (L.) Heynh. | shoot ecotype Columbia (wild-type) | 6-deoxoCT<br>6-deoxoTY<br>CS<br>6-deoxoCS<br>3-dehydro-6-deoxoTE<br>TY<br>6-deoxoTE<br>BL<br>TE<br>28-norCS, 28-norTY | 1.96<br>0.95<br>0.75<br>0.71<br>0.13<br>0.11<br>0.10<br>0.04<br>0.025 | Fujioka et al., 1996, 1997, 2000a; Nomura et al., 2001 |
| | seeds ecotype Columbia (wild-type) | 6-deoxoCS<br>TY<br>BL<br>6-deoxoTY<br>6-deoxoTE<br>CS<br>24-epiBL | 1.5-3<br>1.3<br>0.5-1,9<br>0.5-5.4<br>0.5-1<br>0.4-5<br>0.22 | Fujioka et al., 1998 |
| | seeds (ecotype 24) | CS<br>24-epiBL | 0.36<br>0.22 | Schmidt et al., 1997 |
| | root callus | BL, 3-epiBL | | Konstantinova et al., 2001 |

*Table 6. The occurrence of brassinosteroids in the dicotyledons – the Chloripetalae.*

| *Family / Species* | *Plant parts* | *Brassinosteroid* | *Isolated quantity (μg/kg fresh wt.)* | *References* |
|---|---|---|---|---|
| **Brassicaceae** | | | | |
| *Arabidopsis thaliana* (L.) Heynh. | seedlings | 6-deoxoCT<br>3-epi-6-deoxoCT<br>6-deoxo-28-norCT<br>3-epi-6-deoxo-28-norCT<br>22-OH-4-en-3-one<br>28-nor-22-OH-4-en-3-one<br>22-OH-3-one<br>28-nor-22-OH-3-one<br>22-OHCR<br>28-nor-22-OHCR | | Choe et al., 2001; Fujioka et al., 2002 |
| *Brassica campestris var. pekinensis* L. | seeds | CS<br>28-norBL<br>BL<br>28-norCS<br>28-homoCS | 1600<br>1300<br>940<br>780<br>130 | Abe et al., 1982, 1983; Ikekawa et al., 1984 |
| *Brassica napus* L. | pollen | BL | 100 | Grove et al., 1979 |
| *Raphanus sativus* L. | seeds | CS<br>BL<br>TE, 28-homoTE | 0.8<br>0.3 | Schmidt et al., 1991, 1993b |

*Table 6. The occurrence of brassinosteroids in the dicotyledons – the Chloripetalae – continued.*

| *Family / Species* | *Plant parts* | *Brassinosteroid* | *Isolated quantity (μg/kg fresh wt.)* | *References* |
|---|---|---|---|---|
| **Fabaceae** | | | | |
| *Dolichos lablab* L. | seeds | DL<br>DS<br>28-homoDS<br>28-homoDL<br>BL<br>CS, 6-deoxoCS<br>6-deoxoDS | 160<br>50<br>20<br>12 | Baba et al., 1983;<br>Yokota et al., 1982b, 1983b, 1984 |
| *Robinia pseudo-acacia* L. | pollen | CS, TY, 6-deoxoCS | | Abe et al., 1995b |
| *Vicia faba* L. | seeds | BL<br>24-epiBL<br>CS, 28-norCS | 190<br>5 | Park et al., 1987;<br>Ikekawa et al., 1988 |
| | pollen | 28-norCS<br>DS<br>BL<br>CS | 628<br>537<br>181<br>134 | Gamoh et al., 1989 |
| *Ornithopus sativus* Brot. | seeds | 24-epiCS<br>CS | 25<br>5 | Schmidt et al., 1993a |
| | shoot | CS, 6-deoxoCS, 24-epiCS, 6-deoxo-24-epiCS, 6-deoxo-28-norCS | | Spengler et al., 1995 |

*Table 6. The occurrence of brassinosteroids in the dicotyledons – the Chloripetalae – continued.*

| *Family / Species* | *Plant parts* | *Brassinosteroid* | *Isolated quantity (μg/kg fresh wt.)* | *References* |
|---|---|---|---|---|
| **Fabaceae** | | | | |
| *Phaseolus vulgaris* L. | seeds | BL, CS, 2-epiCS, 3-epiCS, 2,3-diepiCS, 3,24-diepiCS, TY, TE, 6-deoxoCS, 3-epi-6-deoxoCS, 1β-OH-CS, 3-epi-1α-OH-CS, DL, DS, 6-deoxoDS, 6-deoxo-28-homoDS, 25-MeDS, 2-epi-25-MeDS, 2,3-diepi-25-MeDS, 2-deoxy-25-MeDS, 2-epi-2-deoxy-25-MeDS, 3-epi-2-deoxy-25-MeDS, 6-deoxo-25-MeDS, 25-MeDS-Glu, 2-epi-25-MeDS-Glu | | Yokota et al., 1983c, 1987c; Kim et al., 1987, 1988, 2000b; Kim, 1991; Park et al., 2000 |
| *Psophocarpus tetragonolobus* (Stickm.) DC. | seeds | BL, CS, 6-deoxoCS, 6-deoxoDS | | Takatsuto, 1994 |

*Table 6. The occurrence of brassinosteroids in the dicotyledons – the Chloripetalae – continued.*

| *Family / Species* | *Plant parts* | *Brassinosteroid* | *Isolated quantity (μg/kg fresh wt.)* | *References* |
|---|---|---|---|---|
| **Fabaceae** | | | | |
| *Pisum sativum* L. | seeds | BL, CS, TY, 6-deoxoCS, 2-deoxyBL | | Yokota et al., 1996 |
| | shoot | 6-deoxoCS<br>6-deoxoCT<br>TY<br>6-deoxoTY<br>CS<br>BL<br>3-dehydro-6-deoxoTE<br>6-deoxoTE | 5.2<br>3.7<br>1.0<br>0.8<br>0.4-2.4<br>0.2-0.8<br>0.074<br>0.047 | Nomura et al., 1997, 1999, 2001 |
| **Hamamelidaceae** | | | | |
| *Distylium racemosum* Sieb. et Zucc. | galls | CS<br>28-norCS | 2500<br>5 | Ikekawa et al., 1984 |
| | leaves | 28-norBL<br>CS<br>BL<br>28-norCS<br>TE, TY | 0.16<br>0.13<br>0.023<br>0.016 | Ikekawa et al., 1984; Abe et al., 1994 |
| **Myrtaceae** | | | | |
| *Eucalyptus calophylla* R. Br. | pollen | BL | | Takatsuto, 1994 |
| *Eucalyptus marginata* Sn. | pollen | DS | | Takatsuto, 1994 |

*Table 6. The occurrence of brassinosteroids in the dicotyledons – the Chloripetalae – continued.*

| *Family / Species* | *Plant parts* | *Brassinosteroid* | *Isolated quantity (μg/kg fresh wt.)* | *References* |
|---|---|---|---|---|
| **Rosaceae** | | | | |
| *Eriobotrya japonica* (Thunb.) Lindl. | flower buds | CS | | Takatsuto, 1994 |
| ***Rutaceae*** | | | | |
| *Citrus unshiu* Marcov. | pollen | BL, CS, TY, TE | | Abe, 1991 |
| *Citrus sinensis* Osbeck | pollen | BL<br>CS | 36.2<br>29.4 | Motegi et al., 1994 |
| **Theaceae** | | | | |
| *Thea sinensis* L. | leaves | CS<br>TY<br>TE<br>BL<br>28-norCS<br>28-homoCS | 0.1<br>0.06<br>0.02<br>0.006<br>0.002<br>< 0.001 | Abe et al., 1983, 1984a; Morishita et al., 1983; Ikekawa et al., 1984 |
| | seeds | 28-norCS, 6-deoxo-28-norCS, 6-deoxo-28-norTY, 6-deoxo-28-norTE, 6-deoxo-28-norCT, 3-dehydro-6-deoxo-28-norTE | | Kaur et al., 2002 |

*Table 7. The occurrence of brassinosteroids in the dicotyledons – the Sympetalae.*

| *Family / Species* | *Plant parts* | *Brassinosteroid* | *Isolated quantity (μg/kg fresh wt.)* | *References* |
|---|---|---|---|---|
| **Apocynaceae** | | | | |
| *Catharanthus roseus* G. Don. | culture cells | 6-deoxoCT<br>6-deoxoCS<br>CT<br>6-deoxoTY<br>CS<br>BL<br>6-deoxoTE<br>3-epi-6-deoxoCT, 3-DT, TY, TE<br>6-deoxo-28-norCT<br>3-epi-6-deoxo-28-norCT<br>22-OH-4-en-3-one<br>28-nor-22-OH-4-en-3-one<br>22-OH-3-one<br>28-nor-22-OH-3-one<br>22-OHCR<br>28-nor-22-OHCR | 30<br>5.9-18.9<br>2-4<br>0.76<br>0.6-4.5<br>0.4-8.7<br>0.047 | Choi et al., 1993, 1996, 1997; Fujioka et al., 1995, 2000b; Park et al., 1989; Suzuki et al., 1993, 1994a, c, 1995; Yokota et al., 1990; Choe et al., 2001; Fujioka et al., 2002 |
| **Asteraceae** | | | | |
| *Helianthus annuus* L. | pollen | BL<br>28-norCS<br>CS | 106<br>65<br>21 | Takatsuto et al., 1989 |
| *Solidago altissima* L. | shoot | BL | | Takatsuto, 1994 |

*Table 7. The occurrence of brassinosteroids in the dicotyledons – the Sympetalae– continued.*

| *Family / Species* | *Plant parts* | *Brassinosteroid* | *Isolated quantity ($\mu$g/kg fresh wt.)* | *References* |
|---|---|---|---|---|
| **Asteraceae** | | | | |
| *Zinnia elegans* L. | culture cells | CS, TY, 6-deoxoCS, 6-deoxoTY, 6-deoxoTE | | Yamamoto et al., 2001 |
| **Boraginaceae** | | | | |
| *Echium plantagineum* L. | pollen | BL | | Takatsuto, 1994 |
| **Convolvulaceae** | | | | |
| *Pharbitis purpurea* Voigt | seeds | CS<br>28-norCS | 1.1<br>0.2 | Suzuki et al., 1985 |
| **Cucurbitaceae** | | | | |
| *Cucurbita moschata* Duch. | seeds | BL, CS | | Jang et al., 2000 |
| **Lamiaceae** | | | | |
| *Perilla frutescens* (L.) Britt. | seeds | CS | | Park et al., 1994b |
| **Solanaceae** | | | | |
| *Nicotiana tabacum* L. | culture cells | CS | | Park et al., 1994b |

*Table 7. The occurrence of brassinosteroids in the dicotyledons – the Sympetalae– continued.*

| *Family / Species* | *Plant parts* | *Brassinosteroid* | *Isolated quantity (µg/kg fresh wt.)* | *References* |
|---|---|---|---|---|
| **Solanaceae** | | | | |
| *Lycopersicon esculentum* Mill. | shoot | 6-deoxoCS | 1.7 | Yokota et al., 1997d |
| | | CS | 0.2 | |
| | | 28-norCS | 0.03 | |
| - dwarf mutant | shoot | 6-deoxoCS | 52 | Bishop et al., 1999 |
| | | 6-deoxoCT | 1.1 | |
| | | 6-deoxoTY | 0.5 | |
| | | CS | 0.2 | |
| | | 6-deoxoTE | 0.04 | |
| | | 3-dehydro-6-deoxoTE | 0.03 | |
| | | BL | < 0.001 | |
| | | TY | < 0.001 | |
| | | 3-DT | < 0.001 | |
| | | TE | < 0.001 | |
| | | CT | < 0.001 | |
| | | 6α-OH-CS | | |

*Table 8. The occurrence of brassinosteroids in gymnosperms.*

| *Family / Species* | *Plant parts* | *Brassinosteroid* | *Isolated quantity (μg/kg fresh wt.)* | *References* |
|---|---|---|---|---|
| **Cupressaceae** | | | | |
| *Cupressus arizonica* Greene | pollen | 6-deoxoTY<br>3-dehydro-6-deoxoTE<br>6-deoxoCS<br>CS<br>TY<br>TE<br>28-homoCS<br>3-DT<br>BL | 6400<br>2300<br>1200<br>1000<br>460<br>5<br>4<br>2<br>< 1 | Griffiths et al., 1995 |
| **Ginkgoaceae** | | | | |
| *Ginkgo biloba* L. | seeds | TE | 15 | Takatsuto et al., 1996a |
| **Pinaceae** | | | | |
| *Piceae sitchensis* Trantv. ex Mey | shoot | TY<br>CS | 7<br>5 | Yokota et al., 1985 |
| *Pinus silvestris* L. | cambial region | BL, CS | | Kim et al., 1990 |
| *Pinus thunbergii* Parl. | pollen | TY | 89.5 | Yokota et al., 1983a |
| **Taxodiaceae** | | | | |
| *Cryptomeria japonica* D. Don. | pollen<br>anther | TY, DL, 3-DT, 28-homoBL, 28-homoDL, 23-dehydroBL, 2-epi-23-dehydroBL, 3-epi-23-dehydroBL, 2,3-diepi-23-dehydroBL | | Yokota et al., 1998<br>Watanabe et al., 2000 |

*Table 9. The occurrence of brassinosteroids in lower plants.*

| *Family / Species* | *Plant parts* | *Brassinosteroid* | *Isolated quantity (μg/kg fresh wt.)* | *References* |
|---|---|---|---|---|
| **Equisetaceae** | | | | |
| *Equisetum arvense* L. | whole plant | DS<br>28-norCS<br>CS<br>28-norBL | 0.75<br>0.35<br>0.17<br>0.15 | Takatsuto et al., 1990a |
| **Hydrodictyaceae** | | | | |
| *Hydrodictyon reticulatum* (L.) Lager. | whole plant | 28-homoCS<br>24-epiCS | 4.0<br>0.3 | Yokota et al., 1987b |
| **Marchantiaceae** | | | | |
| *Marchantia polymorpha* L. | culture cells | TE<br>3-DT<br>TY | | Park et al., 1999 |

## DISTRIBUTION IN GYMNOSPERMS

The occurrence of BRs in gymnosperms has been reported from six conifers (Table 8). The presence of new 6 BRs was shown in *Cupressus arizonica* and *Cryptomeria japonica*. Among plant species so far reported, the level of BR in the mature pollen of *Cupressus arizonica* is the highest (6.4 mg/kg 6-deoxoTY).

## DISTRIBUTION IN LOWER PLANTS

BRs have been identified in lower plants such as the green alga (*Hydrodictyon reticulatum*), a pteridophyte (*Equisetum arvense*), a bryophyte (*Marchantia polymorpha*) (Table 9). Total quantity of BRs amount 9 various compounds, among them 6-oxo type of BRs is dominated (8 compounds). Furthermore, the occurrence of 24-epiCS in algae and the first time in plants has been demonstrated.

## REFERENCES

Abe, H. (1991). Rice-lamina inclination, endogenous levels in plant tissues and accumulation during pollen development of brassinosteroids. In Brassinosteroids: Chemistry, Bioactivity and Applications, pp. 200-207. Eds H G Cutler, T Yokota and G Adam. American Chemical Society, Washington.

Abe, H., Honjo, C., Kyokawa, Y., Asakawa, S., Narsume, M., Narushima, M. (1994). 3-Oxoteasterone and the epimerization of teasterone: identification in lily anthers and *Distylium racemosum* leaves and its biotransformation into typhasterol. Bioscience, Biotechnology and Biochemistry 58: 986-989.

Abe, H., Morishita, T., Uchiyama, M., Marumo, S., Munakata, K., Takatsuto, S., Ikekawa, N. (1982). Identification of brassinolide-like substances in chinese cabbage. Agricultural and Biological Chemistry 46: 2609-2611.

Abe, H., Morishita, T., Uchiyama, M., Takatsuto, S., Ikekawa, N. (1984a). A new brassinolide-related steroid in the leaves of *Thea sinensis*. Agricultural and Biological Chemistry 48: 2171-2172.

Abe, H., Morishita, T., Uchiyama, M., Takatsuto, S., Ikekawa, N., Ikeda, M., Sassa, T., Kitsuwa, T., Marumo, S. (1983). Occurrence of three new brassinosteroids: brassinone, 24(*S*)-24-ethylbrassinone and 28-norbrassinolide, in higher plants. Experientia 39: 351-353.

Abe, H., Nakamura, K., Morishita, T., Uchiyama, M., Takatsuto, S., Ikekawa, N. (1984b). Endogenous brassinosteroids of the rice plant: castasterone and dolichosterone. Agricultural and Biological Chemistry 48: 1103-1104.

Abe, H., Soeno, K., Koseki, N-N., Natsume, M. (2001). Conjugated and unconjugated brassinosteroids. In Agrochemical Discovery. Insect, Weed, and Fungal Control, pp. 91-101. Eds D R Baker and N K Umetsu. American Chemical Society, Washington.

Abe, H., Takatsuto, S., Nakayama, M., Yokota, T. (1995a). 28-Homotyphasterol, a new natural brassinosteroid from rice (*Oryza sativa* L.) bran. Bioscience, Biotechnology and Biochemistry 59: 176-178.

Abe, H., Takatsuto, S., Okuda, R., Yokota, T. (1995b). Identification of castasterone, 6-deoxocastasterone, and typhasterol in the pollen of *Robinia pseudoacacia* L. Bioscience, Biotechnology and Biochemistry 59: 309-310.

Adam, G., Petzold, U. (1994). Brassinosteroide – eine neue phytohormon-gruppe? Naturwissenschaften 81: 210-217.

Antonchick, A.P., Schneider, B., Zhabinskii, V.N., Konstantinova, O.V., Khripach, V.A. (2003). Biosynthesis of 2,3-epoxybrassinosteroids in seedlings of *Secale cereale*. Phytochemistry, in press.

Arima, M., Yokota, T., Takahashi, N. (1984). Identification and quantification of brassinolide-related steroids in the insect gall and healthy tissues of the chestnut plant. Phytochemistry 23: 1587-1591.

Asakawa, S., Abe, H., Kyokawa, Y., Nakamura, S., Natsume, M. (1994). Teasterone 3-myristate: a new type of brassinosteroid derivative in *Lilium longiflorum* anthers. Bioscience, Biotechnology and Biochemistry 58: 219-220.

Asakawa, S., Abe, H., Nishikawa, N., Natsume, M., Koshioka, M. (1996). Purification and identification of new acyl-conjugated teasterones in lilly pollen. Bioscience, Biotechnology and Biochemistry 60: 1416-1420.

Asami, T., Yoshida, S. (1999). Brassinosteroid biosynthesis inhibitors. Trends in Plant Science 4 (9): 348-353.

Baba, J., Yokota, T., Takahashi, N. (1983). Brassinolide-related new bioactive steroids from *Dolichos lablab* seed. Agricultural and Biological Chemistry 47: 659-661.

Bishop, G. J., Nomura, T., Yokota, T., Harrison, K., Noguchi, T., Fujioka, S., Takatsuto, S., Jones, J. D. G., Kamiya, Y. (1999). The tomato DWARF enzyme catalyses C-6 oxidation in brassinosteroid biosynthesis. Proceeding of the National Academy of Science (USA) 96: 1761-1766.

Bishop, G. J., Yokota, T. (2001). Plant steroid hormones, brassinosteroids: current highlights of molecular aspects on their synthesis/metabolism, transport, perception and response. Plant and Cell Physiology 42 (2): 114-120.

Choe, S., Fujioka, S., Noguchi, T., Takatsuto, S., Yoshida, S., Feldmann, K.A. (2001). Overexpression of *DWARF4* in brassinosteroid biosynthesis pathway results in increased vegetative growth and seed yield in *Arabidopsis*. Plant Journal 26: 573-582.

Choi, Y-H., Fujioka, S., Harada, A., Yokota, T., Takatsuto, S., Sakurai, A. (1996). A brassinolide biosynthetic pathway via 6-deoxocastasterone. Phytochemistry 43: 593-596.

Choi, Y-H., Fujioka, S., Nomura, T., Harada, A., Yokota, T., Takatsuto, S., Sakurai, A. (1997). An alternative brassinolide biosynthetic pathway via late C-6 oxidation. Phytochemistry 44: 609-613.

Choi, Y-H., Inoue, T., Fujioka, S., Saimoto, H., Sakurai, A. (1993). Identification of brassinosteroid-like active substances in plant-cell cultures. Bioscience, Biotechnology and Biochemistry 57: 860-861.

Clouse, S. D., Feldmann, K. A. (1999). Molecular genetics of brassinosteroid action. In Brassinosteroids: Steroidal Plant Hormones, pp. 163-190. Eds A Sakurai, T Yokota and S D Clouse, Springer-Verlag, Tokyo.

Clouse, S. D., Sasse, J. M. (1998). Brassinosteroids: essential regulators of plant growth and development. Annual Review of Plant Physiology and Plant Molecular Biology 49: 427-451.

Friebe, A., Volz, A., Schmidt, J., Voigt, B., Adam, G., Schnabl, H. (1999). 24-Epi-secasterone and 24-epi-castasterone from *Lychnis viscaria* seeds. Phytochemistry 52: 1607-1610.

Friedrichsen, D., Chory, J. (2001). Steroid signaling in plants: from the cell surface to the nucleus. BioEssays 23 (11): 1028-1036.

Fujioka, S. (1999). Natural occurrence of brassinosteroids in the plant kingdom. In Brassinosteroids: Steroidal Plant Hormones, pp. 21-45. Eds A Sakurai, T Yokota and S D Clouse, Springer-Verlag, Tokyo.

Fujioka, S., Choi, Y-H., Takatsuto, S., Yokota, T., Li, J., Chory, J., Sakurai, A. (1996). Identification of castasterone, typhasterol, and 6-deoxotyphasterol from the shoots of *Arabidopsis thaliana*. Plant and Cell Physiology 37: 1201-1203.

Fujioka, S., Inoue, T., Takatsuto, S., Yanagisawa, T., Sakurai, A., Yokota, T. (1995). Identification of a new brassinosteroid, cathasterone, in cultured cells of *Catharanthus roseus* as a biosynthetic precursor of teasterone. Bioscience, Biotechnology and Biochemistry 59: 1543-1547.

Fujioka, S., Li, J., Choi, Y-H., Seto, H., Takatsuto, S., Noguchi, T., Watanabe, T., Kuriyama, H., Yokota, T., Chory, J., Sakurai, A. (1997). The *Arabidopsis deetiolated2* mutant is blocked early in brassinosteroid biosynthesis. Plant Cell 9: 1951-1962.

Fujioka, S., Noguchi, T., Sekimoto, M., Takatsuto, S., Yoshida, S. (2000a). 28-Norcastasterone is biosynthesized form castasterone. Phytochemistry 55: 97-101.

Fujioka, S., Noguchi, T., Watanabe, T., Takatsuto, S., Yoshida, S. (2000b). Biosynthesis of brassinosteroids in cultured cells of *Catharanthus roseus*. Phytochemistry 53: 549-553.

Fujioka, S., Noguchi, T., Yokota, T., Takatsuto, S., Yoshida, S. (1998). Brassinosteroids in *Arabidopsis thaliana*. Phytochemistry 48: 595-599.

Fujioka, S., Takatsuto, S., Yoshida, S. (2002). An early C-22 oxidation branch in the brassinosteroid biosynthetic pathway. Plant Physiology 130: 930-939.

Gamoh, K., Okamoto, N., Takatsuto, S., Tejima, I. (1990). Determination of traces of natural brassinosteroids as dansylaminophenylboronates by liquid chromatography with fluorimetric detection. Analytica Chimica Acta 228: 101-105.

Gamoh, K., Omote, K., Okamoto, N., Takatsuto, S. (1989). High-performance liquid chromatography of brassinosteroids in plants with derivatization using 9-phenanthrene-boronic acid. Journal of Chromatography 469: 424-428.

Griffiths, P.G., Sasse, J. M., Yokota, T., Cameron, D. W. (1995). 6-Deoxotyphasterol and 3-dehydro-6-deoxoteasterone, possible precursors to brassinosteroids in pollen of *Cupressus arizonica*. Bioscience, Biotechnology and Biochemistry 59: 956-959.

Grove, M. D., Spencer, G. F., Rohwedder, W. K., Mandava, N., Worley, J. F., Warthen Jr, J. D., Steffens, G. L., Flippen-Anderson, J. L., Cook Jr, J. C. (1979). Brassinolide, a plant growth-promoting steroid isolated from *Brassica napus* pollen. Nature 281: 216-217.

Ikeda, M., Takatsuto, S., Sassa, T., Ikekawa, N., Nukina, M. (1983). Identification of brassinolide and its analogues in chestnut gall tissue. Agricultural and Biological Chemistry 47: 655-657.

Ikekawa, N., Nishiyama, F., Fujimoto, Y. (1988). Identification of 24-epibrassinolide in bee pollen of the broad bean, *Vicia faba* L. Chemical Pharmaceutical Bulletin 36: 405-407.

Ikekawa, N., Takatsuto, S., Kitsuwa, T., Saito, H., Morishita, T., Abe, H. (1984). Analysis of natural brassinosteroids by gas chromatography and gas chromatography-mass spectrometry. Journal of Chromatography 290: 289-302.

Jang, M-S., Han, K-S., Kim, S-K. (2000). Identification of brassinosteroids and their biosynthetic precursors from seeds of pumpkin. Bulletin of the Korean Chemical Society 21: 161-164.

Kaur, S., Bhardwaj, R., Nagar, P.K. (2002). Isolation and characterization of brassinosteroids from immature seeds of *Camelia sinensis* (L.) O. Kuntze. In: 13th Congress of the Federation of European Societies of Plant Physiology, Hersonissos, Heraklion, Crete, Greece, Abstract 206.

Kim, S-K. (1991). Natural occurrences of brassinosteroids. In Brassinosteroids: Chemistry, Bioactivity and Applications, pp. 26-35. Eds H G Cutler, T Yokota and G Adam, American Chemical Society, Washington.

Kim, S-K., Abe, H., Little, C. H. A., Pharis, R. P. (1990). Identification of two brassinosteroids from the cambial region of scots pine (*Pinus silvestris*) by gas chromatography-mass spectrometry, after detection using a dwarf rice lamina inclination bioassay. Plant Physiology 94: 1709-1713.

Kim, S-K., Akihisa, T., Tamura, T., Matsumoto, T., Yokota, T., Takahashi, N. (1988). 24-Methylene-25-methylcholesterol in *Phaseolus vulgaris* seed: structural relation to brassinosteroids. Phytochemistry 27: 629-631.

Kim, S-K., Chang, S. C., Lee, E. J., Chung, W-S., Kim, Y-S., Hwang, S., Lee, J. S. (2000a). Involvement of brassinosteroids in the gravitropic response of priary root of maize. Plant Physiology 123: 997-1004.

Kim, S-K., Yokota, T., Takahashi, N. (1987). 25-Methyldolichosterone, a new brassinosteroid with tertiary butyl group from immature seed of *Phaseolus vulgaris*. Agricultural and Biological Chemistry 51: 2303-2305.

Kim, T-W., Park, S-H., Han, K-S., Choo, J., Lee, J. S., Hwang, S., Kim, S-K. (2000b). Occurrence of teasterone and typhasterol, and their enzymatic conversion in *Phaseolus vulgaris*. Bulletin of the Korean Chemical Society 21: 373-374.

Konstantinova, O. V., Antonchick, A. P., Oldham, N. J., Zhabinskii, V. N., Khripach, V. A., Schneider, B. (2001). Analysis of underivatized brassinosteroids by HPLC/APCI-MS. Occurrence of 3-epibrassinolide in *Arabidopsis thaliana*. Collection of Czechoslovak Chemical Communications 66: 1729-1734.

Mandava, N. B. (1988). Plant growth-promoting brassinosteroids. Annual Review of Plant Physiology and Plant Molecular Biology 39: 23-52.

Morishita, T., Abe, H., Uchiyama, M., Marumo, S., Takatsuto, S., Ikekawa, N. (1983). Evidence for plant growth promoting brassinosteroids in leaves of *Thea sinensis*. Phytochemistry 22: 1051-1053.

Motegi, C., Takatsuto, S., Gamoh, K. (1994). Identification of brassinolide and castasterone in the pollen of orange (*Citrus sinensis* Osbeck) by high-performance liquid chromatography. Journal of Chromatography A 658: 27-30.

Müssig, C., Altmann, T. (2001). Brassinosteroid signaling in plants. Trends in Endocrinology and Metabolism 12: 398-402.

Nomura, T., Kitasaka, Y., Takatsuto, S., Reid, J. B., Fukami, M., Yokota, T. (1999). Brassinosteroid/sterol synthesis and plant growth as affected by *lka* and *lkb* mutations of pea. Plant Physiology 119: 1517-1526.
Nomura, T., Nakayama, M., Reid, J. B., Takeuchi, Y., Yokota, T. (1997). Blockage of brassinosteroid biosynthesis and sensitivity causes dwarfism in garden pea. Plant Physiology 113: 31-37.
Nomura, T., Sato, T., Bishop, G. J., Kamiya, Y., Takatsuto, S., Yokota, T. (2001). Accumulation of 6-deoxocathasterone and 6-deoxocastasterone in *Arabidopsis*, pea and tomato is suggestive of common rate-limiting steps in brassinosteroid biosynthesis. Phytochemistry 57: 171-178.
Park, K-H., Park, J-D., Hyun, K-H., Nakayama, M., Yokota, T. (1994a). Brassinosteroids and monoglycerides in immature seeds of *Cassia tora* as the active principles in the rice lamina inclination bioassay. Bioscience, Biotechnology and Biochemistry 58: 1343-1344.
Park, K-H., Park, J-D., Hyun, K-H., Nakayama, M., Yokota, T. (1994b). Brassinosteroids and monoglycerides with brassinosteroid-like activity in immature seeds of *Oryza sativa* and *Perilla frutescens* and in cultured cells of *Nicotiana tabacum.* Bioscience, Biotechnology and Biochemistry 58: 2241-2243.
Park, K-H., Saimoto, H., Nakagawa, S., Sakurai, A., Yokota, T., Takahashi, N., Syono, K. (1989). Occurrence of brassinolide and castasterone in crown gall cells of *Catharanthus roseus.* Agricultural and Biological Chemistry 53: 805-811.
Park, K-H., Yokota, T., Sakurai, A., Takahashi, N. (1987). Occurrence of castasterone, brassinolide and methyl 4-chloroindole 3-acetate in immature *Vicia faba* seeds. Agricultural and Biological Chemistry 54: 3081-3086.
Park, S.C., Kim, T-W., Kim, S-K. (2000). Identification of brassinosteroids with 24*R*-methyl in immature seeds of *Phaseolus vulgaris*. Bulletin of the Korean Chemical Society 21: 1274-1276.
Park, S-H., Han, K-S., Kim, T-W., Shim, J-K., Takatsuto, S., Yokota, T., Kim, S-K. (1999). In vivo and in vitro conversion of teasterone to typhasterol in cultured cells of *Marchantia polymorpha*. Plant and Cell Physiology 40: 955-960.
Plattner, R. D., Taylor, S. L., Grove, M. D. (1986). Detection of brassinolide and castasterone in *Alnus glutinosa* (European alder) pollen by mass spectrometry/mass spectrometry. Journal of Natural Product 49: 540-545.
Sakurai, A., Fujioka, S. (1993). The current status of physiology and biochemistry of brassinosteroids. Plant Growth Regulation 13: 147-159.
Sasse, J. M. (1997). Recent progress in brassinosteroid research. Physiologia Plantarum 100: 696-701.
Sasse, J. M. (1999). Physiological actions of brassinosteroids. In Brassinosteroids: Steroidal Plant Hormones, pp. 137-161. Eds A Sakurai, T Yokota and S D Clouse, Springer-Verlag, Tokyo.
Schmidt, J., Altmann, T., Adam, G. (1997). Brassinosteroids from seeds of *Arabidopsis thaliana.* Phytochemistry 45: 1325-1327.
Schmidt, J., Böhme, F., Adam, G. (1996). 24-Epibrassinolide from *Gypsophila perfoliata*. Zeitschrift für Naturforschung 51c: 897-899.
Schmidt, J., Himmelreich, U., Adam, G. (1995a). Brassinosteroids, sterols and lup-20(29)-en-2$\alpha$,3$\beta$,28-triol from *Rheum rhabarbarum*. Phytochemistry 40: 527-531.
Schmidt, J., Kuhnt, C., Adam, G. (1994). Brassinosteroids and sterols from seeds of *Beta vulgaris.* Phytochemistry 36: 175-177.
Schmidt, J., Porzel, A., Adam, G. (1998). Brassinosteroids and a pregnane glucoside from *Daucus carota.* Phytochemical Analysis 9: 14-20.
Schmidt, J., Spengler, B., Yokota, T., Adam, G. (1993a). The co-occurrence of 24-epi-castasterone and castasterone in seeds of *Ornithopus sativus.* Phytochemistry 32: 1614-1615.
Schmidt, J., Spengler, B., Yokota, T., Nakayama, M., Takatsuto, S., Voigt, B., Adam, G. (1995b). Secasterone, the first naturally occurring 2,3-epoxybrassinosteroid from *Secale cereale*. Phytochemistry 38: 1095-1097.
Schmidt, J., Voigt, B., Adam, G. (1995c). 2-Deoxybrassinolide – a naturally occurring brassinosteroid from *Apium graveolens*. Phytochemistry 40: 1041-1043.
Schmidt, J., Yokota, T., Adam, G., Takahashi, N. (1991). Castasterone and brassinolide in *Raphanus sativus* seeds. Phytochemistry 30: 364-365.
Schmidt, J., Yokota, T., Spengler, B., Adam, G. (1993b). 28-Homoteasterone, a naturally occurring brassinosteroid from seeds of *Raphanus sativus*. Phytochemistry 34: 391-392.

Schneider, B. (2002). Pathways and enzymes of brassinosteroid biosynthesis. Progress in Botany 63: 286-306.
Schneider, J. A., Yoshihara, K., Nakanishi, K., Kato, N. (1983). Typhasterol (2-deoxycastasterone): a new plant growth regulator from cat-tail pollen. Tetrahedron Letters 24: 3859-3860.
Shimada, K., Abe, H., Takatsuto, S., Nakayama, M., Yokota, T. (1996). Identification of castasterone and teasterone from seeds of canary grass (*Phalaris canariensis*). Recent Research and Development in Chemistry and Pharmaceutical Sciences 1: 1-5.
Soeno, K., Kyokawa, Y., Natsume, M., Abe, H. (2000). Teasterone-3-O-β-D-glucopyranoside, a new conjugated brassinosteroid metabolite from lily cell suspension cultures and its identification in lily anthers. Bioscience, Biotechnology and Biochemistry 64: 702-709.
Spengler, B., Schmidt, J., Voigt, B., Adam, G. (1995). 6-Deoxo-28-norcastasterone and 6-deoxo-24-epicastasterone – two new brassinosteroids from *Ornithopus sativus*. Phytochemistry 40: 907-910.
Suzuki, H., Fujioka, S., Takatsuto, S., Yokota, T., Murofushi, N., Sakurai, A. (1993). Biosynthesis of brassinolide from castasterone in cultured cells of *Catharanthus roseus*. Journal of Plant Growth Regulation 12: 101-106.
Suzuki, H., Fujioka, S., Takatsuto, S., Yokota, T., Murofushi, N., Sakurai, A. (1994a). Biosynthesis of brassinolide from teasterone via typhasterol and castasterone in cultured cells of *Catharanthus roseus*. Journal of Plant Growth Regulation 13: 21-26.
Suzuki, H., Fujioka, S., Takatsuto, S., Yokota, T., Murofushi, N., Sakurai, A. (1995). Biosynthesis of brassinosteroids in seedlings of *Catharanthus roseus, Nicotiana tabacum* and *Oryza sativa*. Bioscience, Biotechnology and Biochemistry 59: 168-172.
Suzuki, H., Fujioka, S., Yokota, T., Murofushi, N., Sakurai, A. (1994b). Identification of brassinolide, castasterone, typhasterol, and teasterone from the pollen of *Lilium elegans*. Bioscience, Biotechnology and Biochemistry 58: 2075-2076.
Suzuki, H., Inoue, T., Fujioka, S., Takatsuto, S., Yanagisawa, T., Yokota, T., Murofushi, N., Sakurai, A. (1994c). Possible involvement of 3-dehydroteasterone in the conversion of teasterone to typhasterol in cultured cells of *Catharanthus roseus*. Bioscience, Biotechnology and Biochemistry 58: 1186-1188.
Suzuki, Y., Yamaguchi, I., Takahashi, N. (1985). Identification of castasterone and brassinone from immature seeds of *Pharbitis purpurea*. Agricultural and Biological Chemistry 49: 49-54.
Suzuki, Y., Yamaguchi, I., Yokota, T., Takahashi, N. (1986). Identification of castasterone, typhasterol and teasterone from the pollen of *Zea mays*. Agricultural and Biological Chemistry 50: 3133-3138.
Takatsuto, S. (1994). Brassinosteroids: distribution in plants, bioassays and microanalysis by gas chromatography - mass spectrometry. Journal of Chromatography A 658: 3-15.
Takatsuto, S., Abe, H., Gamoh, K. (1990a). Evidence for brassinosteroids in strobilus of *Equisetum arvense* L. Agricultural and Biological Chemistry 54: 1057-1059.
Takatsuto, S., Abe, H., Shimada, K., Nakayama, M., Yokota, T. (1996a). Identification of teasterone and 4-desmethylsterols in the seeds of *Ginkgo biloba* L. Journal of Japanese Oil Chemical Society 45: 1349-1351.
Takatsuto, S., Abe, H., Yokota, T., Shimada, K., Gamoh, K. (1996b). Identification of castasterone and teasterone in seeds of *Cannabis sativa* L. Journal of Japanese Oil Chemical Society 45: 871-873.
Takatsuto, S., Omote, K., Gamoh, K., Ishibashi, M. (1990b). Identification of brassinolide and castasterone in buckwheat (*Fagopyrum esculentum* Moench) pollen. Agricultural and Biological Chemistry 54: 757-762.
Takatsuto, S., Yokota, T., Omote, K., Gamoh, K., Takahashi, N. (1989). Identification of brassinolide, castasterone and norcastasterone (brassinone) in sunflower (*Helianthus annuus* L.) pollen. Agricultural and Biological Chemistry 53: 2177-2180.
Taylor, P. E., Spuck, K., Smith, P. M., Sasse, J. M., Yokota, T., Griffiths, P. G., Cameron, D. W. (1993). Detection of brassinosteroids in pollen of *Lolium perenne* L. by immunocytochemistry. Planta 189: 91-100.
Watanabe, T., Yokota, T., Shibata, K., Nomura, T., Seto, H., Takatsuto, S. (2000). Cryptolide, a new brassinolide catabolite with a 23-oxo group from Japanese cedar pollen / anther and its synthesis. Journal of Chemical Research (S) 18-19.
Yamamoto, R., Fujioka, S., Demura, T., Takatsuto, S., Yoshida, S., Fukuda, H. (2001). Brassinosteroid levels increase drastically prior to morphogenesis of tracheary elements. Plant Physiology 125: 556-563.

Yasuta, E., Terahata, T., Nakayama, M., Abe, H., Takatsuto, S., Yokota, T. (1995). Free and conjugated brassinosteroids in the pollen and anthers of *Erythronium japonicum* Decne. Bioscience, Biotechnology and Biochemistry 59: 2156-2158.

Yokota, T. (1997). The structure, biosynthesis and function of brassinosteroids. Trends in Plant Science 2: 137-143.

Yokota, T. (1999a). Brassinosteroids. In Biochemistry and Moleculat Biology of Plant Hormones, pp. 277-293. Eds P J J Hooykaas, M A Hall and K R Libbenga, Elsevier Science, London.

Yokota, T. (1999b). The history of brassinosteroids: discovery to isolation of biosynthesis and signal transduction mutants. In Brassinosteroids: Steroidal Plant Hormones, pp. 1-20. Eds A Sakurai , T Yokota and S D Clouse , Springer-Verlag, Tokyo.

Yokota, T., Arima, M., Takahashi, N. (1982a). Castasterone, a new phytosterol with plant-hormone potency, from chestnut insect gall. Tetrahedron Letters 23: 1275-1278.

Yokota, T., Arima, M., Takahashi, N., Crozier, A. (1985). Steroidal plant growth regulators, castasterone and typhasterol (2-deoxycastasterone) from the shoots of Sitka spruce (*Picea sitchensis*). Phytochemistry 24: 1333-1335.

Yokota, T., Arima, M., Takahashi, N., Takatsuto, S., Ikekawa, N., Takematsu, T. (1983a). 2-Deoxycastasterone, a new brassinolide-related bioactive steroid from *Pinus* pollen. Agricultural and Biological Chemistry 47: 2419-2420.

Yokota, T., Baba, J., Koba, S., Takahashi, N. (1984). Purification and separation of eight steroidal plant-growth regulators from *Dolichos lablab* seed. Agricultural and Biological Chemistry 48: 2529-2534.

Yokota, T., Baba, J., Takahashi, N. (1982b). A new steroidal lactone with plant growth-regulatory activity from *Dolichos lablab* seed. Tetrahedron Letters 23: 4965-4966.

Yokota, T., Baba, J., Takahashi, N. (1983b). Brassinolide-related bioactive sterols in *Dolichos lablab*: brassinolide, castasterone and a new analog, homodolicholide. Agricultural and Biological Chemistry 47: 1409-1411.

Yokota, T., Higuchi, K., Takahashi, N., Kamuro, Watanabe, T., Takatsuto, S. (1998). Identification of brassinosteroids with epimerized substituents and / or the 23-oxo group in pollen and anthers of Japanese cedar. Bioscience, Biotechnology and Biochemistry 62: 526-531.

Yokota, T., Kim, S. K., Fukui, Y., Takahashi, N., Takeuchi, Y., Takematsu, T. (1987a). Conjugation of brassinosteroids. In Conjugated Plant Hormones. Structure, Metabolism and Function, pp. 288-296. Eds K Schreiber, H R Schuette and G Sembdner, VEB Deutscher Verlag der Wissenschaften, Berlin.

Yokota, T., Kim, S. K., Fukui, Y., Takahashi, N., Takeuchi, Y., Takematsu, T. (1987b). Brassinosteroids and sterols from a green alga, *Hydrodictyon reticulatum*: configuration at C-24. Phytochemistry 26: 503-506.

Yokota, T., Koba, S., Kim, S. K., Takatsuto, S., Ikekawa, N., Sakakibara, M., Okada, K., Mori, K., Takahashi, N. (1987c). Diverse structural variations of the brassinosteroids in *Phaseolus vulgaris*. Agricultural and Biological Chemistry 51: 1625-1631.

Yokota, T., Matsuoka, T., Koarai, T., Nakayama, M. (1996). 2-Deoxybrassinolide, a brassinosteroid from *Pisum sativum* seed. Phytochemistry 42: 509-511.

Yokota, T., Morita, M., Takahashi, N. (1983c). 6-Deoxocastasterone and 6-deoxodolichosterone: putative precursors for brassinolide-related steroids from *Phaseolus vulgaris*. Agricultural and Biological Chemistry 47: 2149-2151.

Yokota, T., Nakayama, M., Wakisaka, T., Schmidt, J., Adam, G. (1994). 3-Dehydroteasterone, a 3,6-diketobrassinosteroid as a possible biosynthetic intermediate of brassinolide from wheat grain. Bioscience, Biotechnology and Biochemistry 58: 1183-1185.

Yokota, T., Nomura, T., Nakayama, M. (1997d). Identification of brassinosteroids that appear to be derived from campesterol and cholesterol in tomato shoots. Plant and Cell Physiology 38: 1291-1294.

Yokota, T., Ogino, Y., Takahashi, N., Saimoto, H., Fujioka, S., Sakurai, A. (1990). Brassinolide is biosynthesized from castasterone in *Catharanthus roseus* crown gall cells. Agricultural and Biological Chemistry 54: 1107-1108.

Yokota, T., Sato, T., Takeuchi, Y., Nomura, T., Uno, K., Watanabe, T., Takatsuto, S. (2001). Roots and shoots of tomato produce 6-deoxo-28-cathasterone, 6-deoxo-28-nortyphasterol and 6-deoxo-28-norcastasterone, possible precursors of 28-norcastasterone. Phytochemistry 58: 233-238.

Zaki,A. K., Schmidt, J., Hammouda, F. M., Adam, G. (1993). Steroidal constituents from pollen grains of *Phoenix dactylifera*. Planta Medica 59: A 613.

CHAPTER 2

JULIE CASTLE, TERESA MONTOYA AND GERARD J. BISHOP

# SELECTED PHYSIOLOGICAL RESPONSES OF BRASSINOSTEROIDS: A HISTORICAL APPROACH

Brassinosteroids are endogenous growth promoting hormones that have a structure similar to steroids in mammals. For over 70 years mammalian steroids have been known to function as hormones and, given that steroidal hormones have also been found in many multicelluar organisms, it was logical to assume a hormonal role for steroids in plants. However, even though the existence of growth stimulating chemicals had been observed in the reproductive tissues of plants in 1849, it was not until 1979 that the first plant steroidal hormone was isolated, characterised and named brassinolide. Brassinosteroids have been found in diverse species of plants and the physiological role they play in growth and development has been considerably researched. Here we review these initial physiological experiments using BRs in the context of their agronomic potential. Initial experiments showed that seeds treated with brassinosteroids increased seed yield and plant size, particularly in the case of smaller and slower growing plants. However, subsequent larger field-test trials in the United States of America resulted in disappointing results with little economic significance. This, in addition to the identification of brassinosteroid mutant plants rescued by the exogenous application of brassinolide, indicated an essential role for brassinosteroids in normal plant development. These physiological studies have shown that although brassinosteroids function in growth and development, their complex interaction with other plant hormones and environmental signals indicate that more detailed studies are needed in order to elucidate more fully the potential agronomic benefit of BRs in crop production. .

## INTRODUCTION

Plants have an inherent ability to sense and respond to stimuli in their environment; this is crucial to their survival as they depend on adaptation to environmental signals for optimal growth and development. Such an adaptation brings about many mechanisms, including altering the levels of endogenous chemicals/hormones. Brassinosteroids (BRs) are steroid hormones that play a crucial role in plant growth and development. The role of BRs in plant growth and development has been covered in many reviews (Hooley, 1996; Yokota, 1997; Clouse and Sasse, 1998; Altmann; 1998a, b, 1999; Li and Chory, 1999; Bishop and Yokota 2001; Bishop and Koncz 2002). The aim of this review is to focus on selected discoveries of the physiological responses of plants to exogenous application of BRs with particular reference to crops.

*S.Hayat and A.Ahmad (eds.), Brassinosteroids,* 45-68.

Due to the necessary space constraint we restrict our coverage of the advances in *Arabidopsis* to those that help in clarifying the defects in mutants and the role of BRs in crops.

The origin of BR research can be considered to have started either in 1968 in Japan (Marumo *et al.,* 1968) or in 1970 in the USA (Mitchell *et al.,* 1970). However, there are several other citations in literature going back as early as 1849 reporting growth stimulating properties of substances that may have been BRs. In 1849 spores of the pteridophyte *Lycopodium* were reported to promote the growth of stigmas (Gaertner, 1849). Other early observations in 1901 and 1902 include the discovery of induction of parthenocarpic development in fruit when live or dead pollen from other species was applied to the stigmas of certain cucurbits, grapes and orchids (Massast, 1902). To develop an understanding of the nature of the growth promoting substance found in pollen, Mitchell and Whitehead (1941) studied growth responses and histological changes of cut and intact plants to a pollen extract. The crude pollen extract was obtained by the addition of ether to pollen collected from *Zea mays* and after the ether was evaporated a 'fatty residue' was obtained. The residue was mixed with lanolin and applied to specific vegetative parts of various species of plants. This resulted in internodal elongation that was shown to be a consequence of increase in cell length rather than an increase in cell number (Mitchell and Whitehead, 1941). In the same bioassay internode elongation induced by the indole acetic acid (IAA) was less and growth occurred for a shorter period of time. Interestingly, it was also found that treated plants grown in light conditions showed the greatest increase in linear growth, whereas treated plants grown in conditions that favoured etiolation showed little or no difference in internodal elongation compared with untreated, control plants. However, this was not developed further until 1983 (Krizek and Mandava, 1983a,b), and a more detailed account of this will be provided later in this review.

Mitchell's subsequent studies involved obtaining bioactive extracts from immature bean seeds using the same method as that to obtain pollen extracts (Mitchell *et al.,* 1951). The extract was applied to the first, second, third, fourth and fifth internodes of bean plants where it was found that the most significant increase in length occurred in the second internode in comparison with untreated plants as shown in Figure 1. This bioassay is referred to as the bean second internode bioassay and used for detection of BR response. However, in 1971 chromatographical tests on hormone extracts from pollen and immature bean seeds indicated that the extracts are qualitatively different (Mitchell *et al.,* 1971). The bioactive extract from pollen was invisible in white light, whereas the extract from seeds was readily visible. Also the extracts collected from pollen were not visible under UV light and in contrast fluorescence was observed in the seed extract (Mitchell *et al.,* 1971).

*Pollen extract – a new plant hormone?*

As mentioned previously evidence for the presence of a new type of plant hormone also came from Japan when Marumo *et al.*(1968) isolated three active fractions from leaves of Distylium racemosum Sieb et Zucc, an evergreen tree known in Japan as

'Isunoki'. Small insect galls of this tree were studied in an attempt to isolate and identify the substance(s) that induce the cell proliferation/enlargement observed in the insect gall (Marumo *et al.*, 1968). The active substance was not only found in the insect galls, but also in young leaves. Extracts were obtained and two fractions A1

*Figure 1. The Bean Second Internode Assay. Initial experiments using 10µg of the brassin complex obtained from Brassica napus pollen and applied to the bean second internode in a lanolin-based liquid resulted in elongation of the bean's second internode. Left, the control plant treated with lanonin minus the brassin complex. Right, plant treated with lanolin and brassin complex. Drawing based on results from Mitchell et al., (1970)*

and A2 were identified. Further purifications were carried out before chromatographical $R_F$ values for each fraction were determined and it was found these values were clearly different from known phytohormones. Biological activities of the Distylium factors were investigated using known bioassays for plant growth promoters. The active fractions were found to be more active in the rice lamina inclination test than IAA. This assay was developed as an assay for auxin in the early 1960s when rice leaves were excised and floated in distilled water or auxin solutions (Maeda, 1960). The inclination is defined as the variation of the angle between laminae and sheathes after treatment with varying concentrations of auxin. Large inclination angles were induced by high concentrations of auxin (Maeda, 1965). In cytokinin bioassays activity of onc factor was found to be almost equal to cytokinin but in GA bioassays the factors showed a negative response. It was concluded that the fractions contained a new type of plant hormone that has both auxin and cytokinin-like activity (Marumo *et al.*, 1968).

In 1970, in the USA, a pollen extract derived from *Brassica napus* was isolated and characterised as a lipophilic compound with hormone activity (Mitchell *et al.,* 1970). The extract was assayed for activity using the bean second internode test and was found to induce elongation of the second internodes of bean seedlings by an average of 155 mm at a dose of 10 μg per plant compared with 12 mm growth for untreated plants. The response was histologically different to that induced by gibberellic acid. The bioactive pollen extract was then referred to as brassin, derived from the name of the plant species in which it was first identified (Mitchell *et al.,* 1970).

Initial experiments using brassins/BRs on young bean plants, *Phaseolus vulgaris* L., Pinto variety, and young Siberian elm trees, *Ulmus pumila* L., resulted in two different types of growth response (Mitchell and Gregory, 1972). An early response is that internodal growth is increased in the treated plant. This is followed by an enhancement of overall plant growth. This had not been demonstrated previously using either plant hormones or synthetic plant growth regulators. In particular, it was noted that smaller, slower growing plants responded to a greater extent than vigorously growing plants. This was also seen in experiments with barley seeds where 50% of seeds were pre-treated by soaking in brassin-containing solution and 50% were left untreated. The pre-treatment significantly reduced the phenotypic variation of seedlings (Gregory, 1981). Agricultural implications of this observation warranted the United States Department of Agriculture (USDA) to initiate a special program to process 227 kg of bee-collected Brassica napus (rape) pollen (Steffens, 1991). This program was not without its critics (Milborrow and Pryce 1973). However, the potential agronomic importance of this research ensured sufficient funding and it was rewarded when Grove *et al*., (1979) isolated and characterised brassinolide (BL) as a plant growth-promoting steroid. The structure of BL was established using X-ray analysis and shown to have similarities with ecdysome, an insect-moulting hormone (Grove *et al*., 1979). BL was tested using the bean second internode assay and it was found that at concentrations as low as 10 ng per plant there was approximately a 200% increase in length of the internode in comparison to control plants (Grove *et al*., 1979). This increased size could be so dramatic that the enhanced cell divisions resulted in extreme curvature and even splitting of the internode (Figure. 2). Identification of BRs was not confined to *Brassica napus*; reports of brassinosteroid-like active substances were being reported in a variety of higher plants, the first of which was *Thea sinensis* (green tea) from where BL and its 6-keto analogue, castasterone, a precursor of BL were subsequently identified (Morishita *et al*., 1983).

*Development of a specific bioassay for BRs*

During this early period, experiments in Japan were aimed at establishing a specific bioassay for BL. Using the lamina inclination of rice plants it was found that even at extremely low concentrations of BL (0.5 ng/ml) the angle attained between lamina and sheath could reach $140^0$. Applications of IAA and GA, at much higher concentrations, only resulted in slight effects of the lamina, thus the rice lamina inclination bioassay for detection of BRs was proposed (Wada et al., 1981). This was corroborated in

Canada where when using two varieties of dwarf rice, auxin was shown to promote the BL-induced bending of the rice lamina (Takeno and Pharis, 1982).

In USA an alternative method was proposed to determine the biological activity of BL. In this bioassay the elongation of epicotyls of young mung beans was measured following the application of compounds either in solution or suspension (Gregory and Mandava, 1982). The mung bean epicotyl was found to be especially

*Figure 2. Splitting of the Bean Second Internode. Brassinolide, isolated from Brassica napus pollen, applied in concentrations as low as 10 ng per plant, resulted in splitting of the second internode. Drawing based on results in (Grove et al. 1979).*

sensitive to BL, a significant elongation of the epicotyl occurred with $10^{-7}$M BL after 1 hour. Responses to BL included epinastic curvature of the epicotyl and petioles, whereas treatment using other plant growth substances and other animal sex hormones did not produce this response (Gregory and Mandava, 1982).

Another assay extensively used to test for BR activity is the root elongation assay where seedlings or cuttings are either germinated on or transferred to medium containing BRs. Initially inhibition of root elongation was observed (Hewitt and Hillman, 1980) and then contradictory reports of growth promotion were made (Romani *et al.* 1983). Guan and Roddick (1988a, b) clarified that BRs inhibited root growth using cuttings, seedlings and roots excised from culture. Clouse *et al.* (1993) observed root inhibition in *Arabidopsis* and also that there were mutants insensitive to this response. They proved that BRs inhibit root growth in the auxin insensitive *Arabidopsis* mutant *axr1*, which meant that they could use it as a screen to select specifically for BR insensitive plants. Since then, this test has been widely used to screen for BR insensitive mutants (Clouse *et al.,* 1996; Koka *et al.,* 2000; Montoya *et al.*, 2002).

## FIELD TRIALS USING BRs

Motivated by the agronomic potential of BRs field trials began using brassins on crops. Initial field trials held in 1973 showed that field-grown vegetables and cereals produced larger seed yields and/or larger plants in response to brassins (Steffens, 1991). However, one major problem was the difficulty in obtaining an adequate supply of brassins, so a principal part of the project was devoted to improving the method of synthesis.

In 1974, a coordinated programme at Beltsville was initiated; it consisted of three parts, purification, bean second internode assay and field tests in the U.S. and Brazil (Steffens, 1991). Between 1974-1975 field trials commenced in Brazil in collaboration with USA. Seeds were incubated in brassin solutions containing 150 ppm, 300 ppm and 450 ppm and solvent only control. After incubation the seeds were dried and sent to Brazil where no significant differences were found in seed yield or plant size. Similar experiments in USA resulted in an increase of 0.5% in yield, which was considered economically insignificant. Similar results were obtained using barley, so it was concluded that yield did not increase after seeds were pre treated with brassins, before germination. Because this method was unsuccessful, therefore, various other methods were tested and the most efficient was found to be spraying the plants with a brassin solution. Field trials on radishes, leafy vegetables and potatoes were performed using this application method resulting significant increase in yield (Steffens, 1991).

The isolation of BL, the most bioactive BR, provided the knowledge required for its chemical synthesis, which was reported, simultaneously by Fung and Siddall (1980) and Ishiguro *et al.,* (1980). As production of BL was now possible it enabled considerable advances in experimentation to be made on agricultural crops. However, results from field trials have not been reported widely, many being in native languages. Reviews summarising this work are available (Steffens, 1991; Ikekawa and Zhao, 1991; Khripach, 1999) and highlights taken from these are provided below.

Experiments conducted in USA, where BRs were sprayed on seedlings at regular intervals resulted in yield improvements in many but not all crops (Steffens, 1991): lettuce (25-32%), radishes (20%), peppers (9%), tomatoes (no differences), beans (6%) and corn (no differences). In Japan, Takematsu conducted extensive field trials on wheat where spraying of epiBL at flowering resulted in increasing panicle weight by ~ 30% and the number of seeds per panicle increasing by 30% (Ikekawa and Zhao, 1991). When flowers were sprayed with epiBL the yield of rice increased by 11 % and in rapeseed and soybean a 10-20% increase in yield was observed (Ikekawa and Zhao, 1991). However, environmental conditions and the method of application had significant effect on the results with BL showing improved crop performance when crops were grown in more stressful conditions thus it was proposed that BL could be called a "stress hormone".

Japan and China collaborated in 1985 to study the practical applications of BL. The results were summarised by Ikekawa and Zhao (1991). Preliminary experiments were carried out at the Shanghai Institute of Plant Physiology. Large and small-scale field tests were also performed at this institute, obtaining very promising results with wheat, corn, watermelon, cucumber, grapes and tobacco. Between 1985-1990, field trial experiments were extended to other stations in Shangai, Henan and Zhejiang. All the results reported were encouraging, for example, in wheat, yield increased between 1% and 18%, depending on environmental conditions. An increase in disease resistance was also observed. In corn, yield increased between 10 and 20%

following BL application. In tobacco, root weight increased between 15% and 90%, the leaf area between 10-20% and the nicotine content between 40 and 70%. Experiments with watermelon showed that BL increased the yield by 10-20% by promoting fruit setting. In cucumber a 10-20% yield increase was detected with the application of BL. Grape yield increased between 66% and 30% (depending on the concentration used) due to an increase in the number of grapes per cluster. In general, reports shown by Chinese scientists were more positive than those from Japan. The reason for these differences could be due to differences in agricultural conditions between the two countries, i.e. differences in soil, plant species, cultural practices and/or climate (Ikekawa and Zhao, 1991).

Other countries have also been carrying out field trials in the eighties and nineties such as Russia, Belarus, Germany, Moldova, Uzbekistan, China, and The Ukraine. Khripach (1999) summarised the results. The report includes crops such as rice, wheat, rye, corn, buckwheat, oats, soybean and pea and a crop enhancement ranging from 6% to 63% depending on the year, the location and the method of application. The typical yield increase was about 17-20%. Better results were obtained in regions where drought conditions predominated, whereas in the Ukraine, where the soil is richer, the yield increases were not so dramatic.

## PHYSIOLOGY

### *Interactions with hormones*

Initial experiments in studying the growth promoting properties and other physiological responses of plants to brassins was complicated by the fact that the extract may be a mixture of hormones. To ascertain whether brassins acted independently of other plant hormones an etiolated bean hypocotyl test was used (Yopp *et al.*, 1979). This test demonstrated that brassins and IAA when applied to excised etiolated bean hypocotyls exhibited similar effects on dark-grown seedlings. Likewise brassins and GA promoted elongation of red light-inhibited hypocotyls. Following the isolation of BL, Yopp *et al.* (1981) and Mandava *et al.* (1981) were able to develop further the uniqueness of BR activity using BRs in auxin, GA and cytokinin bioassays. Here it was shown that BL did elicit certain auxin-like responses in some bioassays, however, it was ineffective in others and could even elicit the completely opposite response to IAA. Likewise, BL could induce GA- or cytokinin-type responses, but what was clearly apparent was that BL treated samples did not respond in the same manner as those treated with any of the other plant hormones used. BL application to plants consistently promoted the elongation of stems responsive to IAA or $GA_3$: maize mesocotyl, pea epicotyl, azuki epicotyl sections (Yopp *et al.*, 1981), bean hypocotyl, cucumber hypocotyl, etiolated pea epicotyl tips, (Mandava *et al.*, 1981) and bean seedlings (Gregory and Mandava, 1982). BL decreased the formation of adventitious roots and retarded the opening of the hypocotyl hook. Certain responses were due to a synergistic action of BL and IAA or in an additive manner with $GA_3$. Some of the effects seen by exogenous application of

BL may reflect ethylene responses, particularly those similar to responses induced by cytokinins, as cytokinins stimulate ethylene synthesis (Mandava, 1988). BL was also shown to inhibit senescence and abscission promoted by abscisic acid (ABA), suggesting that BR and ABA act antagonistically (Mandava, 1988). Although these early experiments showed that BRs exerted physiological effects either synergistically or in an additive manner with other hormones, BRs were not generally accepted as plant hormones until the biosynthesis was widely understood and BR mutants had been discovered. Subsequent research into the mechanism of BR morphological responses revealed that treatment with BL does not have a direct effect on IAA uptake, metabolism or cell-to-cell transport, though clearly it is involved in auxin activity (Cohen and Meudt, 1983). It had previously been demonstrated that auxin might stimulate the production of ethylene (Yu and Yang, 1979). Further evidence for synergistic action of BRs and IAA was gained when it was determined that BRs stimulate auxin-induced ethylene production in mung bean hypocotyls (Schlagnhaufer *et al.,* 1984; Arteca, 1984; Arteca *et al.,* 1984).

In later experiments with pakchoi (*Brassica chinensis*), cell elongation of the hypocotyls induced by BRs was found to occur with little or no change in the mechanical properties of cell walls but with an increase in wall relaxation properties and a dilution of the osmotic pressure of the cell sap (Wang *et al.,* 1993). It was suggested that the molecular mechanisms of BR-induced cell elongation may be different from those of auxin- and GA-induced elongation, which may produce the synergistic and additive effects previously mentioned. However, Tominaga *et al.* (1994) claimed a modification of cell wall properties occurred, induced by BRs in *Cucurbita maxima* hypocotyl inner tissues. They suggested the effect of BR was mediated by some aspects of the metabolism of cell walls in the inner tissues. This was further developed by Mayumi and Shibaoka (1995). They showed that BL regulated cell expansion via at least two processes, a microtubule dependent process and a microtubule independent process. They hypothesised that regulation of cell expansion by BL occurred by the same mechanisms as that of GA. This occurs by arranging cortical microtubules transversely to the cell axis causing the arrangement of newly deposited cellulose microfibrils in the same orientation. It was found that BL, similar to auxin, induced elongation by a microtubule independent process and once elongation is induced, BL acts in a way similar to GAs in a microtubule dependent manner.

Cell expansion is usually associated with acidification of the incubation medium caused by an increase in activity of the $H^+$ pump as a response to endogenous hormones (Cerana *et al.,* 1983). Cell elongation due to exogenous application of BRs was thought to be associated with an increase in the capacity of the tissue to acidify the medium in a manner similar to that of IAA, suggesting the decrease in pH in the wall space allows wall loosening and growth (Cerana *et al.,* 1983). In order to differentiate between the activities of IAA and BRs on the proton pump Romani *et al.* (1983) examined maize root segments, as IAA has an inhibitory effect on root growth and proton extrusion in exact contrast to its effect on stem segments (Pilet *et al., 1979*). It was found that electrogenic proton secretions in roots were at the same concentrations as in stems, and that the mechanism of secretion was the same as in stems when growth was stimulated by the application of BRs. These findings implied that BR cell elongation is dependent on the decrease in pH of the cell wall space inducing loosening of the cell wall. However, root growth was later shown to be inhibited by BRs as stated previously in this review.

In radish the effect of exogenously applied BL on cotyledon growth was compared with that of benzyladenine (BA), a synthetic cytokinin (De Michelis and Lado, 1986). Cytokinin-induced cell expansion does not occur through the acidification of the cell wall, whereas BRs were found to promote acidification of the incubation medium suggesting that the stimulatory effect of BRs could depend on BRs ability to lower the pH in the wall space by stimulating electrogenic $H^+$ extrusion. Dahse *et al.* (1990) further investigated the effects of steroids on membrane potential and found evidence strongly suggesting BRs stimulated the electrogenic pump.

Further evidence to implicate the role of BRs in $H^+$ pump-driven cell elongation was obtained when mutants in a $H^+$-ATPase activity were discovered. The *det*3 *Arabidopsis* mutants are dwarfs and found to develop as light-grown plants when grown in the dark (Schumacher *et al.,* 1999). *DET*3 encodes a subunit of the vacuolar $H^+$-ATPase that has a primary function in the acidification of endomembrane compartments. BRs applied exogenously to *det3* plants do not induce hypocotyl elongation, which is indicative of a role for DET3 in BR signal transduction or as a target in BR signalling. Histological studies of *det*3 mutant plants show they have a reduction in cell elongation.

BRs are also implicated in cell division and expansion of embryonic and post embryonic development in plants. Supporting evidence for this role was supplied by the discovery of the *fackel Arabidopsis* dwarf mutant defective in embryogenesis and post embryonic development. This mutant is deficient in C-14 sterol reductase activity and cannot be restored to wild type phenotype by BR-application suggesting a critical role for BRs in sterol regulation and signalling in developing embryos (Jang *et al.,* 2000)

### *Development and Environmental Interaction*

#### *Pollen tube growth and seedling germination*

The highest concentrations of BRs are found in pollen, the tissue in which it was first identified (Grove *et al.,* 1979) and it is logical to assume a role for BRs in pollen tube growth. One method employed to study pollen tube growth *in vitro* is the multiple hanging-drop technique and this was used to compare the response of pollen tubes between applications of fusicoccin, GA, auxin and BR over a range of concentrations (Hewitt *et al.,* 1985). Pollen collected from *Prunus avium* was suspended in germination medium with the appropriate concentrations of growth regulators, then added to wells of micro-titre trays. The trays were inverted during incubation allowing the cultures to form hanging drops. The effect of plant growth hormones on pollen tube growth was compared between nM and μM concentrations. It was found that 1 nM BR induced the maximum elongation of the pollen tube, 27 % above that of the control, in both light and dark conditions. In comparison with other plant growth promoters, BR was effective an order of magnitude below IAA and $GA_3$, although kinetin and ABA were not effective at all (Hewitt *et al.,* 1985).

BRs are also implicated in seed germination in *Arabidopsis*. Seed dormancy and germination are often regulated by the antagonistic actions of ABA and GA. ABA induces dormancy and inhibits germination and GA breaks dormancy and induces germination (Koornneef and Karssen, 1994). GA mutants often require exogenous applications of GA in order for germination to occur. However, BR has been shown to rescue the germination phenotype of GA biosynthetic mutants and a GA perception mutant (*sleepy* 1) suggesting BRs may play a role in germination (Stebber and McCourt, 2001)

*Formation of vascular tissue*

The first report of a role for BRs in differentiation of vascular tissue came in 1991 (Clouse and Zurek, 1991). Jerusalem artichoke (*Helianthus tuberosus)* cells transferred into xylem differentiation medium in the presence of auxin and cytokinins will differentiate into xylem elements in 72 - 96 hours. Very few vascular elements develop in the first 24 hours following transfer into this medium. However, nanomolar concentrations of BL included in the medium, resulted in a 10-fold increase in xylem differentiation, this was observed in the first 24 hours. Also significant increases in cell numbers were observed, indicating a role for BRs in cell division and differentiation (Clouse and Zurek, 1991).

*Zinnia elegans* has been used extensively to study the formation of xylem/tracheary elements, a process that has three distinct stages (Fukuda, 1997). BRs have been implicated in the transition between Stage II, and Stage III where secondary wall formation and cell death occurs (Fukuda, 1997). It had previously been shown that the effects of uniconazole (a putative BR biosynthesis inhibitor) prevents differentiation of *Zinnia* mesophyll cells into tracheary elements and this inhibition was overcome by exogenous BR application (Iwasaki and Shibaoka, 1991). Uniconazole appears to suppress the transcription of genes involved in the final stages of differentiation and this suppression was recovered by the exogenous application of BL (Yamamoto *et al.,* 1997). This suggests that BRs are synthesised immediately prior to secondary cell wall development and cell death and possibly induces entry into this stage (Yamamoto *et al.,* 1997).

*Interactions of BRs with light responses*

As mentioned previously early experiments showed a relationship between BR and light responses (Mitchell *et al.,* 1941). This was further addressed when Krizek and Mandava (1983a) studied the effect of light on stem elongation and morphogenesis in intact bean plants treated with BRs. They noted that bean plants grown under cool white fluorescent light or incandescent light and treated with BRs (5 μg) showed the characteristic splitting in the treated second internodes. Whereas, treated plants grown under far red light did not show this response. They concluded that BRs act to partially overcome the inhibitory effects of cool white fluorescent light on internode elongation, but the relationship between phytochrome and BR action was unresolved. BR treatment also increased chlorophyll content in the primary leaves of bean plants (Krizek and Mandava, 1983b) suggesting a role for BRs in influencing photosynthesis and the levels of photosynthate.

Symons *et al.* (2002) measured levels of BRs in light- and dark-grown pea seedlings and found that endogenous BR levels in light-grown wild type pea seedlings are increased in comparison with dark grown plants. This suggests that de-etiolation may be a secondary effect of the dwarf phenotype and not due to a role of BRs in etiolation.

*Stress responses*

In early preliminary agricultural trials it had been seen that applications of BRs to crops enhanced tolerance to less favourable environmental conditions. Plants usually respond to stress by synthesising stress related proteins. To demonstrate that this is the method employed by plants in response to exogenously applied BRs, Kulaeva *et al.,* (1991) studied the effect of BRs on total protein synthesis in the first leaves of 14-day-old wheat plants. The leaf segments were incubated in BL for 18 hours and then placed at elevated temperatures. In leaves, pre-treated with BL, protein synthesis at 43ºC was similar to that seen at 23ºC. However, in the untreated control leaf segments, total protein synthesis at 40ºC decreased 2.5 fold. It was concluded that, in wheat, physiologically active BRs activate protein synthesis, modify cell membrane properties and increase stress resistance.

Experiments, under saline stress conditions have been carried out in barley leaves (Kulaeva *et al*., 1991). Barley leaf segments from 10-day-old plants were pre-incubated in BL solution ($10^{-8}$M) for 2 hours, and then transferred to 0.5M NaCl solutions for 24 hours in the presence or absence of BL. Cells of the untreated control leaves had nuclei with condensed and diffused chromatin and well developed lamellae in the chloroplasts. Incubation in 0.5M NaCl solution resulted in the disruption of the chloroplast membrane system and strong chromatin condensation. However, cells incubated with BL had significantly reduced damage and had thus exerted a protective effect on cell ultrastruture under saline conditions. In China experiments were conducted on maize (*Zea mays*) seedlings to determine the effect of BL on growth recovery after chilling (He *et al.,* 1991). Maize seedlings are extremely sensitive to chilling during germination and early development where growth recovery is severely inhibited whilst the temperature is low and also when seedlings are restored to a normal temperature. The effect of BL on the growth and recovery of maize seedlings following chilling was examined. Coleoptiles after 8 days of chilling (0-3ºC) ceased to grow when returned to 25ºC. However, BL-treated seedlings showed growth recovery after the chilling treatment has ceased. Maize radicles are very sensitive to chilling and treatment with BL had a positive effect on re-growth after warming. Other early stress-related experiments have also shown that applications of BRs protects Chinese cabbage from pathogen attack and rice plants from salt injury (Cutler, 1991).

These and other preliminary studies suggested that, potentially, applications of BRs, as a chemical protectant to agricultural crops, could be of significant economic importance in less favourable climates. However, only limited research has been reported elucidating the mechanisms employed by BRs that enable plants to flourish under stressful conditions. Wilen *et al.*,(1995) endeavoured to provide a model system to show changes in gene expression during cold and heat acclimation. Exogenous applications of ABA to bromegrass cell suspension cultures has been shown to enable both cold and thermo tolerance (Chen and Gusta, 1983 and Robertson *et al.,* 1994). Further experiments have demonstrated that cells treated with ABA showed increased freezing and heat tolerance over cells treated with BL and control cells (Wilen *et al.,* 1995). BL-treated cells showed a minimal increase in freezing tolerance but increased cell viability after exposure to high temperature. As BL increases the expression of ABA inducible heat shock proteins it was concluded that the mechanisms by which BRs implement stress tolerance is similar to those implemented by ABA. However, transcript levels of another heat shock protein, hsp90, was increased after treatment with BL and decreased after treatment with ABA suggesting that the mechanisms employed by ABA and BL must only be partially similar (Wilen *et al.,* 1995).

Dhaubhadel *et al.* (1999) developed this further by showing that *Brassica napus* and tomato seedlings grown in the presence of BL are more tolerant to heat. Western blot analysis showed that there was an induction of heat shock proteins in BL-treated plants indicating that BL directly induces the expression of these proteins. Recent molecular studies have suggested BL-treated plants are able to endure heat stress by limiting the loss of some of the components involved in mRNA translation and this enables faster recovery of protein synthesis after heat stress (Dhaubhadel *et al.,* 2002).

## BR MUTANTS

The physiological activities of BRs that were initially studied were insufficiently distinguishable from those of other plant hormones in order to classify BRs as plant hormones. However, the discovery of mutants, defective in either BR biosynthesis or perception confirmed BRs as plant hormones. The typical phenotypic alterations of the BR-deficient mutants due to the loss of various biosynthetic activities are dwarfism (Figure 3) and a de-etiolated phenotype in dark-grown seedlings. Other phenotypic differences include delayed senescence, lack of xylem formation and often male sterility, which married well with the previously reported physiological responses. In the case of BR-sensitive mutants the phenotype can often be restored to wild type by exogenous application of BL (Figure 3), however, this is not possible in the case of BR-insensitive mutants. The phenotypic differences of BR mutants have been validated by the identification of brassinazole (BRZ), an azole-based inhibitor of BR biosynthesis, which can generate dwarfism that is phenotypically similar to BL-related mutants (Asami and Yoshida 1999). These discoveries have led to the acceptance of

BRs' essential role in plant growth and development.The *Arabidopsis* mutants det2 (Li *et al.*, 1996) and cpd (Szekeres *et al.*, 1996) were the first mutants to be shown to be defective in BR biosynthesis however dwarf1 was the first BR-mutant in *Arabidopsis* to be T-DNA tagged (Feldman *et al.* 1989). Although at that time dwarf1 was not characterised as a BR mutant. Arabidopsis BR-insensitive mutants were also identified based on their ability for root elongation in BL-containing medium. The dwarf brassinosteroid insensitive1 (bri1-1) was first characterised by Clouse *et al.* (1996) and many mutant alleles of this gene have been subsequently identified (Li and Chory, 1997; Noguchi *et al.*, 1999; Friedrichsen *et al.*, 2000; Bouquin *et al.*, 2001). BRI1 was isolated by positional cloning (Li and Chory, 1997) and interestingly BRI1 was not a nuclear steroid receptor as had previously been speculated but a leucine-rich repeat receptor kinase. The protein contains a signal peptide, an extra cellular domain, a trans-membrane domain and an intracellular kinase domain (Bishop and Koncz, 2002). More recently other mutants have been found including brassinosteroid insensitive-2, (BIN2) a gain-of-function mutation resulting in reduced BL signalling and ABA hypersensitivity (Li *et al.*, 2001).

*Figure 3. BR biosynthesis mutant (Photograph of 54 day old dumpy tomato plants sprayed with $10^{-6}$M BL. Left, dumpy control; middle dumpy sprayed with BL; right, wt control.)*

*History of mutants in agricultural crops*

Although many of the pioneering advances in BR research and BR mutants have been carried out in *Arabidopsis*, extensive research in many crop species such as tomato, rice, pea, *Vicia faba*, *Brassica napus* has provided insights into conservation of BR biosynthesis and signalling.

*Tomato biosynthesis mutants*

The tomato *dwarf* (*d*) mutation has been known since 1862 (Graham, 1959) and has been used to create varieties such as Dwarf Champion, Tiny Tim, Dwarf Stone and Tom Thumb but in general this mutation has had limited use in the commercial production of tomatoes (Bishop, 1995). *Dwarf* was isolated by transposon tagging using a derivative of the maize transposon *Activator* and is predicted to encode a cytochrome P450 enzyme (Bishop *et al.*, 1996). More recently it was shown that DWARF catalyses synthesis of castasterone, the penultimate step in BL biosynthesis (Figure 4) (Bishop *et al.,* 1999). *dumpy* is a spontaneous dwarf mutant with dark and wrinkled leaves (Hernández Bravo, 1967) (Figure 3). It was recovered in the trisomic F2 of triplo-7 x diploid stock of mixed origin. *dumpy* can be restored to a normal phenotype by the exogenous application of BRs (Figure 3) and has been associated with the C-23 hydroxylation of BRs and the conversion of (Figure 4) (Koka et al., 2000). However, the gene for this mutation has not yet been isolated.

*Tomato signalling mutants*

An extreme dwarf and infertile mutant was found in 1977 in a collection of *Lycopersicon pimpinellifolium* collected in Perú, (Yu, 1982). Due to its resemblence with other *curl* mutations of tomato it was called *curl3* (*cu3*). BR application in the root elongation assay indicated that *cu3* was a BR-insensitive mutant (Koka *et al.*, 2000). Two groups working in seemingly unrelated subject areas isolated the *Curl3* gene simultaneously. One lab isolated the tomato homologue of the *Arabidopsis BRI1* gene using degenerate primers with homology to the kinase region of *BRI1* (Montoya *et al.*, 2002). The other lab isolated *SR160* as the receptor for the peptide hormone systemin that is involved in wound signalling (Scheer and Ryan, 2002). Sequence analysis showed that CURL3 and SR160 are the same, which may provide a very exciting link between wounding and BR perception.

*Mutants of pea and bean*

*lka* and *lkb* are two ethylmethanosulfonate-induced dwarf mutations in pea that were generated in the eighties by Dr. K.K. Sidorova (T. Nomura, personal communication). Initially it was thought that they were GA insensitive mutants (Reid and Ross, 1989). They found that the two mutations were non-allelic and not allelic to known GA-mutations. Application of exogenous gibberellins had little phenotypic effect suggesting that *lka* and *lkb* should be classified as GA-insensitive mutations. It was later proposed that *lka* and *lkb* were involved in IAA metabolism, (Mckay *et al.,* 1994), although the phenotype could not be explained solely on the basis of IAA deficiency. Since neither of the above theories was conclusive it was hypothesised that they could be involved in BR biosynthesis or perception (Nomura *et al.*, 1997).

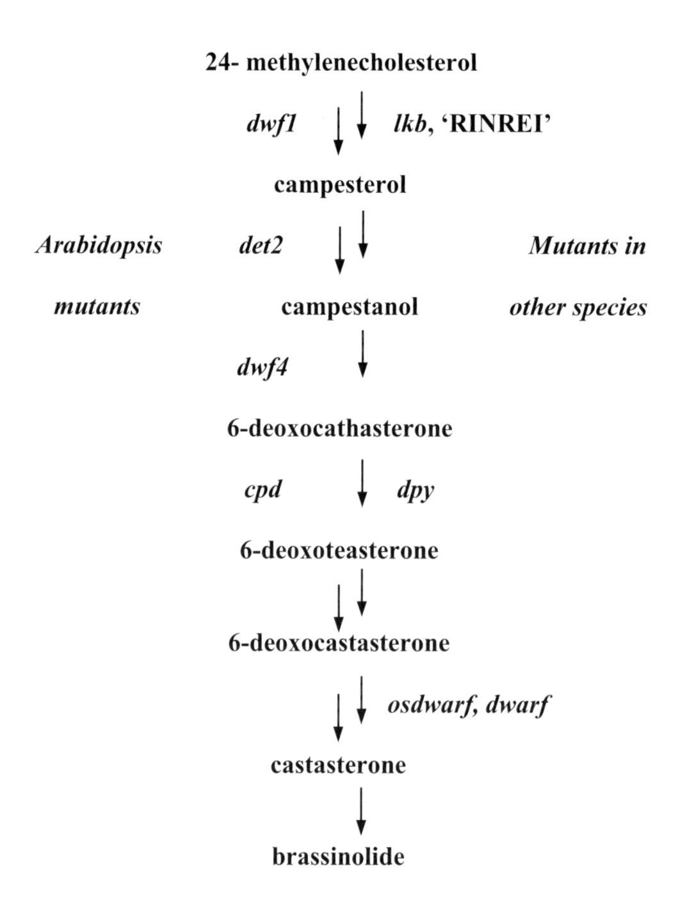


*Figure 4. BR biosynthesis pathway.*

[A simplified scheme of BR biosynthesis pathway including steps of late C-6 oxidation that lead to the synthesis of BL. Mutations in genes coding for key biosynthetic enzymes are highlighted. Only one allele from Arabidopsis has been selected. Tomato, D, DWARF (Bishop *et al.*, 1996; Bishop *et al.*, 1999); DPY, DUMPY (Koka *et al.*, 2000). Pea, LKB (Nomura *et al.*, 1997; Schultz *et al.*, 2001). *Vicia, rinrei*, (Fukuta *et al.*, 2002) Rice, OSDwarf, (Hong *et al.*, 2002 and Mori *et al.*, 2002). *Arabidopsis* DWF1, DWARF1 (Feldmann *et al.*, 1989); DET-2, DEETIOLATED-2, (Li *et al.*, 1996); DWF4, DWARF4 (Choe *et al.*, 1998); CPD, CONSTITUTIVE PHOTOMORPHOGENIC and DWARFISM (Szekeres *et al.*, 1996).]

Different components of the BRs pathway were quantified and the effect of BR application on growth was measured. It was found that the *lkb* phenotype was restored to wild type by the application of BL. This and further studies indicate that *lkb* is a BR biosynthetic mutant equivalent to the *dwarf1* mutant of *Arabidopsis* and is defective in the synthesis of campesterol (Nomura *et al.*, 1997; 1999; Schultz *et al.*, 2001).

The *lka* mutation did not respond to the application of BL and it was suggested that it was a BR insensitive mutant. Recently, the *lka* mutation has been confirmed to be a mutation in the *BRI1* gene of pea called *PsBRI1* (Nomura, *et al.*, 2002; Bishop and Koncz 2002). More recent has been the observation of a gamma radiation induced rinrei dwarf mutant of bean (Fukuta et al., 2002). The mutation is believed to be in a gene homologous to *dwarf*1 and *lkb* that catalyses the synthesis of campesterol.

### *Mutations in rice*

BR mutants in monocots are less well known and the mechanism of action is unclear. Two dwarf rice mutants *d61-1* and *d61-2* are BR-related mutants. These mutations were obtained independently by mutagenesis with N-methyl-N-Nitrosurea. They were initially classified independently as they were different phenotypically. However, crosses between the two mutants revealed they were allelic (Yamamuro *et al.,* 2000), both mutants showed reduced sensitivity to BL. *d61-1* and *d61-2* were found to be defective in OsBRI1 the rice homolog of the Arabidopsis BRI1 gene (Yamamuro *et al.*, 2000).

Recently rice dwarf mutants have been observed that are defective in BR biosynthesis (Hong et al., 2002 and Mori et al., 2002). These mutants have a more severe phenotype than the insensitive mutants and they have been found to be defective in osDWARF the rice equivalent of the tomato Dwarf gene. Further BR-related mutants are likely to be identified in the near future with the aid of the rice genome sequence and associated genome technologies.

## MOLECULAR BIOLOGY OF BRs' ACTION

### *Gene expression studies*

To gain insight into the physiological responses after BL application Clouse *et al.*, (1992) initiated research into BR regulated gene expression. They proposed two different theories for the mode of action of BRs: firstly, as BRs were steroids it was proposed that the mode of action would be similar to that in animals and secondly, BRs would act via the auxin receptor. They subsequently initiated an experiment to analyse the effect of BLs on the expression of auxin inducible genes to try and identify the mutant for the BR receptor. Following BR treatment, *SAUR* increased expression after 12 hours and *GH1* after 24 hours, but *JCW1* and *GH3* showed no differences in the conditions used. This suggested an independent pathway of action for auxins and BL (Clouse *et al.,* 1992).

This investigation was continued by studying how BL affects elongation and gene expression in digeotropica (*dgt*), a tomato mutant defective in auxin signalling (Zurek *et al.,*1994). In this mutant auxin does not induce either elongation or expression of the *SAUR* genes, whereas BRs induce elongation and an increase in *SAUR* gene transcript after 18 hours (Zurek *et al.*,1994). They also found the levels of auxin went down with BL treatment, suggesting that auxin and BL signalling pathways may be interconnected although not in way previously thought.

The first report on the cloning of a BR responsive gene was published in 1994 (Zurek and Clouse, 1994). The gene (*BRU1*) was isolated by cDNA differential library screening from apical epicotyl sections of soybean. The sections were incubated in 0.1 μM BR or buffer for 17 hours. Northern expression analysis revealed that BR increased the level of *BRU1* after 2 hours of incubation, auxin also induced *BRU1* but after 18-24 hours and not to the same extent as BL. Results from other experiments suggested that BRs regulate the expression of *BRU1* at the post-transcriptional level (Zurek and Clouse, 1994). *BRU1* encodes an endo-transglycosylase (XET) a wall-modifying enzyme involved in stem elongation (Zurek and Clouse, 1994): Interestingly *BRU1* expression is higher in actively growing tissues of soybean (Zurek and Clouse, 1994) and an increase in both the expression of *BRU1* and the plastic wall extensibility occurs, 2 hours after BR addition (Zurek *et al.*,1994). The *TCH4* XET gene in *Arabidopsis* appears not to be post transcriptionally but transcriptionally regulated by auxins and BRs (Xu *et al.*, 1995). Other genes have been found to be regulated by BL and could be involved in growth and development. Muñoz *et al.*, (1998) found that BRs promote expression of a new β-tubulin gene from *Cicer arietinum* in parallel with promoted growth of the epicotyls.

Yopp showed that BRs induce ethylene production and act synergistically with auxin to stimulate ethylene synthesis in etiolated mung bean seedlings (Yopp *et al.*, 1979). However, the molecular basis for this mechanism was not addressed until Yi *et al.* (1999) found that auxin and BRs regulate expression of three 1-aminocyclopropane 1-carboxylate (ACC) synthase genes in mung bean (*Vigna radiata* L.). ACC synthase catalyses the rate limiting step in ethylene biosynthesis. The genes are called *VR-ACS-1*, *VR-ACS-6* and *VR-ACS-7*. It was found that BRs interact indirectly with auxins or directly with the three genes in different pathways. They studied gene expression with different hormone combinations and suggested that BRs and auxins control the expression of these genes via interdependent pathways. BRs independently increased the transcript levels of VR-ACS7 and in an interaction with auxin activate expression of *VR-ACS6* whereas BRs inactivate the auxin induction of the gene *VRACS-1*.

However, histochemical analysis with a GUS-VRACS7 promoter fusion revealed that auxin induced GUS expression in the hypocotyls but BRs did not, therefore, three hypothesis were proposed: (1) the induction of *VR-ACS7* by BRs is controlled by post-transcriptional processes, (2) there were negative regulatory sequences in the promoter used or (3) there were some cis-trans acting factors functional only for the induction of the VR-ACS7 gene

*Genome wide transcript analysis*

Physiological effects of BRs have been extensively, studied as has been discussed in this review, however little is known about the cellular machinery that regulates these processes. Recently, the use of new technologies, which combine robotics with molecular biology (array technology), has allowed the development of expression profiles for BL action, which hopefully will allow an understanding of the molecular mechanism of BL action.

Müssig *et al.* (2002) studied the expression profile of 8,200 genes to identify BR-regulated transcripts. They classified the genes into four categories. 1) Genes which were BR-regulated (their expression was affected by BR application and in the dwarf mutant). 2) Genes affected only by BR treatment. 3) Genes affected by BR deficiency (their expression changed in the mutants only). 4) Genes, which were upregulated by BL treatment and down regulated in the wild type compared with the dwarf mutant. The results revealed many genes with a BL dependent pattern of expression, which were involved in functions such as BR biosynthesis, auxin response, nitrogen transport proteins, wax biosynthesis and pathogen defence, cell wall modification, phytohormone responses, stresses chromatin components and transcription factors. Future work from such studies using more complete arrays will provide a comprehensive understanding of how BRs transcriptionally regulate cellular processes.

## PROPSPECTS

What we have attempted to present here is not a complete history of BR research but some of the highlights in the advances in understanding BR-induced physiological changes. It has been shown that BRs are important in normal plant growth and development and that the physiological changes induced by BRs need to be considered in the context of other plant hormones. The agronomic potential resulting from utilising research into BRs cannot be underestimated. Various approaches can be utilised to enhance crop performance. One approach is to enhance plant growth via increasing BR concentrations in the plant by genetic manipulation of BL biosynthesis or signalling. It is possible that by over-expressing BR biosynthetic genes in specific organs, production and yield could be increased. An alternative to using BRs to generate larger plants and increase yield, it is conceivable that by using dwarfs (BR mutants) it could be possible to increase yield by increasing lodging resistance in a similar way in which GA mutants have been used (Evans 1993; Peng *et al.*, 1999; Milach *et al.,* 2002). Wheat production increased dramatically in the 1960s and 1970s with the introduction of new varieties, which were shorter than the old ones. These varieties were more resistant to damage by wind and rain and were later found to be affected in GA synthesis or perception (Peng *et al.*, 1999). It might be possible to use BR mutants to create plants possessing a dwarf habit but in which the wild type gene is over expressed only in the organ of choice. Alternatively antisense or RNAi technologies may be utilised to create 'semi-dwarfs', which would have the good qualities from being a dwarf (i.e. reduced lodging) without the inconvenient (low productivity). Schlüter *et al.* (2002) used this technology to create *Arabidopsis CPD* antisense plants and studied the effect on carbohydrate metabolism. The antisense inhibition of *CPD* gene led to less severe phenotypic alterations than the *cpd* mutant.

To carry out such experiments in a meaningful way it is advantageous to have more detailed knowledge about the physiological responses to BL and the effects of mutations in BR-related genes, as paradoxically the lack of BRs (mutants) and BL application seem to enhance plants' response to stress.

## ACKNOWLEDGEMENTS

We thank T. Nomura for his help and advice and personal communication on certain aspects of this review. For figure preparation we thank Vivian Evans for his expert drawings and Tony Pugh and Gwen Jenkins for photography. The Bishop Lab is currently supported by the BBSRC, HFSP and for JC a UWA/IBS studentship.

## REFERENCES

Altmann, T. (1998a). Recent advances in brassinosteroid molecular genetics. Current Opinion in Plant Biology 1: 378-383.

Altmann, T. (1998b). A tale of dwarfs and drugs: brassinosteroids to the rescue. Trends in Genetics 14(12): 490-495.

Altmann, T. (1999). Molecular physiology of brassinosteroids revealed by the analysis of mutants. Planta 208(1): 1-11.

Arteca, R. N. (1984). $Ca^{+2}$ acts synergistically with brassinosteroid and indole-3- acetic-acid in stimulating ethylene production in etiolated mung bean hypocotyl segments. Physiologia Plantarum 62(1): 102-104.

Arteca, R. N., Schlagnhaufer, C. (1984). The effect of brassinosteroid and 2,4-D-L-amino acid conjugates on ethylene production by etiolated mung bean segments. Physiologia Plantarum 62(3): 445-447.

Asami, T., Yoshida, S. (1999). Brassinosteroid biosynthesis inhibitors. Trends in Plant Science 4(9): 348-353.

Bishop, G. J. (1995). Transposon tagging in *Lycopersicon esculentum*. PhD Thesis University of East Anglia, UK

Bishop, G. J., Harrison, K., Jones, J. D. G. (1996). The tomato Dwarf gene isolated by heterologous transposon tagging encodes the first member of a new cytochrome P450 family. Plant Cell 8(6): 959-969.

Bishop, G. J., Koncz, C. (2002). Brassinosteroids and plant steroid hormone signalling. Plant Cell 14: 97-110.

Bishop, G. J., Nomura, T., Yokota, T., Harrison, K., Noguchi, T., Fujioka, S., Takatsuto, S., Jones, J. D. G., Kamiya, Y. (1999). The tomato DWARF enzyme catalyses C-6 oxidation in brassinosteroid biosynthesis. Proceedings of the National Academy of Sciences (USA) 96(4): 1761-1766.

Bishop G. J., Yokota, T. (2001). Plants steroid hormones, brassinosteroids: Current highlights of molecular aspects on their synthesis/metabolism, transport, perception and response. Plant and Cell Physiology 42(2): 114-120.

Bouquin, T., Meier, C., Foster, R., Nielsen, M. E., Mundy, J. (2001). Control of specific gene expression by gibberellin and brassinosteroid. Plant Physiology 127(2): 450-458.

Cerana, R., Bonetti, A., Marre, M. T., Romani, G., Lada, P., Marre, E. (1983). Effects of brassinosteroid on growth and electrogenic proton extrusion in asuki bean epicotyls. Physiologia Plantarum 59(1): 23 - 27.

Chen, T. H. H., Gusta, L. V. (1983). Abscisic acid induced freezing in cultured plant cells. Plant physiology 73: 71 - 75.

Choe, S. W., Dilkes, B. P., Fujioka, S., Takatsuto, S., Sakurai, A., Feldmann, K. A. (1998). The DWF4 gene of Arabidopsis encodes a cytochrome P450 that mediates multiple 22 alpha-hydroxylation steps in brassinosteroid biosynthesis. Plant Cell 10(2): 231-243.

Clouse, S. D., Hall, A. F., Langford, M., McMorris, T. C., Baker, M. E. (1993). Physiological and molecular effects of brassinosteroids on *Arabidopsis thaliana*. Journal of Plant Growth Regulation 12(2): 61-66.

Clouse, S. D., Sasse, J. M. (1998). Brassinosteroids: Essential regulators of plant growth and development. Annual Review of Plant Physiology and Plant Molecular Biology 49: 427-451.

Clouse S.D., Zurek, D. (1991). Molecular analysis of brassinolide action in plant growth and development. In Brassinosteroids: Chemistry, Bioactivity and Applications, pp 122-140. Eds H G Cuttler, T Yokota and G Adam. American Chemical Society, Washington.

Clouse, S. D., Langford, M., McMorris, T. C. (1996). A brassinosteroid-insensitive mutant in *Arabidopsis thaliana* exhibits multiple defects in growth and development. Plant Physiology 111: 671-678.

Clouse, S. D., Zurek, D. M., McMorris, T. C., Baker, M. E. (1992). Effect of brassinolide on gene-expression in elongating soybean epicotyls. Plant Physiology 100(3): 1377-1383.

Cohen, J. D., Meudt, W. J. (1983). Investigations on the mechanism of the brassinosteroid response. 1. Indole-3-acetic acid metabolism and transport. Plant Physiology 72: 691 - 694.

Cutler, H. G. (1991). Brassinosteroids through the looking glass. In Brassinosteroids: Chemistry, Bioactivity and Applications, pp 334-345. Eds H G Cuttler, T Yokota and G Adam. American Chemical Society, Washington.

Dahse, I., Sack, H., Bernstein, M., Petzold, U., Muller, E., Vorbrodt, H. M., Adam, G. (1990). Effects of (22s, 23s)-homobrassinolide and related compounds on membrane-potential and transport of Egeria leaf-cells. Plant Physiology 93(3): 1268-1271.

De Michelis, M. I., Lado, P. (1986). Effects of a brassinosteroid on growth and $H^+$ extrusion in isolated radish cotyledons: Comparison with the effects of benzyladenine. Physiologia Plantarum 68: 603 - 607.

Dhaubhadel, S., Browning, K. S., Gallie, D. R., Krishna, P. (2002). Brassinosteroid functions to protect the translational machinery and heat-shock protein synthesis following thermal stress. Plant Journal 29(6): 681-691.

Dhaubhadel, S., Chaudhary, S., Dobinson, K. F., Krishna, P. (1999). Treatment with 24-epibrassinolide, a brassinosteroid, increases the basic thermotolerance of *Brassica napus* and tomato seedlings. Plant Molecular Biology 40(2): 333-342.

Evans, L. T. (1993). Crop evolution adaptation and Yield. Cambridge University Press, Cambridge.

Feldmann, K. A., Marks, M. D., Christianson, M. L., Quatrano, R. S. (1989). A dwarf mutant of *Arabidopsis* generated by T-DNA insertion mutagenesis. Science 243: 1351-1354.

Friedrichsen, D. M., Joazeiro, C. A. P., Li, J. M., Hunter, T., Chory, J. (2000). Brassinosteroid-insensitive-1 is a ubiquitously expressed leucine-rich repeat receptor serine/threonine kinase. Plant Physiology 123(4): 1247-1255.

Fukuda, H. (1997). Tracheary element differentiation. Plant Cell 9: 1147 - 1156.

Fukuta, N., Fujioka, S., Takatsuto, S., Yoshida, S., Nakayama, M. (2002). A new brassinosteroid-deficient mutant of faba bean (*Vicia faba* L.). Plant and Cell Physiology 43:184-184.

Fung, S., Siddall, J. B. (1980). Stereoselective synthesis of brassinolide: a plant promoting steroidal lactone. Journal of the American Chemistry society, 102: 6580-6581.

Gaertner K.F., Versuche und Beobachtungen über die Bastardzeugung im Planzenreich. Stuttgart, 1849.

Graham, T. O. (1959). Impact of recorded mendelian factors on the tomato 1929-1959. Report Tomato Genetics Cooperative, 9, 37.

Gregory, L. E. (1981). Acceleration of plant growth through seed treatment with brassins. American Journal of Botany 68: 586 - 588.

Gregory, L. E., Mandava, N. B. (1982). The activity and interaction of brassinolide and gibberellic acid in mung bean epicotyls. Physiologia Plantarum 54: 239 - 243.

Grove, M. D., Spencer, G. F., Rohwedder, W. K., Mandava, N., Worley, J. F., Jr, J. D. W., Steffens, G. L., Flippen-Anderson, J. L., Cook, Jr. J. C. (1979). Brassinolide, a plant growth-promoting steroid isolated from *Brassica napus* pollen. Nature 281: 216-217.

Guan, M., Roddick, J. G. (1988a). Comparison of the effects of epibrassinolide and steroidal estrogens on adventitious root-growth and early shoot development in mung bean cuttings. Physiologia Plantarum 73: 426-431.

Guan, M., Roddick, J. G. (1988b). Epibrassinolide-inhibition of development of excised, adventitious and intact roots of tomato (*Lycopersicon esculentum*) - Comparison with the effects of steroidal estrogens. Physiologia Plantarum 74: 720-726.

He, R., Wang, G., Wang, X. (1991). Effects of brassinolide on growth and chilling resistance of maize seedlings. In Brassinosteroids: Chemistry, Bioactivity and Applications, pp 220-230. Eds H G Cuttler, T Yokota and G Adam. American Chemical Society, Washington.

Hernández-Bravo, G. (1967). Two new dwarf mutants and their linkage relations. Report Tomato Genetics Cooperative, 17, 30-31.
Hewitt, S., Hillman, J. R. (1980). Steroidal estrogens and adventitious root formation in *Phaseolus* cuttings. Annals of Botany 46: 153-164.
Hewitt, F. R., Hough, T., Oneill, P., Sasse, J. M., Williams, E. G., Rowan, K. S. (1985). Effect of brassinolide and other growth-regulators on the germination and growth of pollen tubes of *Prunus avium* using a multiple hanging-drop assay. Australian Journal of Plant Physiology 12(2): 201-211.
Hong, Z., Ueguchi-Tanaka, M., Shimizu-Sato, S., Inukai, Y., Fujioka, S., Shimada, Y., Takatsuto, S., Agetsuma, M., Yoshida, S., Watanabe, Y., Uozu, S., Kitano, H., Ashikari, M., Matsuoka, M. (2002). Loss-of-function of a rice brassinosteroid biosynthetic enzyme, C-6 oxidase, prevents the organized arrangement and polar elongation of cells in the leaves and stem. The Plant Journal 32, 4: 495-508.
Hooley, R. (1996). Plant steroid hormones emerge from the dark. Trends in Genetics 12(8): 281-283.
Ikekawa, N., Zhao, Y. (1991). Application of 24-epibrassinolide in agriculture. In Brassinosteroids: Chemistry, Bioactivity and Applications, pp 280-291. Eds H G Cuttler, T Yokota and G Adam. American Chemical Society, Washington.
Ishiguro M., Takatsuto, S., Morisaki, M. (1980). Synthesis of brassinolide, a steroidal lactone with plant-growth promoting activity. J.Chem. Soc. Chemical Communication 1980: 962-964.
Iwasaki, T., Shibaoka, H. (1991). Brassinosteroids act as regulators of tracheary-element differentiation in isolated zinnia mesophyll-cells. Plant and Cell Physiology 32: 1007 - 1014.
Jang, J. C., Fujioka, S., Tasaka, M., Seto, H., Takatsuto, S., Ishii, A., Aida, M., Yoshida, S., Sheen, J. (2000). A critical role of sterols in embryonic patterning and meristem programming revealed by the fackel mutants of *Arabidopsis thaliana*. Genes and Development 14(12): 1485-1497.
Khripach, V.A. Zhabinskii, V.N., Groot, A.E.(1999). Practical applications and toxicology. In Brassinosteroids: A new class of plant hormones. pp.325-346. Eds V.A. Khripach, V.N. Zhabinskii, and A.E. Groot. Academic Press. London.
Koka, C. V., Cerny, R. E., Gardner, R. G., Noguchi, T., Fujioka, S., Takatsuto, S., Yoshida, S., Clouse, S. D. (2000). A putative role for the tomato genes *DUMPY* and *CURL-3* in brassinosteroid biosynthesis and response. Plant Physiology 122(1): 85-98.
Koornneef M., Karssen CM., 1994. Seed dormancy and germination. In Arabidopsis Eds C R Somerville and E M Meyerowitz. pp. 313-334. Cold Spring Harbor Laboratory Press, Cold Spring Harbor, NY.
Krizek, D. T., Mandava, N. B. (1983a). Influence of spectral quality on the growth-response of intact bean-plants to brassinosteroid, a growth-promoting steroidal lactone .1. Stem elongation and morphogenesis. Physiologia Plantarum 57(3): 317-323.
Krizek, D.T., Mandava, N.B. (1983b). Influence of spectral quality on the growth response of intact bean plants to brassinosteroid, a growth-promoting steroidal lactone. 2. Chlorophyll content and partitioning of assimilate. Physiologia Plantarum 57: 324-329.
Kulaeva, O. N., Burkhanova, E. A., Fedina, A. B., Khokhlova, V. A., Bokebayeva, G. A., Vorbrodt, H. M., Adam, G. (1991). Effects of brassinosteroids on protein synthesis and plant-cell ultrastructure under stress conditions. In Brassinosteroids: Chemistry, Bioactivity and Applications. Eds H G Cutler, T Yokota and G Adam, pp. 141-155. American Chemical Society, Washington.
Li, J. M., Nagpal, P., Vitart, V., McMorris, T. C., Chory, J. (1996). A role for brassinosteroids in light-dependent development of *Arabidopsis*. Science 272(5260): 398-401.
Li, J. M., Chory, J. (1997). A putative leucine-rich repeat receptor kinase involved in brassinosteroid signal transduction. Cell 90(5): 929-938.
Li, J. M., Chory, J. (1999). Brassinosteroid actions in plants. Journal of Experimental Botany 50: 275-282.
Li, J. M., Nam, K. H., Vafeados, D., Chory, J. (2001). BIN2, a new brassinosteroid-insensitive locus in *Arabidopsis*. Plant Physiology 127(1): 14-22.
Maeda, E. (1960). Interaction of gibberellin and auxins in lamina joints of excised rice leaves. Physiologia Plantarum 13: 214 - 226.
Maeda, E. (1965). Rate of lamina inclination in excised rice leaves. Physiologia Plantarum 18: 813 - 827.
Mandava, N. B. (1988). Plant growth-promoting brassinosteroids. Annual Review of Plant Physiology and Plant Molecular Biology 39: 23-52.
Mandava, N. B., Sasse, J.M., Yopp, J.H. (1981). Brassinolide, a growth promoting steroidal lactone 2. Activity in selected gibberellin and cytokinin bioassays. Physiologia Plantarum 53(4): 453 - 461.

Marumo, S., Hattori, H., Abe, H., Nonoyama, Y., Munakata, K. (1968). The presence of novel plant growth regulators in leaves of *Distylium racemosum* Sieb. et Zucc. Agricultural and Biological Chemistry 32: 528 529.

Mayumi, K., Shibaoka, H. (1995). A possible double role for brassinolide in the reorientation of cortical microtubules in the epidermal cells of Azuki bean epicotyls. Plant and Cell Physiology 36: 173 - 181.

McKay, M. J., Ross, J. J., Lawrence, N. L., Cramp, R. E., Beveridge, C. A., Reid, J. B. (1994). Control of internode length in *Pisum sativum* - Further evidence for the involvement of indole-3-acetic-acid. Plant Physiology 106(4): 1521-1526.

Milach, S. C. K., Rines, H. W., Phillips, R. L. (2002). Plant height components and gibberellic acid response of oat dwarf lines. Crop Science 42(4): 1147-1154.

Milborow, B. V., Pryce, R. J. (1973). The brassins. Nature 243: 46.

Mitchell, J. D., Gregory, L. E. (1972). Enhancement of overall plant growth, a new response to brassins. Nature New Biology 239: 253 - 254.

Mitchell, J. D., Mandava, N. B., Worley, J. F., Plimmer, J. R., Smith, M. V. (1970). Brassins - a new family of plant hormones from rape pollen. Nature 225: 1065 - 1066.

Mitchell, J. D., Skaggs, D. P., Anderson, W. P. (1951). Plant growth-stimulating hormones in immature bean seeds. Science 114: 159 - 161.

Mitchell, J. D., Whitehead, M. R. (1941). Responses of vegetative parts of plants following application of extract of pollen from zea mays. Botanical Gazette 102: 770 - 791.

Mitchell J.W., Mandava N., Worley J.F., Drowne M.E. (1971). Fatty horomones in pollen and immature seeds of bean. Journal of Agriculture and Food Chemistry 19: 91-393

Montoya, T., Nomura, T., Farrar, K., Kaneta, T., Yokota, T., Bishop, G.J.(2002). Cloning the tomato *Curl3* gene highlights the putative dual role of the leucine-rich repeat receptor kinase tBRI1/SR160 in plant steroid hormone and peptide hormone signaling. Plant Cell 14: 3163-3176.

Mori, M., Nomura, T., Ooka, H., Ishizaka, M., Yokota, T., Sugimoto, K., Okabe, K., Kajiwara, H., Satoh, K., Yamamoto, K., Hirochika, H., Kikuchi, S. (2002). Isolation and characterization of a rice dwarf mutant with a defect in brassinosteroid biosynthesis. Plant Physiology 130(3): 1152-1161.

Morishita, T., Abe, H., Uchiyama, M., Marumo, S., Takatsuto, S., Ikekawa, N. (1983). Evidence for plant-growth promoting brassinosteroids in leaves of *Thea sinensis*. Phytochemistry 22(4): 1051-1053.

Munoz, F. J., Labrador, E., Dopico, B. (1998). Brassinolides promote the expression of a new *Cicer arietinum* beta-tubulin gene involved in the epicotyl elongation. Plant Molecular Biology 37(5): 807-817.

Mussig, C., Fischer, S., Altmann, T. (2002). Brassinosteroid-regulated gene expression. Plant Physiology 129(3):1241-1251.

Noguchi, T., Fujioka, S., Choe, S., Takatsuto, S., Yoshida, S., Yuan, H., Feldmann, K. A., Tax, F. E. (1999). Brassinosteroid-insensitive dwarf mutants of *Arabidopsis* accumulate brassinosteroids. Plant Physiology 121(3): 743-752.

Nomura, T., Bishop, G., Reid, J., Yokota, T. (2002). Regulation of expression of the brassinosteroid receptor *lka* gene in pea. Plant and Cell Physiology 43: 186-186.

Nomura, T., Kitasaka, Y., Takatsuto, S., Reid, J. B., Fukami, M., Yokota, T. (1999). Brassinosteroid/sterol synthesis and plant growth as affected by *Ika* and *Ikb* mutations of pea. Plant Physiology 119(4): 1517-1526.

Nomura, T., Nakayama, M., Reid, J. B., Takeuchi, Y., Yokota, T. (1997). Blockage of brassinosteroid biosynthesis and sensitivity causes dwarfism in garden pea. Plant Physiology 113: 31-37.

Peng, J. R., Richards, D. E., Hartley, N. M., Murphy, G. P., Devos, K. M., Flintham, J. E., Beales, J., Fish, L. J., Worland, A. J., Pelica, F., Sudhakar, D., Christou, P., Snape, J. W., Gale, M. D., Harberd, N. P. (1999). 'Green revolution' genes encode mutant gibberellin response modulators. Nature 400(6741): 256-261.

Pilet, P. E., Elliott, M. C., Moloney, M. M. (1979). Endogenous and exogenous auxin in the control of root growth. Planta 146: 405 - 408.

Reid, J. B., Ross, J. J. (1989). Internode length in *Pisum* - 2 Further gibberellin-insensitivity genes, *Lka* and *Lkb*. Physiologia Plantarum 75(1): 81-88.

Robertson, A. J., Ishikawa, M., Gusta, L. V., MacKenzie, S. L. (1994). Abscisic acid induced heat tolerance in *Bromus inermis* Leyss cell-suspension cultures. Heat stable, abscisic acid responsive polypeptides in combination with sucrose confer enhanced thermostability. Plant Physiology 105: 181 - 190.

Romani, G., Marre, M. T., Bonetti, A., Cerana, R., Lado, P., Marre, E. (1983). Effects of a brassinosteroid on growth and electrogenic proton extrusion in maize root segments. Physiologia Plantarum 59(4): 528 - 532.

Scheer, J. M., Ryan, C. A. (2002). The systemin receptor SR160 from *Lycopersicon peruvianum* is a member of the LRR receptor kinase family. Proceedings of the National Academy of Sciences (USA) 99(14): 9585-9590.

Schlagnhaufer, C., Arteca, R. N., Yopp, J. H. (1984). Evidence that brassinosteroid stimulates auxin-induced ethylene synthesis in mung bean hypocotyls between S-adenosylmethionine and 1-aminocyclopropane-1-carboxylic acid. Physiologia Plantarum 61(4): 555-558.

Schluter, U., Kopke, D., Altmann, T., Mussig, C. (2002). Analysis of carbohydrate metabolism of CPD antisense plants and the brassinosteroid-deficient *cbb1* mutant. Plant Cell and Environment 25(6): 783-791.

Schultz, L., Kerckhoffs, L. H. J., Klahre, U., Yokota, T., Reid, J. B. (2001). Molecular characterization of the brassinosteroid-deficient *lkb* mutant in pea. Plant Molecular Biology 47(4): 491-498.

Schumacher, K., Vafeados, D., McCarthy, M., Sze, H., Wilkins, T., Chory, J. (1999). The *Arabidopsis det3* mutant reveals a central role for the vacuolar $H^+$-ATPase in plant growth and development. Genes and Development 13(24): 3259-3270.

Steber, C. M., McCourt, P. (2001). A role for brassinosteroids in germination in *Arabidopsis*. Plant Physiology 125(2): 763-769.

Steffens, G. L. (1991). U.S. Department of agriculture brassins project: 1970-1980. In Brassinosteroids: Chemistry Bioactivity and Applications, pp.3-17. Eds. H G Cutler, T Yokota and G Adam, American Chemical Society, Washington.

Symons, G. M., Schultz, L., Kerckhoffs, L. H. J., Davies, N. W., Gregory, D., Reid, J. B. (2002). Uncoupling brassinosteroid levels and de-etiolation in pea. Physiologia Plantarum 115(2): 311-319.

Szekeres, M., Nemeth, K., KonczKalman, Z., Mathur, J., Kauschmann, A., Altmann, T., Redei, G. P., Nagy, F., Schell, J., Koncz, C. (1996). Brassinosteroids rescue the deficiency of CYP90, a cytochrome P450, controlling cell elongation and de-etiolation in arabidopsis. Cell 85(2): 171-182.

Takeno, K., Pharis, R. P. (1982). Brassinosteroid induced bending of the leaf lamina of dwarf rice seedlings and auxin-mediated phenomenon. Plant and Cell Physiology 23: 1275 - 1281.

Tominaga, R., Sakurai, N., Kuraishi, S. (1994). Brassinolide induced elongation of inner tissues of segments of squash *Cucurbit-maxima Duch* hypocotyls. Plant and Cell Physiology 35 :1103 - 1106.

Wada, K., Marumo, S., Ikekawa, N., Morisaki, M., Mori, K. (1981). Brassinolide and homo-brassinolide promotion of lamina inclination of rice seedlings. Plant and Cell Physiology 22: 323 - 325.

Wang, T. W., Cosgrove, D. J., Arteca, R. N. (1993). Brassinosteroid stimulation of hypocotyl elongation and wall relaxation in pakchoi *Brassica-chinensis* cv Lei-Choi hypocotyls. Plant Physiology 101: 965 - 968.

Wilen, R., Sacco, M., Gusta, L. V., Krishna, P. (1995). Effects of 24-epibrassinolide on freezing and thermotolerance of bromegrass (*Bromus-inermis*) Cell cultures. Physiologia Plantarum 95: 195 - 202.

Xu, W., Purugganan, M. M., Polisensky, D. H., Antosiewicz, D. M., Fry, S. C., Braam, J. (1995). *Arabidopsis* Tch4, regulated by hormones and the environment, encodes a xyloglucan endotransglycosylase. Plant Cell 7(10): 1555-1567.

Yamamoto, R., Demura, T., Fukuda, H. (1997). Brassinosteroids induce entry into the final stage of tracheary element differentiation in cultured Zinnia cells. Plant and Cell Physiology 38(8): 980-983.

Yamamuro, C., Ihara, Y., Wu, X., Noguchi, T., Fujioka, S., Takatsuto, S., Ashikari, M., Kitano, H., Matsuoka, M. (2000). Loss of function of a rice brassinosteroid insensitive1 homolog prevents internode elongation and bending of the lamina joint. Plant Cell 12(9): 1591-1605.

Yi, H. C., Joo, S., Nam, K. H., Lee, J. S., Kang, B. G., Kim, W. T. (1999). Auxin and brassinosteroid differentially regulate the expression of three members of the 1-aminocyclopropane-1- carboxylate synthase gene family in mung bean (*Vigna radiata* L.). Plant Molecular Biology 41(4): 443-454.

Yokota, T. (1997). The structure, biosynthesis and function of brassinosteroids. Trends in Plant Science 2(4): 137-143.

Yopp, J., Colclasure, G. C., Mandava, N. B. (1979). Effects of brassin-complex on auxin and gibberellin mediated events in the morphogenesis of etiolated bean hypocotyl. Physiologia Plantarum 46: 247 - 254.

Yopp, J., Mandava, N. B., M, S. J. (1981). Brassinolide, a growth promoting steroidal lactone. 1-Activity in selected auxin bioassays. Physiologia Plantarum 53(4): 445 - 452.

Yu, M. H. (1982). The dwarf curly leaf tomato. Journal of Heredity 73: 270-272.
Yu, Y. B., Yang, S. F. (1979). Auxin induced ethylene production and its inhibition by amino ethoxy vinylglycine and cobalt ion. Plant Physiology 64: 1074 - 1077.
Zurek, D. M., Clouse, S. D. (1994). Molecular-cloning and characterization of a brassinosteroid regulated gene from elongating Soybean (*Glycine-Max* L.) epicotyls. Plant Physiology 104(1): 161-170.
Zurek, D. M., Rayle, D. L., McMorris, T. C., Clouse, S. D. (1994). Investigation of gene expression, growth-kinetics and wall extensibility during brassinosteroid-regulated stem elongation. Plant Physiology 104(2): 505-513.

CHAPTER 3

MARTIN FELLNER

# RECENT PROGRESS IN BRASSINOSTEROID RESEARCH: HORMONE PERCEPTION AND SIGNAL TRANSDUCTION

Brassinosteroids (BRs) are cholestane derivatives that show structural similarity to insect, animal and human steroid hormones. Like steroids, brassinosteroids are signaling molecules in plants, which play important roles in normal growth, development and differentiation. Continued genetic screening and analysis of *Arabidopsis* mutants has provided new insight into brassinosteroid signaling, and today study of brassinosteroid action is the priority of many research laboratories around the world. Molecular genetic studies have led to the cloning and characterization of a BR receptor complex, BRI1/BAK1, a transmembrane receptor serine/threonine kinase pair. Brassinosteroid binding via a BR-binding protein stabilizes heterodimmer formation, activates intrinsic kinase activities and initiates a BR signaling cascade. This results in regulation of the activity of downstream elements of the signal transduction pathways. Although downstream components of BR action are largely unknown, research effort in last two years has led to the identification of a negative cytoplasmic regulator BIN2, a GSK3-like kinase. In the absence of BRs, BIN2 phosphorylates two positive regulators, BES1 and BZR1, resulting in their degradation and preventing their accumulation in the nucleus. In turn, initiation of a BR signaling cascade inhibits BIN2 kinase activity by an unknown mechanism(s). BIN2 is not able to phosphorylate BES1 and BZR1 proteins in the presence of BRs and unphosphorylated BES1 and BZR1 accumulate in the nucleus triggering expression of a variety of BR-induced genes involved in regulation of plant growth.

## INTRODUCTION

Brassinosteroids (BRs) are plant hormones that show structural similarity to insect, animal and human steroid hormones (Evans, 1988). Unlike many steroids identified in plants, only BRs are widely distributed over the plant kingdom (Mandava, 1988). The first brassinosteroid, called brassinolide, was identified in 1979 as the effective compound in a hydrophobic extract called brassin from pollen of *Brassica napus* (Mitchell *et al.*, 1970; Grove *et al.*, 1979). Brassinolide (BL) is the most active among BRs, which elicit various physiological responses such as stimulation of cell elongation and are essential for male fertility and xylem differentiation (Müssig and Altmann, 1999; Altmann, 1999). Relative to study of BR biosynthesis, investigation of BR signaling is a pressing research priority in many laboratories in recent years. Even very recent and excellent plant physiology or biochemistry books and textbooks are still lacking chapters dealing intensively with brassinosteroid signaling.

Using genetic techniques many BR-related mutants were selected and

*S.Hayat and A.Ahmad (eds.), Brassinosteroids,* 69-86.

characterized. BR-insensitive mutants often show phenotypic traits similar to mutants affected in BR biosynthesis, such as extreme dwarfism, altered leaf morphology, reduced fertility or male sterility, delayed senescence, and altered vascular development (Clouse and Feldmann, 1999). However, unlike BR-deficient mutants, the phenotype of BR-insensitive mutants cannot be rescued by exogenous BR application, suggesting that the mutated genes play a role in BR signaling. BR-insensitive mutants have been isolated in rice (*Os-bri1*), pea (*lka*) and tomato (*curl-3*) (Nomura *et al.*, 1999; Yamamuro *et al.*, 2000; Koka *et al.*, 2000). In *Arabidopsis*, numerous alleles of the BR-insensitive mutant *bri1* were identified using a screen based on the inhibition of root growth by BR or on phenotypic similarities to BR-deficient mutants (Clouse *et al.*, 1996; Kauschmann *et al.*, 1996; Li and Chory, 1997; Noguchi *et al.*, 1999). *BRI1* was cloned and has many of the properties expected of a BR receptor (Li and Chory, 1997). Although *BRI1* homologues with high sequence similarity exist in the *Arabidopsis* genome (The *Arabidopsis* Genome Initiative, 2000), BRI1 appears to be the major *Arabidopsis* brassinosteroid receptor. Following some recent excellent reviews dealing with brassinosteroid perception and signal transduction (Friedrichsen and Chory, 2001; Müssig and Altmann, 2001; Bishop and Koncz, 2002), here I summarize advances that have been made in study of BR perception and in identifying novel downstream components in BR signal transduction.

## CELL SURFACE RECEPTOR AND BRASSINOSTEROID PERCEPTION

Structural similarities of BRs to insect and animal steroids led to predictions of similar signal transduction pathways, nongenomic and genomic. In animal systems, the term nongenomic steroid action is used for rapid steroid effects, i.e. taking seconds or minutes. The effects are not blocked by inhibitors of transcription or translation (actinomycin D, cyclohexamine) and are mediated by steroids transported into the cell via carrier proteins or in addition using membrane-binding sites (Wehling, 1997; Watson and Gametchu, 1999; Fleet, 1999; Schmidt *et al.*, 2000). Steroid signals for genomic action are generally mediated by receptors inside the cell, and genomic steroid effects are characterized by delayed responses (i.e. taking minutes or hours) because of time required for transcription and protein synthesis. Therefore, they can be affected by inhibitors of transcription or translation (Beato and Klug, 2000; Mangelsdorf *et al.*, 1995). In plants, direct experimental approaches and analysis of the completed *Arabidopsis* genome sequence have failed to provide any evidence for the intracellular receptor family (The *Arabidopsis* Genome Initiative, 2000), and brassinosteroids are believed to be perceived by plasmamembrane receptors, triggering genomic and nongenomic actions (He *et al.*, 2000; Becraft, 2001).

Positional cloning revealed that *BRI1* encodes a Leu-rich repeat (LRR) transmembrane receptor-like kinase (RLK) (Li and Chory, 1997; Friedrichsen *et al.*, 2000). BRI11 is a typical plasmamembrane associated LRR-RLK, whose localization at the plasmamembrane was demonstrated by expression of the fusion protein BRI1-GFP (green fluorescent protein) (Friedrichsen *et al.*, 2000). BRI1 consists of several typical domains (Fig. 1). It carries a N-terminal signal peptide followed by a putative Leu-zipper motif and cysteine pair. Then, the extracellular LRR domain of 25 imperfect Leu-rich repeats follows. Interestingly, in the LRR domain of BRI1 a 70-amino acid island was found between repeats 21 and 22. Several *bri1* mutations resulting in single amino

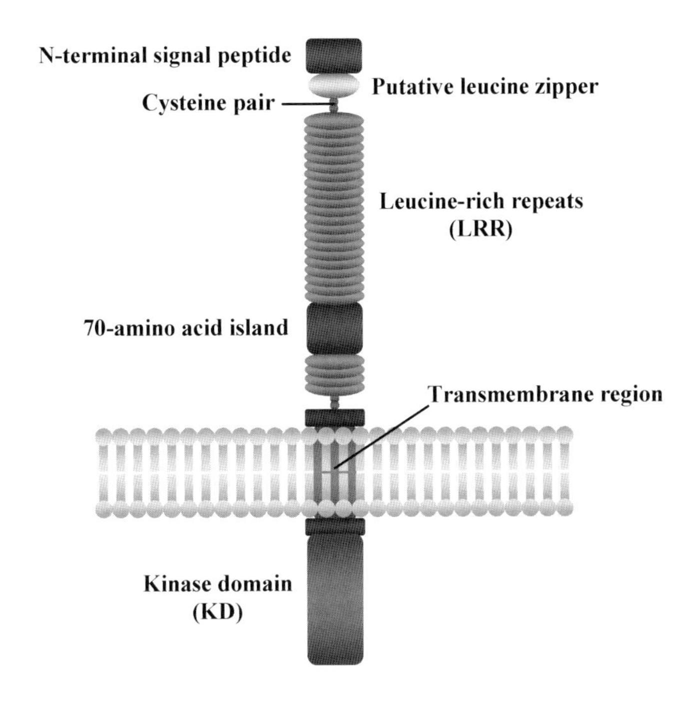


***Figure 1.*** *Schematic structure of Arabidopsis BRI1. Source of primary sequence information: Li and Chory, 1997 (adapted from Bishop and Koncz, 2002)*

acid substitutions were found in the island and highlight its importance for the function of BRI1 (Noguchi *et al.*, 1999; Friedrichsen *et al.*, 2000; Friedrichsen and Chory, 2001). A predicted transmembrane region links extracellular domains with the intracellular Ser/Thr-kinase domain (KD) (Fig. 1). Several *bri1* missense mutations were also mapped to the KD (Friedrichsen and Chory, 2001; Bishop and Koncz, 2002).

A unique feature of BRI1's LRR-RLK is that it contains both the 70-amino acid island and a cytosolic KD. The 70-amino acid island is characteristic of a specific family of LRR receptor like proteins (LRR-R). It is observed in the Cf-9 (Jones *et al.*, 1994), CLAVATA2 (Jeong *et al.*, 1999), and TOLL LRR receptors (Hashimoto *et al.*, 1998). These LRR-Rs, however, lack a cytosolic KD (Bishop and Koncz, 2002). BRI1 also shows significant sequence identity with LRR-RLK ERECTA (Torii *et al.*, 1996) involved in regulation of organ size, to CLAVATA1 (Clark *et al.*, 1997) controlling meristem proliferation, or to Xa21 involved in pathogen resistance (Song *et al.*, 1995). These LRR-RLKs have cytosolic KDs, but in contrast to BRI1, they lack the 70-amino acid island

(Bishop and Koncz, 2002).

Several experimental results strongly support a critical role of *BRI1* in BR signaling. In rice, the product of the *Xa21* gene, conferring resistance to the bacterial plant pathogen *Xanthomonas oryzae* pv. *oryzae* race 6 *(Xoo)*, shows a structure similar to BRI1, contains 23 LRR, a transmembrane domain and Ser/Thr kinase domain (Song *et al.*, 1995). He *et al.* (2000) showed that after inoculation with an incompatible *Xanthomonas Xoo* strain, Xa21 could trigger a hypersensitive response in a rice cell-culture system. The authors then constructed a chimeric protein containing both the extracellular and transmembrane domain plus a short stretch of the intracellular domain of BRI1 fused to the Xa21 kinase domain. They showed that the chimeric protein was also able to elicit the hypersensitive response in rice cells after treatment with BL in the physiologically relevant concentrations of 10 nM to 2 mM. By contrast, the authors also showed that if the 70-amino acid island domain in BRI1 or Xa21 kinase domain was mutated, the hypersensitive response was blocked after treatment with BL (He *et al.*, 2000). By analysis of BL-binding to BRI1, further pieces of evidence were provided indicating that the extracellular domain of BRI1 perceives BR (Wang *et al.*, 2001). Binding of radio labeled BL to microsomal fractions was compared in wild type (WT) and transgenic plants overexpressing BRI1-GFP. The binding activity was dramatically increased in the membrane fractions of BRI1-GFP plants, and the number of BL-binding sites correlates with the amount of BRI1 protein. Unlike the nonspecific competitor ecdysone, the immediate and bioactive BL precursor, castasterone (Yokota, 1997; Bishop *et al.*, 1999), reduced BL binding. The authors also showed that BL-binding activity was abolished when the 70-amino acid island was mutated. In contrast, mutations on the KD of BRI1 have no effects on BL binding. These observations strongly indicate that the 70-amino acid island in the extracellular domain of BRI1 perceives brassinosteroids (Wang *et al.*, 2001).

Brassinolide is a steroid hormone, which classically bind to soluble ligand-activated transcription factors. As showed, BRI1 is an LRR-RLK. LRRs are known to recognize protein ligands. Extracellular LRR of G-protein-coupled receptors in animals are involved in the recognition of peptide hormones, such as gonodotropin, nerve growth factors, and thyroid-stimulating hormones (Kobe and Deisenhofter, 1994). Despite the large numbers of LRR-RLKs in plants, only a few of them have known biological function. For example, LRR-RLKs play a role in hormone perception, meristem signaling, or pathogen responses (Shiu and Bleecker, 2001). Even fewer LRR-RLKs have known ligands. Various peptide ligands were defined for receptors CLV1, FLS2, or PSK (Fletcher *et al.*, 1999; Gomez-Gomez *et al.*, 2001; Matsubayashi *et al.*, 2002). Very recently, a systemin receptor SR160 was identified, closely related to BRI1 (Sheer and Ryan, Jr., 2002). However, LRRs have not been shown to interact with organic compounds such as steroids, and BRI1 was the first LRR-RLK described to be involved in steroid signaling in plants. Therefore, BRI1 structure raised a new question about steroid signaling mechanisms and established a new paradigm for hormonal responses in plants. Although, Li and Chory (1997) admitted that BRs may fit in the cavity formed by the island of 70-amino acid, they proposed that BRI1 may only recognize protein ligands, such as putative steroid carriers, and thus additional factors may mediate BL binding.

In order to learn more about mechanism of BR perception and signaling, Li *et al.* (2001a) performed a gain-of-function suppressor screen with a weak *bri1* allele (*bri1-5*). The authors identified a single dominant *bri1 suppressor dominant* (*brs1-1D*) mutation that resulted in overexpression of a type II serine carboxypeptidase-like protein. However, the *brs1-1D* mutation suppresses only *bri1* alleles with a mutation in the extracellular LRR domain but cannot suppress the *bri1* phenotype due to mutation in cytoplasmic KD (Li *et al.*, 2001a). Also, the *brs1* did not suppress *clavata1* and *erecta* mutant phenotypes, which indicates specificity of BRS1 to BL signaling. Li *et al.* (2001a) suggested that BRS1 acts on a protein that is required for BL perception. The substrates of BRS1 might be steroid-binding proteins represented by genes, which have been identified in the *Arabidopsis* genome sequence (The *Arabidopsis* Genome Initiative, 2000). Suggested from the observation that the BRI1 product may potentially be processed (Wang *et al.*, 2001), BRS1 may alternatively use BRI1 to generate an activated BL receptor (Fig. 2). However, since BRS1 does not process functional BRI1-Xa21 chimera (He *et al.*, 2000) further investigation is necessary to define role of BRS1 in BR signaling (Li *et al.*, 2001a).

When a protein extract from transgenic BRI1-GFP seedlings treated by BL was immunoblotted with antibodies against the N-terminal region of BRI1 or GFP, the mobility of the BRI1 KD protein band in SDS-PAGE was reduced (Wang *et al.*, 2001). The authors suggest that this shift is likely a result of autophosphorylation of BRI1 in the presence of BL. This idea was supported by the observation that the intracellular cytoplasmic KD of BRI1 autophosphorylates on serine/threonine residues (Oh *et al.*, 2000). Li and Chory (1997) proposed that binding of BRs, either directly or via a steroid binding protein to the BRI1 receptor may dimerize BRI1 with other LRR receptors. Dimerization could mediate cell-to-cell interactions through LRRs, leading to activation kinase activity and triggering a phosphorylation cascade inside the plant cell. This was a logical proposal since receptor dimerization is a general mechanism for ligand-induced activation of receptor kinases in animals (Schlessinger, 2000). Although BRI1 is a component of a multiprotein BR receptor complex (Wang *et al.*, 2000), indirect evidence against BRI1 homodimerization as a mechanism for receptor activation has been provided (Clifford and Schupbach, 1994; Simin *et al.*, 1998; Friedrichsen and Chory, 2001; Clark, 2001; Nam and Li, 2002). Instead, it is hypothesized that BR binding results in formation of heterodimer (Fig. 2)(Schumacher and Chory, 2000; Friedrichsen and Chory, 2001).

Very recently, it has been confirmed that BRI1 receptor functions as a heterodimer. Results from two laboratories independently provided evidence that BRI1 forms heterodimer with distinct receptor kinase (Nam and Li, 2002; Li *et al.*, 2002). Using activation tagging, Li *et al.* (2002) identified a dominant genetic suppressor of *bri1*, *bak1-1D* (*bri1-associated receptor kinase1-1Dominant*), which encodes an LRR-RLK, distinct from BRI1. The BAK1 protein was independently identified in a yeast two-hybrid screen where the BRI1 kinase domain was used as bait (Nam and Li, 2002). Both groups showed that overexpression of BAK1 suppresses a weak *bri1* phenotype and results in elongated organ phenotypes, while a null *bak1* mutant is semidwarf and reduced in sensitivity to BRs. Li *et al.* (2002) showed that BAK1 and BRI1 have the same expression pattern, both are plasmamembrane localized, they can phosphorylate one another, and the autophosphorylation of BAK1 is stimulated by BRI1. Both groups further showed that

BAK1 and BRI1 proteins directly interact with one another *in vitro* as well as *in vivo*. On the basis of their genetic analyses, Li *et al.* (2002) and Nam and Li (2002) independently suggest that BAK1 is specifically involved in the BRI1-mediated BR signaling pathway. Li *et al.* (2002) hypothesize that when a ligand binds to BRI1, BRI1 activates BAK1, and the activated BAK1 then phosphorylates other downstream components. Nam and Li (2002) propose that BRI1 and BAK1 exist as inactive monomers (Fig. 2) that are in equilibrium with active dimers. BR binding via a BR-binding protein (BRBP) stabilizes or promotes active heterodimer formation, leading, via transphosphorylation, to activation of both receptor kinases and to initiation of a BR signaling cascade (Fig. 2).

Dephosphorylation by phosphatases plays an important role in the downregulation of receptor kinases in animals (Ostman and Bohmer, 2001). It has been observed that the 2C phosphatase, KAPP (kinase-associated protein phosphatase), can interact with a number of LRR-RLK, and may thus oppose the action of BRI1 KD (Fig. 2) (Schumacher and Chory, 2000). Overexpression of KAPP may thus lead to a *bri1*-like phenotype. So far, this type of experiment has been performed on a *clavata* mutant. The KAPP protein binds the phosphorylated form of LRR-RLK CLV1, and the KAPP overexpression results in a floral phenotype similar to that caused by weak *clv1* alleles (Williams *et al.*, 1997; Trotochaud *et al.*, 1999). Conversely, reduction of the KAPP transcript in the *clv1* mutant suppressed the mutant phenotype (Stone *et al.*, 1998). It suggests that KAPP could function as a negative regulator in the CLV1 signal-transduction pathway. KAPP was cloned and further interactions with RLKs were shown, indicating that KAPP may function in RLK-initiated signaling pathways (Stone *et al.*, 1994; Braun *et al.*, 1997), including BR signaling.

## BR SIGNAL TRANSDUCTION

Until recently, nothing was known about steps which mediate transduction of the BR signal further into the cell. The inability to identify BR-signaling components encoded by genes different from *BRI1* was thought to be due to redundancy in downstream components or due to lethality of mutants with loss-of-function mutation(s) (Clouse, 2002). Screening of 150,000 EMS-mutagenized seeds led to isolation of only two alleles of a new mutant *bin2* (*brassinosteroid-insensitive 2*) (Li *et al.*, 2001b). The *bin2* is dwarf and semidominant. Homozygous *bin2* seedlings are extremely dwarf looking like *bri1* and lack transcriptional downregulation of the *CPD* gene (involved in BR biosynthesis) in the presence of BL. In the heterozygous configuration *bin2* is semi dwarf and shows approx. 50% reduction in CPD transcription level in comparison with the WT (Li *et al.*, 2001b). BIN2 has been cloned by a map-based approach, and was found to encode a cytoplasmic serine/threonine kinase (Li and Nam, 2002). Interestingly, BIN2 is an *Arabidopsis* ortholog of the *Drosophila* SHAGGY protein kinase and human glycogen synthase kinase-3 (GSK-3) (Li and Nam, 2002). SHAGGY/GSK-3-like kinases are widely distributed over eukaryotes and function often like negative regulators of signal transduction pathways (Kim and Kimmel, 2000). Similar to *bin2*, *Arabidopsis* mutant *ultracurvata1* (*Ucu1*), displaying aberrant leaf morphology, was mutated in the same kinase gene (Perez-Perez *et al.*, 2002). In both *bin2* and *Ucu1*, missense mutations resulted in a semidominant phenotype, leading to the suggestion that, as for mammalian SHAGGY/GSK-3 kinases, BIN2 may function as a negative regulator of BR signaling (Li *et al.*, 2001b). Moreover, activity of the mutant *bin2* kinase is increased by approximately 33% in comparison to WT, and seedlings overexpressing BIN2 shows more pronounced dwarfism (Li and Nam, 2002). This suggests that the *bin2* mutation is hypermorphic. Also, overexpression of BIN2 in a weak *bri1* mutant background generated a more pronounced dwarf phenotype, whereas co-suppression of BIN2 transcription resulted in partial suppression of the weak *bri1* phenotype (Li and Nam, 2002). These data indicate that BIN2 is a negative regulator of BR signaling, and the authors speculate that BRI1 interacts with BIN2, phosphorylates and thus inactivates it (Li and Nam, 2002). Oh *et al.* (2000) showed that recombinant BRI1 KD could phosphorylate *in vitro* a conserved peptide motif and the authors deduced putative consensus sequence for peptide-substrate recognition by BRI1 KD. However, BIN2 lacks close homology to the consensus sequence, which would be likely phosporylated by BRI1. Therefore, BRI1 does not, most likely, interact directly with the BIN2 kinase. Instead, Li and Nam (2002) propose that, after BL binding to the BRI1, a signaling cascade is initiated and BRI1-containing complex interacts with and inactivates BIN2 (Fig. 2). Then, BIN2 is not able to phosphorylate and inactivate other pathway components such as positive regulators. In turn, in the absence of BRs, BIN2 is constitutively active and therefore phosphorylates and inactivates positive BR signaling proteins (Fig. 2).

## DOWNSTREAM EVENTS OF BR SIGNALING: IDENTIFICATION OF THE POSITIVE REGULATORS

To elucidate new components of downstream BR signaling Yin *et al.* (2002a) screened EMS-mutagenized *bri1* seeds for mutant plants with suppressed *bri1* phenotypes. The authors isolated a semidominant mutant *bes1-D* (*bri-EMS-suppressor1-D*) which, along with the suppression of the *bri1* dwarf phenotype, also exhibited constitutive BR response phenotypes including long and bending petioles, curly leaves, and constitutive expression of BR-responsive genes encoding cell wall-modifying enzymes. The *BES1* gene [also called *BZR2* (Yin *et al.*, 2002a)] was cloned and found to encode phosphoproteins with nuclear localization signal and multiple consensus sites (S/TXXXS/T) for phosphorylation by GSK-3 type kinases (Yin *et al.*, 2002a). The *bes1-D* mutant phenotype and its gain-of-function nature suggests that BES1 acts downstream of BRI1 as a positive component in the BR signaling pathway. In WT, the BES1 protein was found in low levels in cytoplasm and nucleus, and its nuclear localization was rapidly and specifically enhanced by BR treatment. In contrast, the *bes1-D* mutant protein was found in the nucleus at high levels and BL had apparently no effect on nuclear bes1-D localization (Yin *et al.*, 2002a). The authors further showed that` BL treatment resulted in nuclear accumulation of BES1 in unphosphorylated form suggesting that the protein is regulated by a negative acting kinase. They provided evidence that BES1 interacts with BIN2, so that in the absence of BRs BIN2 phosporylates BES1 in vivo, resulting in low level of BES1 in the nucleus (Yin et al., 2002a).

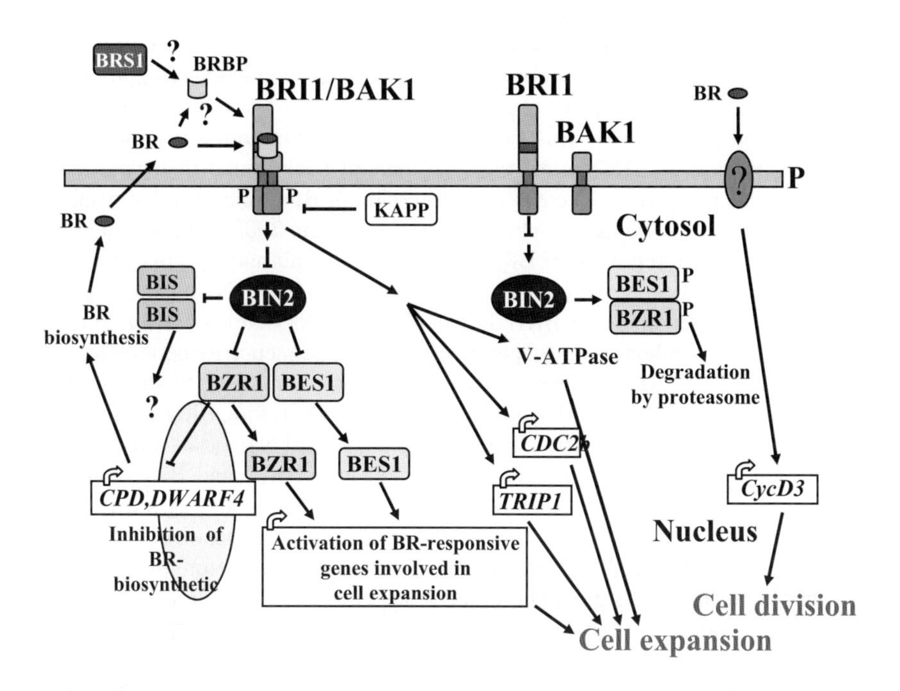


*Figure 2 Model for BR signaling. BRI1 is the major BR receptor and is located on the plasma membrane. BRI1 and BAK1 (BRI1-associated receptor kinase) exist mainly as inactive monomers. BR binding through a putative BR-binding protein (BRBP) or directly to the amino acid island stabilizes or promotes active heterodimer formation, leading to activation of both receptor kinases via transphosphorylation. BRS1 may process BRBP and modulate the BRI1signaling pathway. KAPP (kinase-associated phosphatase) is a potential suppressor of receptor complex. Receptor activation leads to inhibition of the negative regulator BIN2 kinase. It allows stabilization and accumulation of unphosphorylated positive regulators BES1 and BZR1 in the nucleus (suppression of the BIN2 may also stabilize new positive regulators BIS1 and BIS2 recently identified with yet unknown functions). BES1 and BZR1 then activate BR-inducible genes encoding cell wall-modifying enzymes and other genes whose products are involved in regulation of cell expansion. In light, BZR1 also activates a negative feedback pathway that inhibits the BR-biosynthetic genes such as CPD and DWARF4, which leads to reduced cell elongation. BR-mediated activation of the BRI1/BAK1 complex leads to activation of other BR-induced genes, such as TRIP1 and CDC2b. These BR-induced genes work together with the V-ATPase pathway to promote cell growth. In the absence of BR, BRI1 is inactive. Negative regulator BIN2 is able to phosphorylate BES1 and BZR1 and target them for degradation by proteasome. Cyclin gene CycD3 is potentially activated through a separate BR pathway using unidentified BR receptor and affects the cell cycle to promote cell division. This model is adapted from Friedrichsen and Chory, 2001; Müssig and Altman, 2001; Bishop and Koncz, 2002; Clouse, 2002; He et al., 2002; Nam and Li, 2002, and Yin et al., 2002a.*

Screening for mutants resistant to an inhibitor of BR biosynthesis, brassinazole, the research group of J. Chory isolated bzr1-D (brassinazole resistant1-D), a dominant mutation, and identified a second positive regulator of the BR signaling pathway (Wang et al., 2002). Like bes1-D, bzr1-D suppresses BR-deficient and BR-insensitive (bri1) phenotypes, and both *bzr1-D* and *bes1-D* dark-grown seedlings show resistance to brassinazole. The authors therefore suggested that *BZR1* might encode BES1-like protein. The BZR1 has been cloned and it shows 88% sequence identity to BES1, including the nucleus localization signal and multiple sites for phosphorylation by GSK3 kinases (Wang *et al.*, 2002; Yin *et al.*, 2002a). The authors found that in response to BRs, BZR1 accumulates in the nucleus of elongating cells in dark-grown hypocotyls and is stabilized by BR signaling and the *bzr1-D* mutation (Wang *et al.*, 2002; Yin *et al.*, 2002a). Consistent with this, in experiments using a proteasome inhibitor, MG132, He *et al.* (2002) revealed that in the presence of BRs BIN2 kinase activity is inhibited. As for BES1, BZR1 is not phosphorylated and accumulates in nucleus. In turn, in the absence of BRs, BZR1 is phosphorylated and destabilized (degraded) by the negative regulator BIN2 by the proteasome machinery.

Based on their results, Yin *et al.* (2002a) and He *et al.* (2002) proposed a model for BR signal transduction (Fig. 2). In the absence of BRs, BIN2 phosphorylates two positive regulators, BES1 and BZR1, resulting in their degradation and preventing their accumulation in the nucleus (Fig. 2). In the presence of BRs, the hormone binding to BRI1 via a BR-binding protein stabilizes or promotes active heterodimer formation, leading to transphosphorylation and activation of both receptor kinases (Nam and Li, 2002; Li *et al.*, 2002) and other signaling events (Fig. 2). Initiation of a BR signaling cascade inhibits BIN2 kinase activity by an unknown mechanism(s). BIN2 is not able to phosphorylate BES1 and BZR1 proteins and unphosphorylated BES1 and BZR1 accumulate in the nucleus. Nuclear localization of BES1 and BZR1 then activates expression of BR-induced genes such as those encoding cell-wall modifying enzymes required for cell elongation (Fig. 2).

In contrast to *bes1-D*, light-grown *bzr1-D* seedlings exhibit semi-dwarf phenotypes and reduced BR content (Yin *et al.*, 2002a). Therefore, even though BES1 and BZR1 have overlapping functions, they are not completely redundant since *bzr1-D* seems in addition to activate a BR feedback inhibition pathway (Wang *et al.*, 2002;Yin *et al.*, 2002a). The difference in phenotypes between *bes1-D* and *bzr1-D* mutants in light suggests that these two genes or gene products are differently regulated by light. BES1 and BZR1 therefore provide potential targets for crosstalk between the light and BR signaling pathways (Yin *et al.*, 2002a). There is evidence for regulation of BR levels by the phototransduction pathways via *BAS1* gene (Neff *et al.*, 1999), or for the crosstalk between light and BRs through a dark-induced small G-protein (Kang *et al.*, 2001). However, the function of *BES1* and *BZR1* genes in the control of photomorphogenesis and light-regulated gene expression is unknown.

Recently, two novel nuclear proteins BIS1 (BIN2 SUBSTRATE1) and its closest homologue BIS2 (BIN2 SUBSTRATE2), were identified and found to be phosphorylated by BIN2 *in vitro* by a novel phosphorylation mechanism (Fig. 2) (Peng *et al.*, 2002). The authors propose that BIS1/BIS2 are two nuclear components of BR signaling that are, in addition to BES1 and BZR1, negatively regulated by BIN2. Functions of BIS1/BIS2 in BR-mediated responses have yet to be elucidated.

In addition to the BRI1 pathway, which results in targeting of genes involved in cell elongation and expansion, isolation and analysis of the *det3* mutant revealed other cellular components involved in BR-induced cell elongation. Interestingly, this component regulates cell elongation independently of gene expression, since mutations in *DET3* affect cell expansion in the absence of alterations of gene induction (Cabrera y Poch *et al.*, 1993). The *DET3* gene was found to encode a subunit of the vacuolar type $H^+$-ATPase (V-ATPase) (Schumacher *et al.*, 1999). It was reported earlier that BR-induced hypocotyl elongation in cucumber is dependent on membrane-bound ATPase activity, whereas hypocotyl elongation induced by gibberellin does not involve the $H^+$-ATPase (Mandava, 1988). Schumacher *et al.* (1999) propose that BR action triggers V-ATPase activity to initiate the uptake of water into the vacuole, and promote BR-induced cell elongation (Fig. 2). This is coordinated with changes in cell wall properties and changes in gene expression necessary to sustain growth responses.

## BR-INDUCED GENES

Major brassinosteroid effects such as BR-induced growth are mediated through the genomic pathway. A limited number of BR-regulated genes have been identified hitherto. Most of them encode BR biosynthesis enzymes or affect cell expansion and proliferation.

A possible biological significance of the nuclear localization of BES1 is suggested by the fact that several genes encoding cell wall-modifying enzymes and involved in cell expansion are overexpressed in *bes1-D* mutant. Using a commercially available *Arabidopsis* Affymetrix Gene Chip, Yin *et al.* (2002a) revealed that 30 genes were induced by BL in WT and *bes1-D* seedlings, but not significantly changed in the *bri1* mutant. A time course of their BR-regulated expression correlated with BES1 nuclear accumulation. Among these 30 genes, 19 of them showed 2-fold basal expression in *bes1-D* in comparison to WT, but not greater induction by exogenous BL relative to WT. In a second group, 11 genes showed basal expression similar in *bes1-D* and WT, but they were hyper-responsive to BL treatment in *bes1-D* (Yin *et al.*, 2002a). Of the 30 BR-induced genes 7 encode putative cell wall-associated proteins, such as xyloglucan endotransglycosylases (XETs), endo-1,4-β-glucanases (EGases), polygalacturonase, pectin methylesterase, or expansin. All of these proteins are implicated in cell elongation or expansion (Nicol *et al.*, 1998; Darley *et al.*, 2001; Friedrichsen and Chory, 2001; Lamport, 2001). The increased expression of BL-induced genes in *bes1-D* or their hyperinduction by BL in the mutant provide molecular evidence for the constitutive BL-response phenotypes of the *bes1-D* mutant and suggests that the *bes1-D* phenotypes result from changes in gene expression.

The relatively small changes (2- to 4-fold) in expression of BR-induced genes possibly can be explained by tightly controlled BR biosynthesis by a feedback inhibition mechanism (Yin *et al.*, 2002a). It was found that expression of the important BL biosynthetic gene *CPD* (Szekeres *et al.*, 1996) is inhibited by exogenous BRs, and such feedback inhibition requires functional BRI1 (Fig. 2) (Mathur *et al.*, 1998). BL-mediated downregulation of *CPD* was inhibited by the protein biosynthesis inhibitor cyclohexamine, indicating that *de novo* protein synthesis needs genomic effect of BR signaling (Mathur *et al.*, 1998). Similarly, another BL biosynthetic gene, *DWF4*, is negatively controlled by functional BR signaling pathway (Fig. 2) (Noguchi *et al.*, 2000). This possibility was also supported by an observation of Choe *et al.* (2001) who found that *DWF4* and *CPD* show derepressed expression in the *bri1-5* mutant. This is consistent with the accumulation of BR intermediates in *bri1* mutants (Noguchi *et al.*, 1999; Nomura *et al.*, 1999; Yamamuro *et al.*, 2000).

Interestingly, the onset of *CPD* gene expression in the cotyledons of etiolated *Arabidopsis* seedlings (Mathur *et al.*, 1998) correlated temporally with low expression of *BZR1* in cotyledons in seedlings of the same stage (Wang *et al.*, 2002) and, in contrast, with high accumulation of BZR1-GFP in the most actively expanding zone of the etiolated hypocotyls (Wang *et al.*, 2002). It seems that low BZR1 expression allows the high expression of CPD and BR biosynthesis in cotyledons. This mutually exclusive expression pattern of CPD and BZR1 is consistent with the role of BZR1 in BR-regulated cell elongation and feedback regulation of BR biosynthesis.

As mentioned above, BES1 and BZR1 appear to have overlapping functions, but they are not completely redundant since *bes1-D* and *bzr1-D* mutants differ in their-light grown phenotypes. Unlike *bes1-D* seedlings, which have long petioles, *bzr1-D* light-grown plants are semi-dwarf (Yin *et al.*, 2002a). Such differences in light-grown phenotypes are consistent with their different effects on feedback regulation of BR biosynthetic genes, since in contrast to *bes1-D*, the *bzr1-D* mutant has reduced BR levels and *CPD* gene expression (Yin *et al.*, 2002a). These results indicate that BES1 mediates downstream growth responses, but in contrast to BZR1, does not cause feedback regulation of BR biosynthesis (Fig. 2). The physiological significance of two such overlapping pathways seems to lie in the ability of the plant to fine-tune the levels of BR biosynthesis and sensitivity in different tissues and cells (Wang *et al.*, 2002).

Among other BR-induced genes, the expression of CDC2b cyclin-dependent kinase is upregulated by BR in dark (Fig. 2) (Yoshizumi *et al.*, 1999). CDC2b binds to its cyclin partner and promotes progression through the G2-M transition (D'Agostino and Kieber, 1999) suggesting its role in cell division. However, since CDC2b does not contain the conserved PSTAIRE motif, it is not believed to be involved in the cell cycle. Instead, CDC2b is rather important for cell elongation since *cdc2b* seedlings with antisense RNA are de-etiolated in the dark, reduced in cell elongation, but have WT cell number (Yoshizumi *et al.*, 1999).

None of the BR-response genes identified so far participate directly in signaling processes. Jiang and Clouse (2001) have identified a gene encoding the *Arabidopsis* homologue of TRIP1 (TGFβ receptor-interacting protein), which is rapidly induced by BL (Fig. 2). Interestingly, similarly to BRI1, TGFβ receptors are serine/threonine kinases, and in animals, TRIP1 is phosphorylated by TGFβ (Massagué, 1998). Antisense *trip1* *Arabidopsis* plants exhibit a broad range of developmental defects, including some that resemble the phenotype of BR-deficient and -insensitive mutants. TRIP in animals, yeast and plants is also a subunit of the eukaryotic translation factor eIF3 (Burks *et al.*, 2001;

Jiang and Clouse, 2001). This suggests that TRIP1 may mediate some of the molecular mechanisms underlying the regulation of plant growth and development by BRs (Fig. 2), and might establish a link between BR signaling and developmental pathways controlled by homologs of the eukaryotic translation initiation factor elF3 in plants (Jiang and Clouse, 2001; Bishop and Koncz, 2002).

Brassinosteroids have been found to regulate expression of other known or unknown genes in *Arabidopsis* or other plants, such as rice or tomato (Goetz *et al.*, 2000; Friedrichsen and Chory, 2001; Hu *et al.*, 2001; Bouquin *et al.*, 2001; Müssig *et al.*, 2002) but the upstream components which induce them and which are triggered by BRs have yet to be identified.

## PERSPECTIVES

Significant progress has been made over the last few years in elucidating the BR signaling pathway. The discovery of BR-insensitive mutants led to isolation of *BRI1* gene. Molecular characterization of the BRI1 product revealed it to be a receptor-like trans-membrane kinase (RLK) that transduces steroid signals across the plasma membrane. Discovery of *OsBRI1*, a *BRI1* homologue in rice (Yamamuro *et al.*, 2000), suggests that BR action through RLKs is a conserved mechanism in plant steroid signaling. In addition, a calcium-dependent protein kinase (CDPK) was identified in rice as another potential protein involved in BR action (Yang and Komatsu, 2000). Interestingly, using transgenic rice plants expressing the antisense strand of OsBRI1 transcript (Sharma *et al.*, 2001) revealed that the CPDK signaling cascade is parallel and independent to that via activation of BRI1 receptor. The authors speculate that there is involvement of additional receptor(s) other than BRI1 for BR responses in plants. The idea of an alternative BR receptor appeared also from work of Hu *et al.* (2000). The authors found that the cyclin gene *CycD3* can be induced by BRs, but they showed that the induction of CycD3 does not require functional BRI1. Hu *et al.* (2000) speculate that one of the several BRI1 homologues revealed in the *Arabidopsis* genome may serve as an additional BR receptor (Fig. 2) (The *Arabidopsis* Genome Initiative, 2000). In fact, in animals, hormone receptors are released from the cytoplasmic complexes after hormonal stimuli and are translocated to the nucleus (Mangelsdorf *et al.*, 1995). The CPDK activity observed in the study of Sharma *et al.* (2001) might involve a class of receptors similar to that reported in animals (Mangelsdorf *et al.*, 1995). So far, no evidence of a nuclear receptor has been found in plants, including the complete sequencing of the *Arabidopsis* genome (The *Arabidopsis* Genome Initiative, 2000). However, analysis of the rice genome, recently sequenced (Yu *et al.*, 2002; Goff *et al.*, 2002) could possibly reveal candidates of a putative nuclear receptor. It would be interesting to follow the idea of an alternative BR receptor(s), which could function in parallel with the BRI1 pathway. A newly identified receptor might mediate similar plant responses like BRI1, such as cell elongation, or it could trigger different responses, such as cell division (Fig. 2), or alternatively, both pathways may crosstalk, as in the system proposed for auxin in the control of cell elongation and cell division in tobacco leaves (Chen *et al.*, 2001; Chen, 2001). In addition, the proposed BR signaling shares many similarities to the Wnt signaling pathway (Cadigan and Nusse,

1997; Miller *et al.*, 1999; Polakis 2000; Clouse, 2002; Yin *et al.*, 2002a). It will, therefore, be interesting to determine whether other components in the BR pathway share homology to Wnt signaling elements, including a nuclear receptor, or if the two pathways split somewhere for some logical reason.

High homology among the BRI1, the systemin receptor SR160 (Scheer and Ryan, Jr., 2002), and PSK receptor (Matsubayashi *et al.*, 2002), and the fact that systemin and PSK are small peptides, whereas BRs are organic compounds provides unique opportunity to determine how these different ligands interact with very similar receptors. Also, identification of more downstream components for BR, and especially for systemin and PSK signaling will answer how specificity (BRs versus systemin) or similarity (BRs versus PSK?) of the signaling pathways are achieved.

Complete physiological and genetic characterization of already available BR-affected mutants (Yin *et al.*, 2002b), or further genetic screens for mutations that suppress or enhance the above described mutations, are expected to reveal new elements of BR signaling. Finally, because of existence of crosstalk among hormone signaling (Reid and Howell, 1995), we should pay attention to mutants recovered from genetic screens for other hormone-mutants. For example, some mutants screened originally for affected responses to cytokinin (Jang *et al.*, 2000) or auxin (Ephritikhine *et al.*, 1999), were later found to be altered in BR physiology. Physiological, genetic, and molecular studies of such mutants might result in identification of missing components, which link BR signaling with various hormone- or nonhormone-signaling pathways.

## ACKNOWLEDGEMENT

I thank Prof. Elizabeth Van Volkenburgh (University of Washington of Seattle) for her comments on the manuscript.

## REFERENCES

Altman, T. (1999). Molecular physiology of brassinosteroids revealed by the analysis of mutants. Planta 208: 1-11.

Beato, M., Klug J. (2000). Steroid hormone receptors: an update. Human Reproduction Update 6: 225-236

Becraft, P. W. (2001). Plant steroids recognized at the cell surface. Trends Genet 17: 60-62

Bishop, G. J., Koncz, C. (2002). Brassinosteroids and plant steroid hormone signaling. Plant Cell Suppl. S97-S110.

Bishop, G. J., Nomura, T., Yokota, T., Harrison, K., Noguchi, T., Fujioka, S., Takatsuto, S., Jones, J. D. G., Kamiya, Y. (1999). The tomato DWARF enzyme catalyses C-6 oxidation in brassinosteroid biosynthesis. Proceeding of National Academy of Sciences (USA) 96: 1761-1766.

Bouquin, T., Meier, C., Foster, R., Nielsen, M. E., Mundy, J. (2001). Control of specific gene expression by gibberellin and brassinosteroid. Plant Physiology 127: 450-458.

Braun, D. M., Stone, J. M., Walker, J. C. (1997). Interaction of the maize and *Arabidopsis* kinase interaction domains with a subset of receptor-like protein kinases: implications for transmembrane signaling in plants. Plant Journal 12: 83-95.

Burks, E. A., Bezerra, P. P., Le, H., Gallie, D. R., Browning, K. S. (2001). Plant initiation factor 3 subunit composition resembles mammalian initiation factor 3 and has a novel subunit. Journal of Biological Chemistry 276: 2122-2131.

Cabrera y Poch, H. L., Peto, C. A., Chory, J. (1993). A mutation in the *Arabidopsis DET3* gene uncouples photoregulated leaf development from gene expression and chloroplast biogenesis. Plant Journal 4: 671-682.

Cadigan, K. M., Nusse, R. (1997). Wnt signaling: a common theme in animal development. Genes Development 11: 3286-3305.

Chen, J. G. (2001). Dual auxin signaling pathways control cell elongation and division. Journal of Plant Growth Regulation 20: 255-264.

Chen, J. G., Shimomura, S., Sitbon, F., Sandberg, G., Jones, A. M. (2001). The role of auxin-binding protein 1 in the expansion of tobacco leaf cells. Plant Journal 28: 607-617.

Choe, S. W., Fujioka, S., Noguchi, T., Takatsuto, S., Yoshida, S., Feldmann, K. A. (2001). Overexpression of *DWARF4* in the brassinosteroid biosynthetic pathway results in increased vegetative growth and seed yield in *Arabidopsis*. Plant Journal 26: 573-582.

Clark, S. E. (2001). Cell signaling at the short meristem. National Review of Molecular and Cell Biology 2:276-284

Clark, S. E., Williams, R. W., Meyerowitz, E. M. (1997). The *CLAVATA1* gene encodes a putative receptor kinase that controls shoot and floral meristem size in *Arabidopsis*. Cell 89: 575-585.

Clifford, R., Schupbach, T. (1994). Molecular analysis of the *Drosophila* EGF receptor homolog reveals that several genetically defined classes of alleles cluster in subdomains of the receptor protein. Genetics 137: 531-550.

Clouse, S. D. (2002). Brassinosteroid signaling: Novel downstream components emerge. Current Biology 12: R485-R487.

Clouse, S. D., Feldmann, K. (1999). Molecular genetics of brassinosteroid action. In Brassinosteroids: Steroidal plant hormones, pp. 163-190. Eds A Sakurai, T Yokota and S D Clouse, Springer-Verlag, Tokyo.

Clouse, S. D., Langford, M., McMorris, T. C. (1996). A brassinosteroid-insensitive mutant in *Arabidopsis thaliana* exhibits multiple defects in growth and development. Plant Physiology 111: 671-678.

Darley, C. P., Forrester, A. M., McQueen-Mason, S. J. (2001). The molecular basis of plant cell wall extension. Plant Molecular Biology 47: 179-195.

D'Agostino, I. B., Kieber, J. J. (1999). Molecular mechanisms of cytokinin action. Current Opinion in Plant Biology 2: 359-364.

Ephritikhine, G., Fellner, M., Vannini, C., Lapous, D., Barbier-Brygoo, H. (1999). The *sax1* dwarf mutant of *Arabidopsis thaliana* shows altered sensitivity of growth responses to abscisic acid, auxin, gibberellins and ethylene and is partially rescued by exogenous brassinosteroid. Plant Journal 18: 303-314.

Evans, R. M. (1988). The steroid and thyroid hormone receptor superfamily. Science 240: 889-895.

Fleet, J. C. (1999). Vitamin D receptors: not just in the nucleus anymore. Nuture Review 57: 60-62.

Fletcher, J. C., Brand, U., Running, M. P., Simon, R., Meyerowitz, E. M. (1999). Signaling of cell fate decisions by CLAVATA3 in *Arabidospsis* shoot meristems. Science 283: 1911-1914.

Friedrichsen, D., Joazeiro, C. A. P., Li, J. M., Hunter, T., Chory, J. (2000). Brassinosteroid-insensitive-1 is a ubiquitously expressed leucine-rich repeat receptor serine/threonine kinase. Plant Physiology 123: 1247-1255

Friedrichsen, D., Chory, J. (2001). Steroid signaling in plants: from the cell surface to the nucleus. Bioessays 23: 1028-1036.

Goetz, M., Godt, D. E., Roitsch, T. (2000). Tissue-specific induction of the mRNA for an extracellular invertase isoenzyme of tomato by brassinosteroids suggests a role for steroid hormones in assimilate partitioning. Plant Journal 22: 515-522.

Goff, S. A., Ricke, D., Lan, T-H., Presting, G., Wang, R., Dunn, M., Glazebrook, J., Sessions, A., Oeller, P., Varma, H., Hadley, D., Hutchison, D., Martin, C., Katagiri, F., Lange, B. M., Moughamer, T., Xia, Y., Budworth, P., Zhong, J., Miguel, T., Paszkowski, U., Zhang, S., Colbert, M., Sun, W., Chen, L., Cooper, B., Park, S., Wood, T. C., Mao, L., Quail, P., Wing, R., Dean, R., Yu, Y., Zharkikh, A., Shen, R., Sahasrabudhe, S., Thomas, A., Cannings, R., Gutin, A., Pruss, D., Reid, J., Tavtigian, S., Mitchell, J., Eldredge, G., Scholl, T., Miller, R. M., Bhatnagar, S., Adey, N., Rubano, T., Tusneem, N., Robinson, R., Feldhaus, J., Macalma, T., Oliphant, A., Briggs, S. (2002). A draft sequence of the rice genome (*Oryza sativa* L. spp. *Japonica*). Science 296: 92-100.

Gomez-Gomez, L., Bauer, Z., Boller, T. (2001). Both the extracellular leucine-rich repeat and the kinase activity of FSL2 are required for flagellin binding and signaling in *Arabidopsis*. Plant Cell 13: 1155-1163.

Grove, M. D., Spencer, G. F., Rohwedder, W. K., Mandava, N. B., Worley, J. F., Warthen Jr, J. D., Steffens, G. L., Flippen-Anderson, J. L., Cook Jr, J. C. (1979). Brassinolide, a plant growth-promoting steroid isolated from *Brassica napus* pollen. Nature 281: 216-216.

Hashimoto, C., Hudson, K.L., Anderson, K. V. (1998). The *Toll* gene of *Drosophila*, required for dorsal-ventral

embryonic polarity, appears to encode a transmembrane protein. Cell 52: 269-279.
He, Z. H., Gendrom, J. M., Yang, Y., Li, J. K., Wang, Z. Y. (2002). The GSK-like kinase BIN2 phosphorylates and destabilizes BZR1, a positive regulator of brassinosteroid signaling pathway in *Arabidopsis*. Proceeding of National Academy of Sciences (USA) 99: 10185-10190.
He, Z. H., Wang, Z. Y., Li, J. K., Zhu, Q., Lamb, C., Ronald, P., Chory, J. (2000). Perception of brassinosteroids by the extracellular domain of the receptor kinase BRI1. Science 288: 2360-2363.
Hu, Y., Bao, F., Li, J. (2000). Promotive effect of brassinosteroids on cell division involves a distinct CycD3-induction pathway in *Arabidopsis*. Plant Cell 24: 693-701.
Hu, Y., Wang, Z., Wang, Y., Bao, F., Li, N., Peng, Z., Li, J. (2001). Identification of brassinosteroid responsive genes in *Arabidopsis* by cDNA array. Science China 44: 637-643.
Jang, J. C., Fujioka, S., Tasaka, M., Seto, H., Takatsuto, S., Ishii, A., Aida, M., Yoshida, S., Sheen, J. (2000). A critical role of sterols in embryonic pattering and meristem programming revealed by the *fackel* mutants of *Arabidopsis thaliana*. Genes Development 14: 1485-1497.
Jeong, S., Trotochaud, A. E., Clark, S. E. (1999). The *Arabidopsis CLAVATA2* gene encodes a receptor-like protein required for the stability of the CLAVATA1 receptor-like kinase. Plant Cell 11: 1925-1933.
Jones, D. A., Thomas, C. M., Hammondkosack, K. E., Balintkurti, P. J., Jones, J. D. G. (1994). Isolation of the tomato *Cf-9* gene for resistance to *Cladosporium fulvum* by transposon tagging. Science 266: 789-793.
Jiang, J., Clouse, S. D. (2001). Expression of a plant gene with sequence similarity to animal TGF-β receptor interacting protein is regulated by brassinosteroid and required for normal develoment. Plant Journal 26: 35-45.
Kang, J. G., Yun, J., Kim, D. H., Chung, K. S., Fujioka, S., Kim, J. I., Dae, H. W., Yoshida, S., Takatsuto, S., Song, P. S., Park, C. M. (2001). Light and brassinosteroid signals are integrated via a dark-induced small G protein in etiolated seedlings growth. Cell 105: 625-636.
Kauschmann, A., Jessop, A., Koncz, C., Szekeres, M., Willmitzer, L., Altmann, T. (1996). Genetic evidence for an essential role of brassinosteroids in plant development. Plant Journal 9: 701-713.
Kim, L., Kimmel, A. R. (2000). GSK3, a master switch regulating cell-fate specification and tomorigenesis. Current Opinion in Genetic Development 10: 508-514.
Kobe, B., Deisenhofer, J. (1994). The leucine-rich repeat: a versatile binding motif. Trends in Biochemical Science 19: 415-421.
Koka, C. V., Cerny, R. E., Gardner, R. G., Noguchi, T., Fujioka, S., Takatsuto, S., Yoshida, S., Clouse, S. D. (2000). A putative role of the tomato genes *DUMPY* and *CURL-3* in brassinosteroid biosynthesis and response. Plant Physiology 122: 85-98.
Lamport, D. T. (2001). Life behind cell walls: paradigm lost, paradigm regained. Cellular and Molecular Life Sciences 58: 1363-1385.
Lease, K. A., Ingham, E., Walker, J. C. (1998). Challenges in understanding RLK function. Current Opinion in Plant Biology 1: 388-392.
Li, J., Chory, J. (1997). A putative leucine-rich repeat receptor kinase involved in brassinosteroid signal transduction. Cell 90: 929-938.
Li, J., Lease, K. A., Tax, F. E., Walker, J. C. (2001a). BRS1, a serine carboxypeptidase, regulates BRI1 signaling in *Arabidopsis thaliana*. Proceedings of National Academy of Sciences (USA) 98: 5916-5921.
Li, J., Nam, K. H., Vafeados, D., Chory, J. (2001b). *BIN2*, a new brassinosteroid-insensitive locus in *Arabidopsis*. Plant Physiology 127: 14-22.
Li, J., Nam, K. H. (2002). Regulation of brassinoseroid signaling by a GSK3/SHAGGY-like kinase. Science 295: 1299-1301.
Li, J., Wen, J., Lease, K. A., Doke, J. T., Tax, F. E., Walker, J. C. (2002). BAK1, an *Arabidopsis* LRR receptor-like protein kinase, interacts with BRI1 and modulates brassinosteroid signaling. Cell 110: 213-222.
Mandava, N. B. (1988). Plant growth-promoting brassinosteroids. Annual Review of Plant Physiology and Plant Molecular Biology 39: 23-52.
Mangelsdorf, D. J., Thummel, C., Beato, M., Herrlich, P., Schultz, G., Umesono, K., Blumberg, B., Kastner, P., Mark, M., Chambon, P., Evans, R. M. (1995). The nuclear receptor superfamily: the second decade. Cell 83: 835-839.
Mathur, J., Molnár, G., Fujioka, S., Takatsuto, S., Sakurai, A., Yokota, T., Adam, G., Voigt, B., Nagy, F., Mass, C., Schell, J., Koncz, C., Szekeres, M. (1998). Transcription of the Arabidopsis CPD gene, encoding a steroidogenic cytochrome P450, is negatively controlled by brassinosteroids. Plant Journal 14: 593-602.

Massagué, J. (1998). TGF-β signal transduction. Annual Review of Biochemistry 67: 753-791.

Matsubayashi, Y., Ogawa, M., Morita, A., Sakagami, Y. (2002). An LRR receptor kinase involved in perception of a peptide plant hormone, phytosulfokine. Science 296: 1470-1472.

Miller, J. R., Hocking, A. M., Brown, J. D., Moon, R. T. (1999). Mechanism and function by the Wnt/β-catenin and Wnt/$Ca^{2+}$ pathways. Oncogene 18: 7860-7872.

Mitchell, J. W., Mandava, N. B., Worley, J. F., Plimmer, J. R., Smith, M. V. (1970). Brassins: a new family of plant hormones from rape pollen. Nature 225: 1065-1066.

Molnár, G., Bancoş, S., Nagy, F., Szekeres, M. (2002). Characterization of BRH1, a brassinosteroid-responsive RING-H2 gene from *Arabidopsis thaliana*. Planta 215: 127-133.

Müssig, C., Altmann, T. (2001). Brassinosteroid signaling in plants. Trends in Endocrinology Metabolism 12: 398-402.

Müssig, C., Altmann, T. (1999). Physiology and molecular mode of action of brassinosteroids. Plant Physiol and Biochemistry 37: 363-372.

Müssig, C., Fischer, S., Altmann, T. (2002). Brassinosteroid-regulated gene expression. Plant Physiology 129: 1241-1251.

Nam, K. H., Li, J. (2002). BRI1/BAK1, a receptor kinase pair mediating brassinosteroid signaling. Cell 110: 203-212.

Neff, M. M., Nguyen, S. M., Malancharuvil, E. J., Fujioka, S., Noguchi, T., Seto, H., Tsubuki, M., Honda, T., Takatsuto, S., Yoshida, S., Chory, J. (1999). BAS1: A gene regulating brassinoseroid levels and light responsiveness in *Arabidopsis*. Proceeding of National Academy of Sciences (USA) 96: 15316-15323.

Nicol, F., His, I., Jauneau, A., Vernhettes, S., Canut, H., Hofte, H. (1998). A plasma membrane-bound putative endo-1,4-β-D-glucanase is required for normal wall assembly and cell elongation in *Arabidopsis*. EMBO Journal 17: 5563-5576.

Noguchi, T., Fujioka, S., Choe, S., Takatsuto, S., Tax, F. E., Yoshida, S., Feldmann, K. A. (2000). Biosynthetic pathways of brassinolide in Arabidopsis. Plant Physiology 124: 201-209.

Noguchi, T., Fujioka, S., Choe, S., Takatsuto, S., Yoshida, S., Yuan, H., Feldmann, K. A., Tax, F. E. (1999). Brassinosteroid-insensitive dwarf mutants of Arabidopsis accumulate brassinosteroids. Plant Physiology 121: 743-752.

Nomura, T., Kitasaka, Y., Takatsuto, S., Reid, J. B., Fukami, M., Yokota, T. (1999). Brassinosteroid/Sterol synthesis and plant growth as affected by lka and lkb mutations of Pea. Plant Physiology 119: 1517-1526.

Oh, M. H., Ray, W. K., Huber, S. C., Asara, J. M., Gage, D. A., Clouse, S. D. (2000). Recombinant brassinosteroid insensitive 1 receptor-like kinase autophosporylates on serine and threonine residues and phosphorylates a conserved peptide motif in vitro. Plant Physiology 124: 751-765.

Ostman, A., Bohmer, F. D. (2001). Regulation of receptor tyrosine kinase signaling by protein tyrosine phosphatases. Trends in Cell Biology 11: 258-266.

Peng, P., Zhao, J., Li, J., Schmitz, R. J., Tax, F. E. (2002). Two putative GSK3 substrates are nuclear components of plant steroid signaling. Plant Biology 2002, Abstract #512.

Pérez-Pérez, J. M., Ponce, M. R., Micol, J. L. (2002). The UCU1 Arabidopsis gene encodes a SHAGGY/GSK3-like kinase required for cell expansion along the proximodistal axis. Developmental Biology 242: 161-173.

Polakis, P. (2000). Wnt signaling and cancer. Genes and Development 14: 1837-1851.

Reid, J. B., Howell, S. H. (1995). Hormone mutants and plant development. In Plant Hormones. Physiology, Biochemistry and Molecular Biology, pp. 448-485. Eds P J Davies. Kluwer Academic Publishers, Dordrecht.

Scheer, J. M., Ryan, Jr. C. A. (2002). The systemin receptor SR160 from *Lycopersicon peruvianum* is a member of the LRR receptor kinase family. Proceeding of Natational Academy of Sciences (USA) 99: 9585-9590.

Schlessinger, J. (2000). Cell signaling by receptor tyrosine kinase. Cell 103: 211-225.

Schmidt, B. M. W., Gerdes, D., Feuring, M., Falkenstein, E., Christ, M., Wehling, M. (2000). Rapid, nongenomic steroid actions: a new age? Front Neuroendocrinology 21: 57-94.

Schumacher, K., Chory, J. (2000). Brassinosteroid signal transduction: Still casting the actors. Current Opinion in Plant Biology 3: 79-84.

Schumacher, K., Vafeados, D., McCarthy, M., Sze, H., Wilkins, T., Chory, J. (1999). The Arabidopsis det3 mutant reveals a central role for the vacuolar $H^+$-ATPase in plant growth and development. Genes and Development 13: 3259-3270.

Sharma, A., Matsuoka, M., Tanaka, H., Komatsu, S. (2001). Antisense inhibition of a BRI1 receptor reveals additional protein kinase signaling components downstream to the perception of brassinosteroid in rice. FEBS Letter 507: 346-350.

Shiu, S. H., Bleecker, A. B. (2001). Plant receptor-like kinase family: Diversity, function, and signaling. Sci STKE 2001, http://stke.sciencemag.org/cgi/content/full/OC_sigtrans; 2001/113/re22.

Simin, K., Bates, E. A., Horner, M. A., Letsou, A. (1998). Genetic analysis of punt, a type II Dpp receptor that functions throughout the Drosophila melanogaster life cycle. Genetics 148: 801-813.

Song, W. Y., Wang, G. L., Chen, L. L., Kim, H. S., Pi, L. Y., Holsten, T., Gardner, J., Wang, B., Zhai, W. X., Zhu, L. H., Fauquet, C., Ronald, P. (1995). A receptor kinase-like protein encoded by the rice disease resistance gene Xa21. Science 270: 1804-1806.

Stone, J. M., Trotochaud, A. E., Walker, J. C., Clark, S. E. (1998). Control of meristem development by CLAVATA1 receptor kinase and kinase-associated protein phosphatase interactions. Plant Physiology 117: 1217-1225.

Stone, J. M., Collinge, M. A., Smith, R. D., Horn, M. A., Walker, J. C. (1994). Interaction of a protein phosphatase with an Arabidopsis serine/threonine receptor kinase. Science 266: 793-795.

Szekeres, M., Németh, K., Koncz-Kálmán, Z., Mathur, J., Kauschmann, A., Altmann, T., Rédei, G.P., Nagy, F., Schell, J., Koncz, C. (1996). Brassinosteroids rescue the deficiency of CYP90, a cytochrome P450, controlling cell elongation and de-etiolation in *Arabidopsis*. Cell 85: 171-182.

The Arabidopsis Genome Initiative. 2000. Analysis of the genome sequence of the flowering plant *Arabidopsis thaliana*. Nature 408: 796-815.

Torii, K. U., Mitsukawa, N., Oosumi, T., Matsuura, Y., Yokoyama, R., Whittier, R. F., Komeda, Y. (1996). The Arabidopsis ERECTA gene encodes a putative receptor protein kinase with extracellular leucine-rich repeats. Plant Cell 8: 735-746.

Trotochaud, A. E., Hao, T., Wu, G., Yang, Z. B., Clark, S. E. (1999). The CLAVATA1 eceptor-like kinase requires CLAVATA3 for its assembly into a signaling complex that includes KAPP and Rho-related protein. Plant Cell 11: 393-405.

Wang, Z. Y., Nakano, T., Gendron, J., He, J., Chen, M., Vafeados, D., Yang, Y., Fujioka, S., Yoshida, S., Asami, T., Chory, J. (2002). Nuclear-localized BZR1 mediates brassinosteroid-induced growth and feedback suppression of brassinosteroid biosynthesis. Development in Cell 2: 505-513.

Wang, Z. Y., Seto, H., Fujioka, S., Yoshida, S., Chory, J. (2001). BRI1 is a critical component of a plasma-membrane receptor for plant steroids. Nature 410: 380-383.

Wang, Z. Y., Vafeados, D., Cheong, H., Redfern, J., Friedrichsen, D., Chory, J. (2000). Molecular and genetic analysis of brassinosteroid signaling pathways. Plant Biology 2000, Abstract #726.

Watson, C. S., Gametchu, B. (1999). Membrane-initiated steroid actions and the proteins that mediate them. Experimental Biology and Medicine 220: 9-19.

Wehling, M. (1997). Specific, nongenomic actions of steroid hormones. Annual Review of Physiology 59: 365-393.

Williams, R. W., Wilson, J. M., Meyerowitz, E. M. (1997). A possible role for kinase-associated protein phosphatase in the Arabidopsis CLAVATA1 signaling pathway. Proceeding of National Academy of Science (USA) 94: 10467-10472.

Yang, G., Komatsu, S. (2000). Involvement of calcium-dependent protein kinase in rice (Oryza sativa L.) lamina inclination caused by brassinolide. Plant Cell Physiology 41: 1243-1250.

Yamamuro, C., Ihara, Y., Wu, X., Noguchi, T., Fujioka, S., Takatsuto, S., Ashikari, M., Kitano, H., Matsuoka, M. (2000). Loss of function of a rice brassinosteroid insensitive1 homolog prevents internode elongation and bending of the lamina joint. Plant Cell 12: 2591-1606.

Yokota, T. (1997). The structure, biosynthesis and function of brassinosteroids. Trends in Plant Science 2: 137-143.

Yoshizumi, T., Nagata, N., Shimada, H., Matsui, M. (1999). An *Arabidopsis* cell cycle-dependent kinase-related gene, CDC2b, plays a role in regulating seedling growth in darkness. Plant Cell 11: 1883-1896.

Yin, Y., Cheong, H., Friedrichsen, D., Zhao, Y., Hu, J., Mora-Gaarcia, S., Chory, J. (2002b). A crucial role for the putative Arabidopsis topoisomerase VI in plant growth and development. Proceeding of National Academy of Science (USA) 99: 10191-10196.

Yin, Y., Wang, Z. Y., Mora-Gaarcia, S., Li, J., Yoshida, S., Asami, T., Chory, J. (2002a). BES1 accumulates in the nucleus in response to brassinosteroids to regulate gene expression and promote stem elongation. Cell 109: 181-191.

Yu, J., Hu, S., Wang, J., Wong, G. K-S., Li, S., Liu, B., Deng, Y., Dai, L., Zhou, Y., Zhang, X., Cao, M., Liu, J., Sun, J., Tang, J., Chen, Y., Huang, X., Lin, W., Ye, C., Tong, W., Cong, L., Geng, J., Han, Y., Li, L., Li, W., Hu, G., Huang, X., Li, W., Li, J., Liu, Z., Li, L., Liu, J., Qi, Q., Liu, J., Li, L., Li, T., Wang, X., Lu, H., Wu, T., Zhu, M., Ni, P., Han, H., Dong, W., Ren, X., Feng, X., Cui, P., Li, X., Wang, H., Xu, X., Zhai, W., Xu, Z., Zhang, J., He, S., Zhang, J., Xu, J., Zhang, K., Zheng, X., Dong, J., Zeng, W., Tao, L., Ye, J., Tan, J., Ren, X., Chen, X., He, J., Liu, D., Tian, W., Tian, C., Xia, H., Bao, Q., Li, G., Gao, H., Cao, T., Wang, J., Zhao, W., Li, P., Chen, W., Wang, X., Zhang, Y., Hu, J., Wang, J., Liu, S., Yang, J., Zhang, G., Xiong, Y., Li, Z., Mao, L., Zhou, C., Zhu, Z., Chen, R., Hao, B., Zheng, W., Chen, S., Guo, W., Li, G., Liu, S., Tao, M., Wang, J., Zhu, L., Yuan, L., Yang, H.(2002). A draft sequence of the rice genome (*Oryza sativa* L. spp. *indica*). Science 296: 79-92.

CHAPTER 4

MIRIAM NÚÑEZ VÁZQUEZ, CARIDAD ROBAINA RODRÍGUEZ AND FRANCISCO COLL MANCHADO

# SYNTHESIS AND PRACTICAL APPLICATIONS OF BRASSINOSTEROID ANALOGS

Brassinosteroids, a new class of plant hormones of steroidal nature, show intriguing biological properties being now recognized as useful compounds in growth regulation, increasing crop yield and tolerance to abiotic stresses. Therefore, an extensive research work was planned on chemistry, biochemistry, physiology, molecular action and the synthesis of these compounds and their analogs, as the basis for further practical applications. Consequently, in Cuba, for fifteen years, the Laboratory of Natural Products from the University of Havana has been searching for new brassinosteroid analogs with biological activity that may be successfully applied to agriculture. This chapter summarizes the synthesis of several brassinosteroid analogs with structural variations in different parts of the molecules, starting from available steroids such as diosgenin, hecogenin, solasodine, solanidine and bile acids. Besides, the main results of practical applications of some formulations, which have spirostanic analogs as active ingredients, to agricultural fields and plant biotechnology are discussed.

## INTRODUCTION

Brassinosteroids, a new class of plant hormones of steroidal nature, show intriguing biological properties being now recognized as useful compounds in growth regulation, increasing crop yield and tolerance to abiotic stresses (Adam and Marquardt, 1986; Mandava, 1988; Khripach *et al*., 1991; Marquardt and Adam, 1991; Iglesias *et al*., 1996a; Kamuro and Takatsuto, 1999; Khripach *et al*., 1999; Núñez, 1999; Núñez and Robaina, 2000; Núñez and Mazorra, 2001).

Since the discovery of brassinolide [2α, 3α, 22 (R), 23 (R)-tetrahydroxy-24 (S)-methyl-B-homo-7-oxa-5α-cholestan-6-one; 1] (Figure1) in 1979 by Grove *et al*., an extensive research work has been done in the field of brassinosteroids, such as isolation methods, structural elucidation studies, chemistry, biochemistry, physiology and molecular action as well as the synthesis of these compounds and several related bioactive steroids (the last one called brassinosteroid analogs) as the basis for further biological investigations and practical applications.

*S.Hayat and A.Ahmad (eds.), Brassinosteroids,* 87-117.

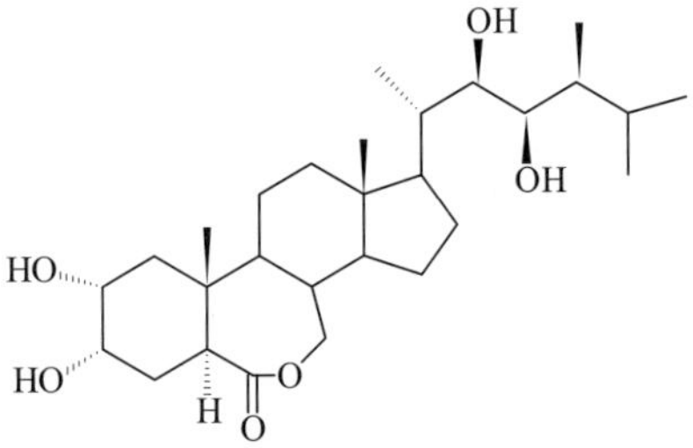

*Fig 1. Structure of brassinolide*

The synthesis of new analogs is a very important task for scientific research, as it enables the studies of structure-activity relationships, biosynthesis and metabolism of the natural brassinosteroids. Therefore, many analogs have been prepared with several variations of the steroidal skeleton and the side chain. Some excellent reviews about this subject until 1999 are available (Adam *et al.*, 1991; Khripach *et al.*, 1991; Kohout *et al.*, 1991; Khripach *et al.*, 1999).

On the other hand, different authors have shown that brassinosteroids have a strong influence on plant growth; since 1981, Maugh considered them promising compounds for application in agriculture, because they show various kinds of regulatory activity in plant growth and development, besides their economic value as yield-promoting agents was predicted by Cutler early in 1990s (Khripach *et al.*, 2000).

In field trials, three natural brassinosteroids (brassinolide, 24-epibrassinolide and 28-homobrassinolide) have been tested to determine their influence on crop yield, under natural conditions; however, even though the activity of brassinolide is higher in bioassays (Adam and Marquardt, 1986), most of the results have been obtained using 24-epibrassinolide and 28-homobrassinolide, because they can not be only more easily synthesised than brassinolide but they also present a higher stability under field conditions.The practical use of brassinosteroids as yield promoters has been previously confirmed by studies in some countries such as Japan, China and USSR and the results summarised in different reviews (Fujita, 1985; Abe, 1989; Ikekawa and Zhao, 1991; Khripach *et al.*, 1997; Kamuro and Takatsuto, 1999). Thus, a preparation with 24-epibrassinolide as active ingredient was extensively investigated in Japan in the middle of the 80's (Nippon Kayaku Co., 1988) whereas another one, which is mainly a mixture of 24-epibrassinolide and its 22S, 23S isomer, was developed in China as a consequence of a joint research program between Japanese and Chinese scientists (Ikekawa and Zhao, 1991). Furthermore, in China, (22S, 23S)-28-homobrassinolide was registered as a plant growth regulator to increase the yields of tobacco, sugarcane, tea and some fruits (Kamuro and Takatsuto, 1999), whereas in Russia and Belarus, 24-epibrassinolide is the active ingredient of the plant growth promoter Epin®, which has been officially registered since 1992 and its application has been recommended to tomato, potato, pepper, cucumber and barley crops (Khripach *et al.*, 1999). In other countries, like India, Ramraj *et al.* (1997)

demonstrated that foliar sprays of 28-homobrassinolide significantly increased wheat, rice and mustard grain yield, groundnut pod yield, potato tuber yield and cotton seed yield.

In spite of all above described, the extent of effectiveness and the stability of results of the applications of natural brassinosteroids have decreased, so giving rise to trials with these compounds were suspended in Japan and European countries during the 90's (Kamuro and Takatsuto, 1999). Therefore, Sasse (1997) pointed out that it was necessary to evaluate the application of structurally related compounds of long-lasting activity, which can be transformed by plants in active brassinosteroids.

From this point of view, Takatsuto *et al.* (1996) and Kamuro *et al.* (1997) reported a brassinosteroid analog termed TS 303, [(22R, 23R)-epoxy-HBl 2,3-diacetate], viable for agricultural application, since it has shown to be better than brassinolide, 24-epibrassinolide and (22S, 23S)-28-homobrassinolide under a wide variety of field trial conditions. Besides, the effects of combining with other plant growth regulators have been explored (Kamuro *et al.*, 1996) and the use of this brassinosteroid precursor seems to be very promising in rice crop (Takeuchi *et al.*, 1996). However, more recent investigations, in Eastern Europe, have demonstrated that natural brassinosteroids were not only able to increase crop yield but also to enhance plant resistance against diseases and viral infections (Bobrick *et al.*, 1998; Korableva *et al.*, 1999; Roth *et al.*, 2000). This poorly investigated characteristic together with known abiotic stress-protective properties of brassinosteroids (Sairam, 1994a,b; Xu *et al.*, 1994; Sairam *et al.*, 1996; Kumawat *et al.*, 1997; Vardhini and Rao, 1997; Hotta *et al.*, 1998; Li and Van Staden, 1998; Dbaubhadel *et al.*, 1999) open new possibilities for the application of these compounds and their analogs in agriculture. This is because the use of very small amounts of environment-friendly substances may increase crop yield, under stressed environments and protects plants against diseases, reducing or substituting the use of traditional pesticides, which are often environmentally unfriendly.

Consequently, in Cuba, since the last fifteen years, the Laboratory of Natural Products from the University of Havana has been searching for new brassinosteroid analogs with biological activity that may be successfully applied to agriculture.

## SYNTHESIS OF BRASSINOSTEROID ANALOGS

### *Analogs from steroidal sapogenins*

Steroidal sapogenins constitute a well-known family of oxygenated steroids, which are widely distributed as glycosides in plant kingdom. These types of compounds show the spiroketalic side chain with oxygenated function at the C-16 and C-22, which can be used for preparing a wide variety of brassinosteroid analogs.

*Starting with diosgenin*

*a) Spirostanic analogs*: Diosgenin **2** is a commercially available steroid that has suitable functional group in the steroidal nucleus for preparing the corresponding analogs.

Synthetic route to spirostanic analogs of the naturally occurring brassinosteroids, castasterone and brassinolide, from diosgenin **2** is shown in Scheme 1. The formation of 2α,3α-dihydroxy-6-ketone compound **5** is carried out by classical methods via mesylation, isosteroidal rearrangement with potassium bicarbonate, oxidation with Jones reagent, isomerization with lithium bromide/lithium carbonate and dihydroxylation with osmium tetroxide. Baeyer-Villiger oxidation of 6-oxosteroid **5** with trifluoroperoxyacetic acid gave mainly the 7-oxa-6-oxo-lactone **6** (Marquardt *et al.,* 1989; Alonso, 1990; Adam *et al.*, 1991; Tian, 1992; Iglesias *et al.,* 1998a).

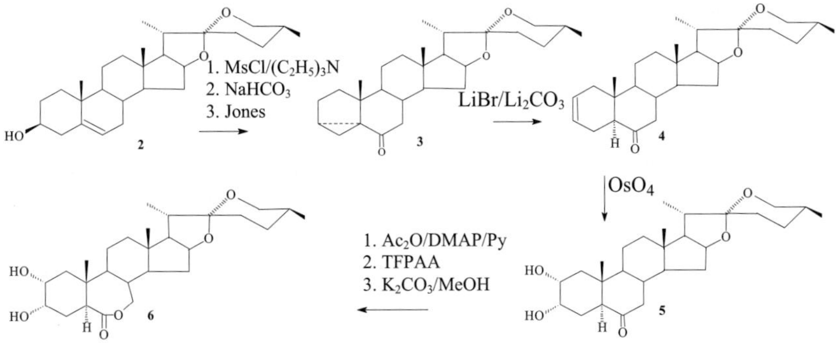

1. $MsCl/(C_2H_5)_3N$
2. $NaHCO_3$
3. Jones

$LiBr/Li_2CO_3$

$OsO_4$

1. $Ac_2O/DMAP/Py$
2. TFPAA
3. $K_2CO_3/MeOH$

*Scheme1*

A number of analogs containing hydroxyl group at C-5 have been prepared (Kovganko and Ananich, 1991; Brosa *et al.,* 1996, 1998; Ramírez *et al.,* 2000a,b). Structures of some of these analogs are shown in Figure 2 which are applicable to agriculture.

The synthesis of two 5α-hydroxy-6-ketone spirostanic analogs from diosgenin **2** is shown in Scheme 2. The $\Delta^5$-double bond of acetate derivatives is epoxidized, followed by oxidative opening of the epoxy ring and basic hydrolysis to afford the 3α, 5α-dihydroxy-6-ketone **7**. Functionalization of the cyclic part as usual manner, afforded the triol **8** (Coll, *et al.,* 1995; Jomarrón, 1995; Dago *et al.,* 1997; Hechavarría, 1998).

Other studies indicate that the introduction of a carbonyl group at C-23 of synthetic spirostanic sapogenins with brassinosteroid-like rings A and B, produces biological activity, compared with the C-23, non-functionalized compound (Iglesias, 1996; Iglesias *et al.*, 1997a,b). It is natural to prepare the corresponding analogs with this group in the spiroketalic side chain, for evaluating their plant growth promoting activity.

Several spirostanic analogs with oxygenated function at C-23 have been

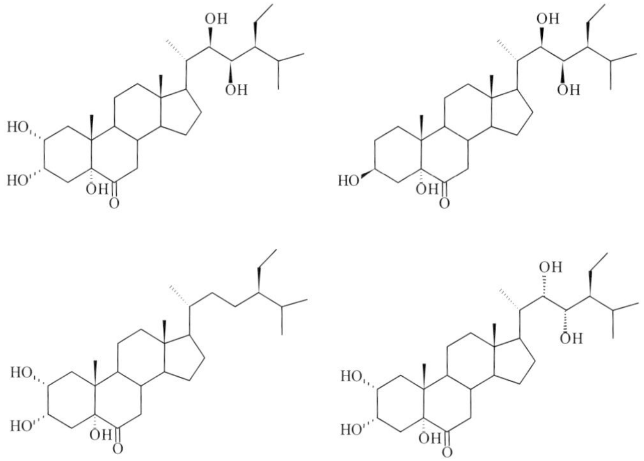

*Fig2. Examples of 5α-hydroxy analogs*

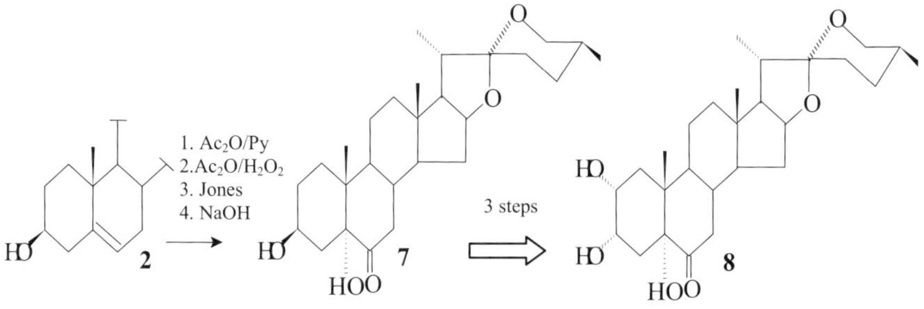

1. $Ac_2O/Py$
2. $Ac_2O/H_2O_2$
3. Jones
4. NaOH

3 steps

*Scheme 2*

prepared by Iglesias (1996) and Iglesias *et al*. (1998b, 2002). Opening of the cyclopropane ring of intermediate **3** with hydrobromic acid, followed by reduction of the 6-ketone with sodium borohydride and further dehydrobromination provided the $\Delta^2$-6β-hydroxy-steroid **9** (Scheme 3). Hydroxylation with osmium tetroxide and acetylation led to the triacetate **10**, which upon treatment with sodium nitrite and boron trifluoride etherate in acetic acid, followed by column chromatography in alumina gave 23-ketocompound **11**. The alkaline hydrolysis of the latter, followed by transketalization, led to acetonide **12**, which upon oxidation of the 6β-hydroxyl group with pyridinium chlorochromate and further acid hydrolysis of the protecting ketal afforded the 2α,3α-dihydroxy-23-ketosteroid **13**.

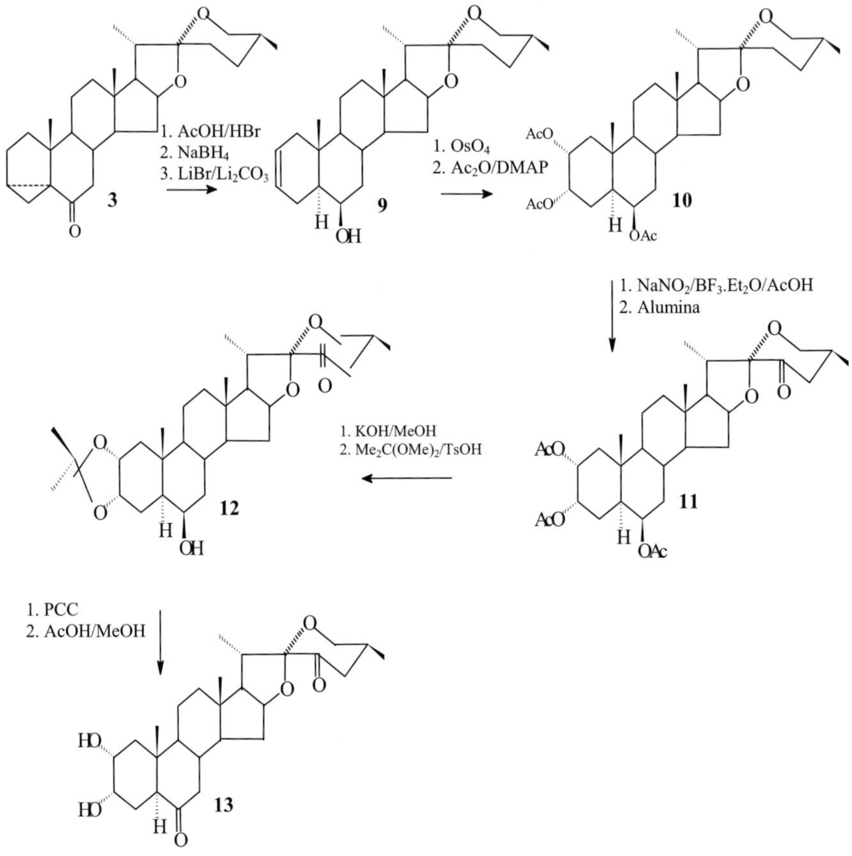

3

1. AcOH/HBr
2. $NaBH_4$
3. $LiBr/Li_2CO_3$

9

1. $OsO_4$
2. $Ac_2O$/DMAP

10

1. $NaNO_2/BF_3.Et_2O$/AcOH
2. Alumina

11

1. KOH/MeOH
2. $Me_2C(OMe)_2$/TsOH

12

1. PCC
2. AcOH/MeOH

13

*Scheme 3*

*b) Furostanic analogs*: A synthetic route for analogs with furostanol skeleton was adopted by Iglesias *et al.* (1996b, 1998c,d). Catalytic hydrogenation of **10** over palladium oxide on acetic acid with a few drops of perchloric acid provided the furostanol **14** (Scheme 4). Saponification of furostanol **14**, followed by transketalization, selective oxidation of 6β-hydroxyl group with n-bromo succinimide and further acid hydrolysis gave furostanic analog of castasterone **15**.

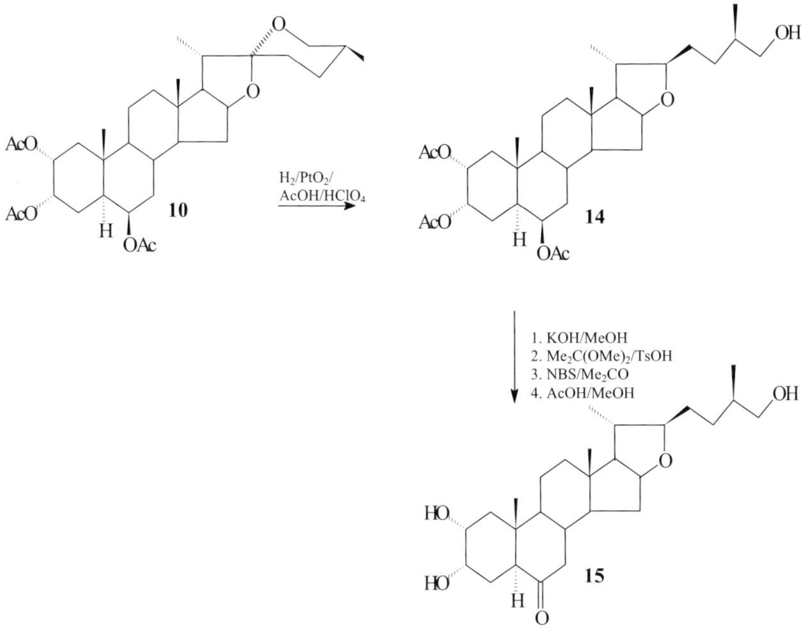

AcO, AcO, H, OAc — **10**

$H_2/PtO_2/$
$AcOH/HClO_4$

AcO, AcO, H, OAc, OH — **14**

1. KOH/MeOH
2. $Me_2C(OMe)_2/TsOH$
3. $NBS/Me_2CO$
4. AcOH/MeOH

HO, HO, H, O, OH — **15**

*Scheme4*

*c) Pregnanic and androstanic analogs*: Many analogs with pregnane and androstane skeleton have been prepared by Kohout *et al*. (1991, 1996, 1998) (Figure 3).

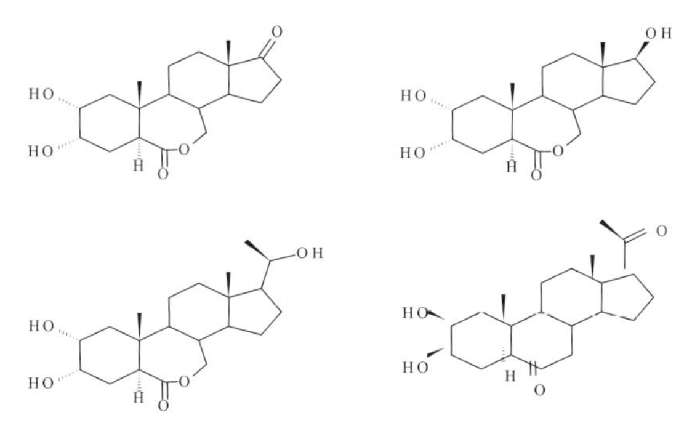

*Figure 3.Examples of analogs of the androstane and pregnane series*

Synthesis of analogs of these types of skeleton with 5α hydroxyl group starting from 16-dehydropregnenolone acetate **16** are shown in Scheme 5 (García, 2002).

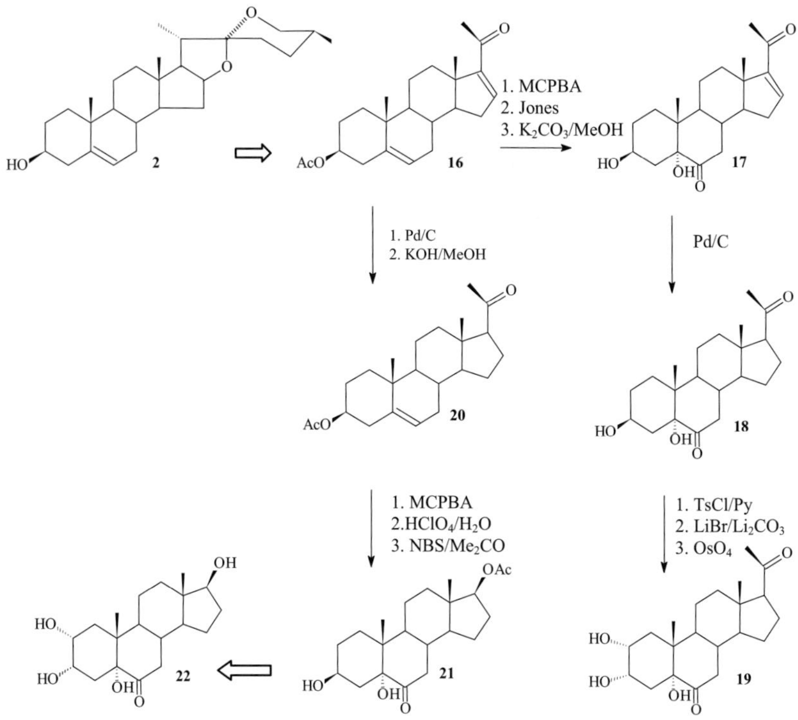
HO
2
AcO
16
1. MCPBA
2. Jones
3. $K_2CO_3$/MeOH
HO
OH
17
1. Pd/C
2. KOH/MeOH
Pd/C
AcO
20
HO
OH
18
1. MCPBA
2.$HClO_4/H_2O$
3. NBS/$Me_2CO$
1. TsCl/Py
2. LiBr/$Li_2CO_3$
3. $OsO_4$
OH
HO
HO
OH
22
OAc
HO
OH
21
HO
HO
OH
19

*Scheme 5*

*d) Other analogs*: Other brassinosteroid analogs, starting from diosgenin has been prepared. Structures of some of them are summarized in Figures 4, 5 and 6.

*Starting with hecogenin*

*Spirostanic analogs*: Till this date, the isolation of natural brassinosteroids with oxygenated function in ring C has not been reported. However, the synthesis of this type of analogs is a very important for structure - activity studies of this kind of phytohormone.

Hecogenin **23** is an abundantly available steroidal sapogenin, which is used as raw material in the production of corticosteroids for Pharmaceutical Industry.

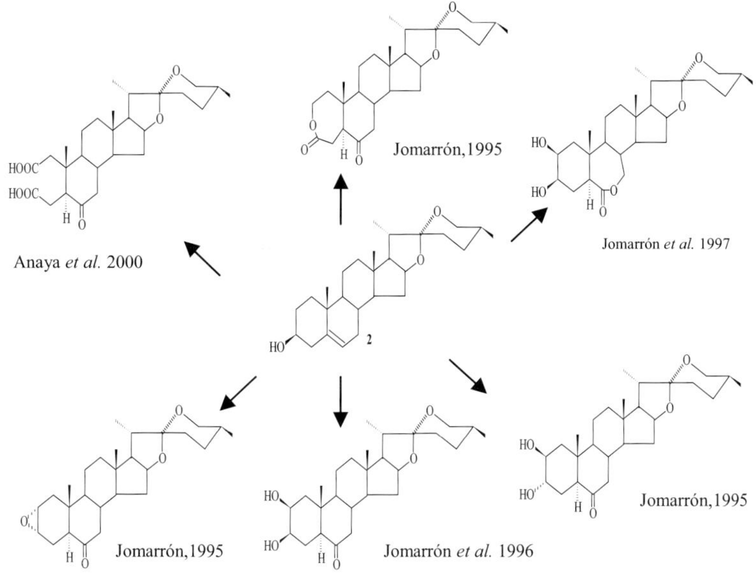


*Figure 4. Examples of analogs with spiroketal side chain*

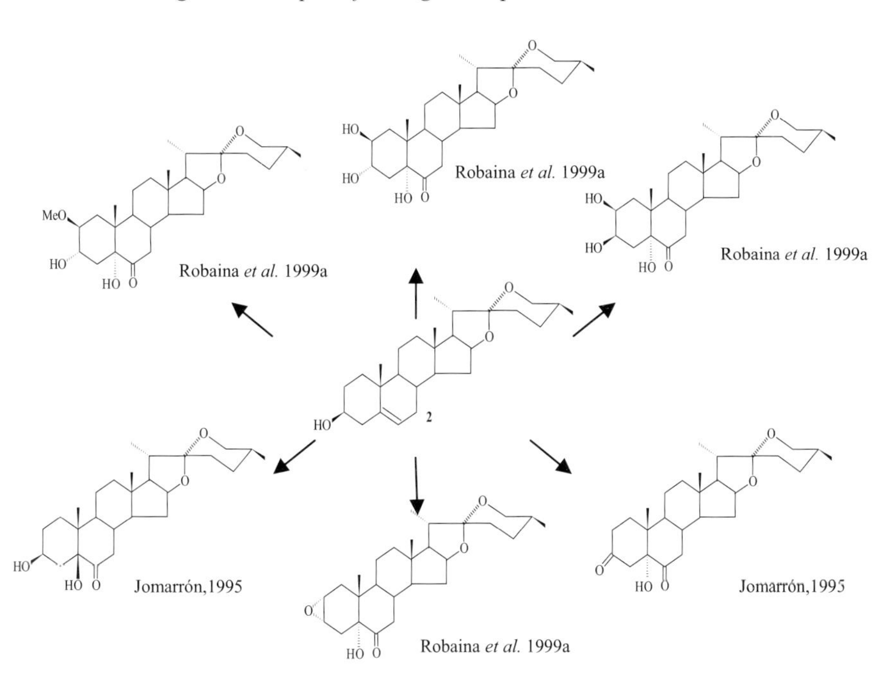


*Figure 5. Examples of analogs with spiroketal side chain and hydroxyl group at C-5*

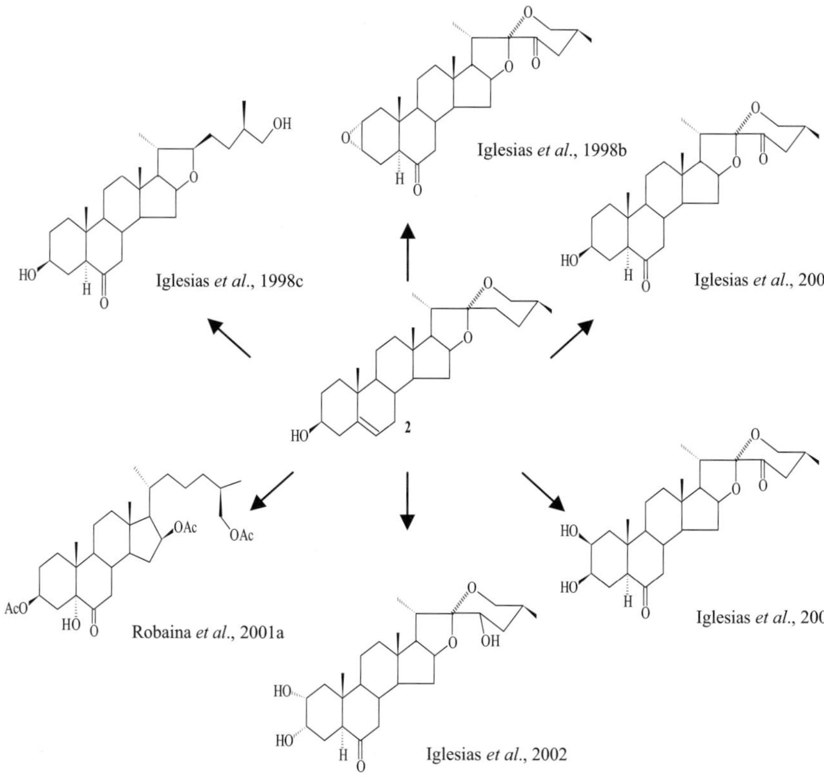

Iglesias *et al.*, 1998b

Iglesias *et al.*, 1998c

Iglesias *et al.*, 2001a

Robaina *et al.*, 2001a

Iglesias *et al.*, 2001b

Iglesias *et al.*, 2002

*Figure 6.Examples of analogs with modifications in the side chain*

The presence of carbonyl group at C-12 in the cyclic part of hecogenin molecule makes it an interesting substrate for the synthesis of brassinosteroid analogs.Synthesis of spirostanic analogs with oxygenated function in ring C starting from hecogenin **23**, is a relatively simple task from a chemical point of view.

A number of analogs containing different oxygenated function in ring C have been prepared by Robaina (1994) and Robaina *et al.* (1995a,b, 1996, 2001b,c). Some of them are shown in Scheme 6. The $\Delta^{2}$-olefin **24** is the key intermediate in this approach. Treatment with m-chloroperbenzoic acid led to the epoxy ketone **25** and epoxy lactone **26**. The Prêvost - Woodward oxidation of alkene **24** followed by basic hydrolysis gave 2β,3β,-dihydroxy-12-ketosteroid **27**.

A 23-oxygenated analog **31** has been synthesized by Iglesias *et al.* (2000). Reduction of 12-ketone with sodium borohydride in methanol led to a mixture of epimeric alcohols at C-12, which were acetylated by standard procedure (Scheme 7). Fractional crystallization of the mixture afforded the desired 3β, 12β-diacetate

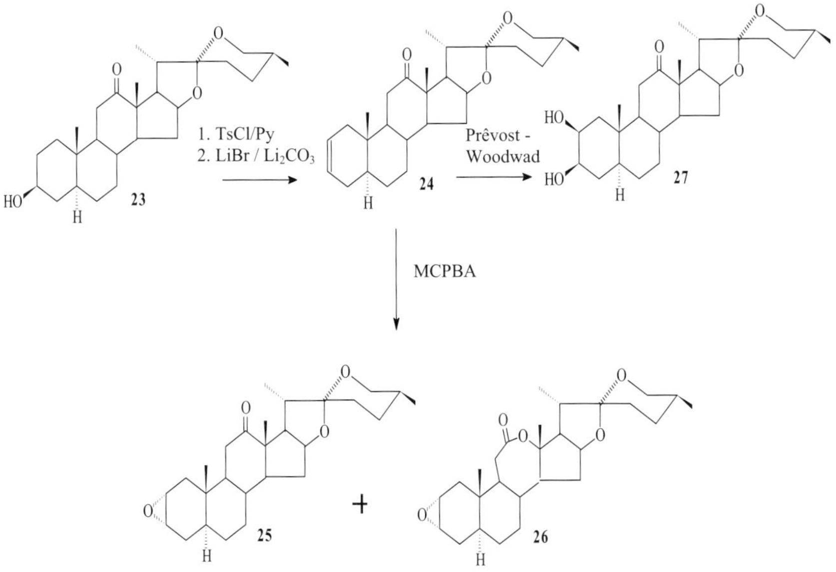

Scheme 6

epimer **28**, which was converted into diacetylated ketone **29** by the method described for compound **11** (see Scheme 3). Partial saponification of 29 led to monoacetate **30**. Protection of the 3β-hydroxyl group as tetrahydropyranil ether and saponification of the 12β-acetate followed by oxidation with pyridinium chlorochromate and further acid hydrolysis of the protecting group afforded the 12-23-dione **31**.

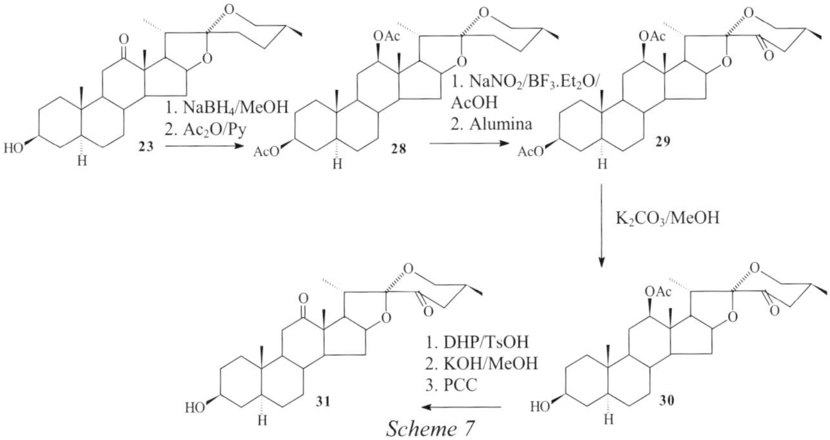

Scheme 7

Some other analogs starting from hecogenin **23** are summarized in Figure 7.

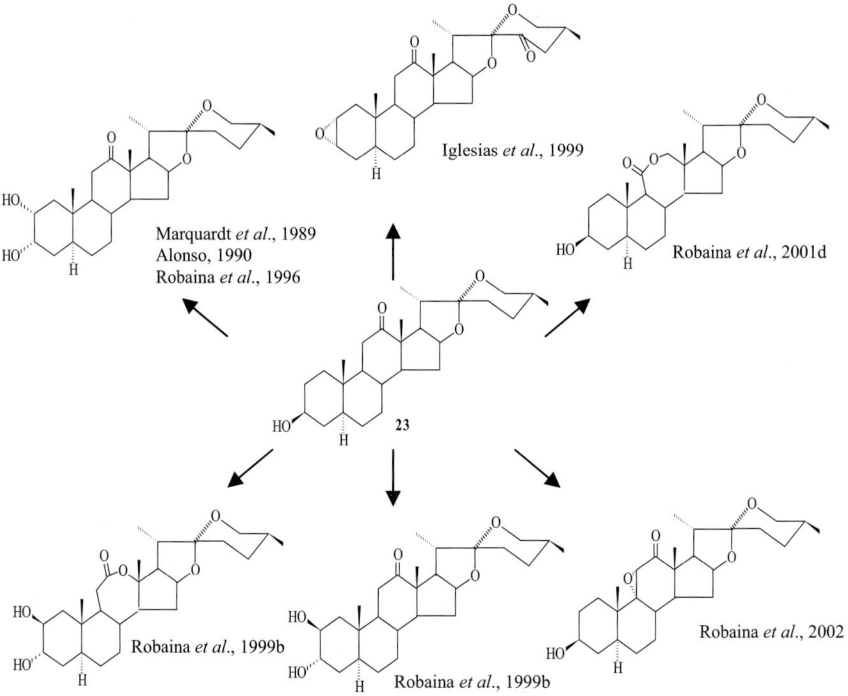

Iglesias *et al.*, 1999

Marquardt *et al.*, 1989
Alonso, 1990
Robaina *et al.*, 1996

Robaina *et al.*, 2001d

23

Robaina *et al.*, 1999b

Robaina *et al.*, 1999b

Robaina *et al.*, 2002

*Figure 7. Structures of other analogs starting from hecogenin (23)*

*Analogs from steroidal alkaloids*

*Starting with solasodine and solanidine*

Some nitrogenous brassinosteroid analogs have been prepared from *Solanum* steroidal alkaloids such as solasodine and solanidine (Quyen *et al.*, 1994a,b). Structure of some of them with epiminocholestane, spirosolane and solanidane skeleton are shown in Figure 8.

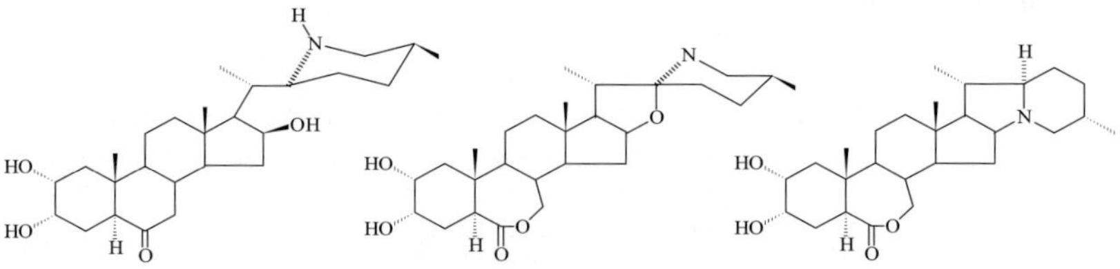

*Figure 8. Examples of nitrogenous analogs with 22, 26-epiminocholestane, spirosolane and solanidane side chain*

Also, a new type of analog containing a lactone in ring C was prepared starting from solasodine **32** by Mola and Magalhães (1998) and Novoa *et al.* (2000). The synthesis of one of them is shown in Scheme 8. Vespertiline **33** was obtained in four steps from solasodine **32** by standard procedure. The remaining steps were traditional for brassinosteroid synthesis (see Scheme 1) and led to the castasterone analog **34**.

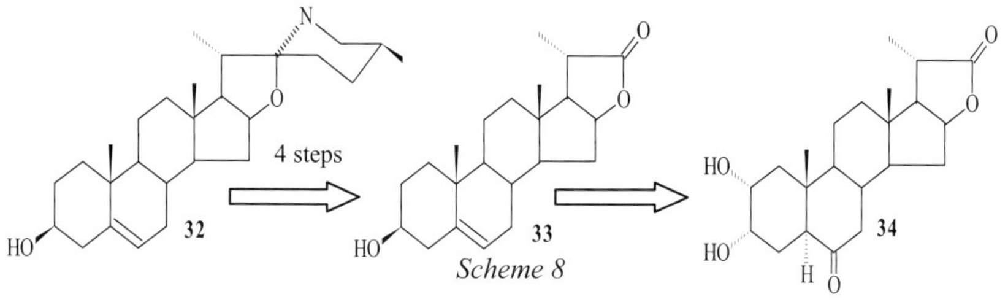

*Scheme 8*

*Analogs from bile acids*

The bile acids are the most important end products of cholesterol metabolism in higher animals. In human bile, the principal bile acids are chenodeoxycholic acid **35**, deoxycholic acid **36** and cholic acid **37**. These compounds are suitable raw materials for the preparation of 5β-cholanic analogs with drastic structural modifications in the side chain and for the cyclic part.

*Starting with chenodeoxycholic acid*

The synthesis of analogs with 7-oxa-7-keto functionality in ring B and 5β-cholanic skeleton was reported by Pérez *et al.* (1998a,b). The synthesis of one of these compounds is shown in Scheme 9. Selective cathylation of methyl chenodeoxycholate with ethyl chloroformate in dioxane and pyridine led to the monocathylate **38**, which upon oxidation of the 7α-hydroxyl group with Jones reagent afforded the ketone **39**. Baeyer-Villiger oxidation of the latter with trifluoroperoxyacetic acid and further saponification furnished the lactone **40**.

*Starting with deoxycholic acid*

Several brassinosteroid analogs with different oxygenated functions in ring C starting from deoxycholic acid **36** have been prepared by Espinoza *et al.* (2000) as shown in Scheme 10. Bromination of ketone **41** followed by alkaline hydrolysis of the intermediate bromide, led to hydroxy ketones **43a**, **43b** and **44**. Periodate cleavage of the mixture of **43a** and **43b** followed by reduction with sodium borohydride and further lactonization in acid medium and subsequent re-esterification afforded the lactone **45**.

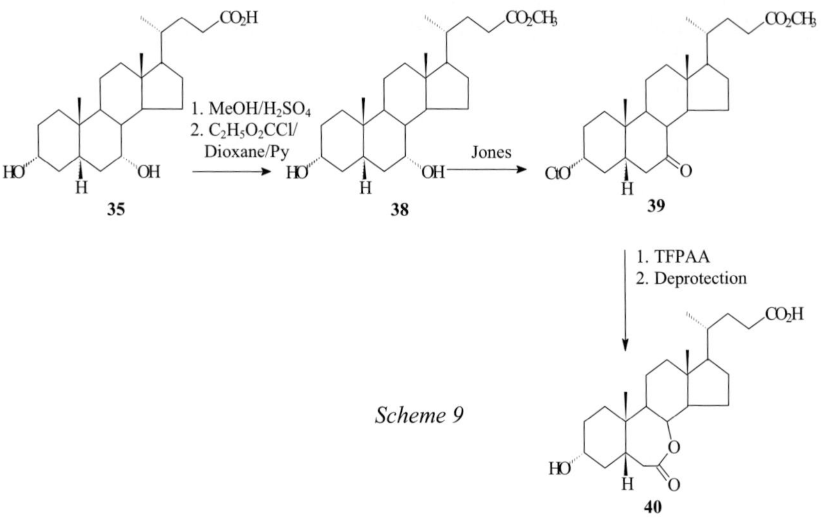
CO2H
HO
OH
H
35
1. MeOH/H2SO4
2. C2H5O2CCl/
Dioxane/Py
CO2CH3
HO
OH
H
38
Jones
CO2CH3
CtO
O
H
39
1. TFPAA
2. Deprotection
CO2H
HO
O
O
H
40

*Scheme 9*

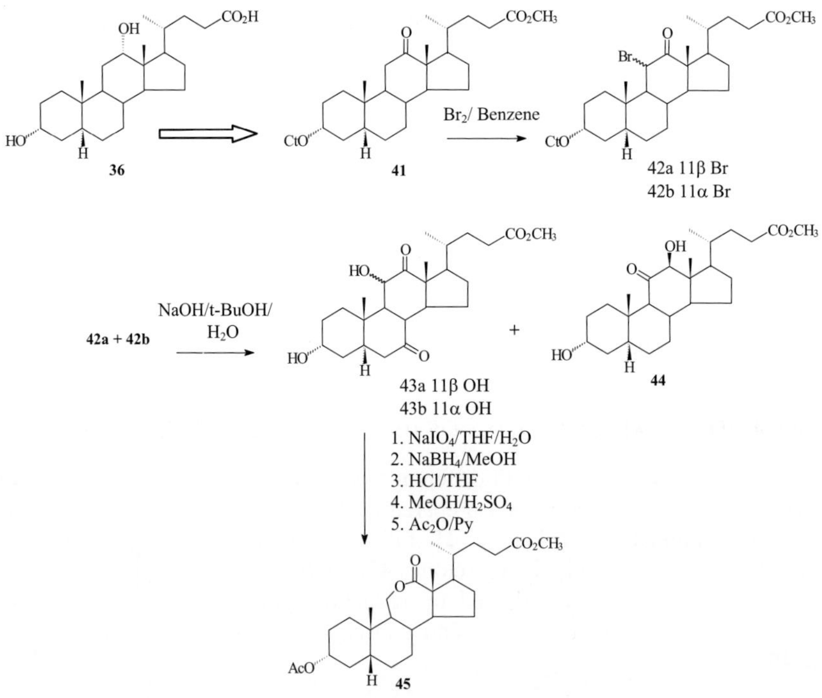
OH
CO2H
HO
H
36
O
CO2CH3
CtO
H
41
Br2/ Benzene
Br
O
CO2CH3
CtO
H
42a 11β Br
42b 11α Br
42a + 42b
NaOH/t-BuOH/
H2O
HO
O
CO2CH3
HO
O
H
43a 11β OH
43b 11α OH
+
OH
O
CO2CH3
HO
H
44
1. NaIO4/THF/H2O
2. NaBH4/MeOH
3. HCl/THF
4. MeOH/H2SO4
5. Ac2O/Py
O
O
CO2CH3
AcO
H
45

*Scheme 10*

*Starting with cholic acid*

Some 12-oxo cholanic analogs have been prepared by Robaina *et al.* (2001e) from cholic acid **37** (Scheme 11). Selective acetylation of methyl ester derivative with acetic anhydride in benzene and pyridine led to 3α, 7α-diacetate **46**, which was oxidized with Jones reagent to give ketone **47**. Saponification of the latter afforded the compound **48**. Baeyer-Villiger oxidation of the ketone **47** with m-chloroperbenzoic acid furnished the lactone **49**.

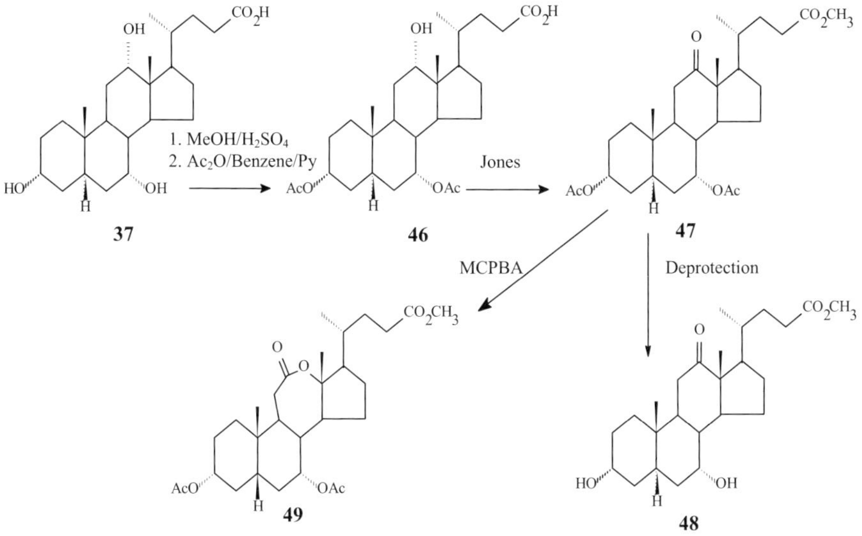

1. $MeOH/H_2SO_4$
2. $Ac_2O$/Benzene/Py

Jones

MCPBA

Deprotection

**37** **46** **47** **49** **48**

*Scheme 11*

## PRACTICAL APPLICATIONS OF BRASSINOSTEROID ANALOGS

*Applications to agricultural fields*

Taking in account the potentialities of brassinosteroids and their analogs as bio-regulators in agriculture, in Cuba, early in the 1990's, investigations on practical applications of brassinosteroid analogs begun, as a result of a joint research program among the Laboratory of Natural Products from the University of Havana and different scientific institutions. Thus, firstly a formulation termed DAA-6 or BIOBRAS-6 (BB-6), which has the compound **5** as active ingredient (Marquardt *et al.*, 1989; see Scheme 1), was tested as yield stimulator in different crops. Later, another formulation known as DI-31 or BIOBRAS-16 (BB-16), which has an isomer of compound **5** as active substance (Coll *et al.*, 1995), was also tested and the most important results of both formulations are presented in this section.

*a) Cereals*

Among cereals, rice and corn were mainly investigated. Concerning rice, an important crop for Cuban people's feeding, Franco *et al.* (2002) studied the effect of BIOBRAS-6 foliar sprays on yield in two rice varieties, grown during two seasons. They tested different doses (10, 50, 100, 250, 500, 750 and 1000 $mg.ha^{-1}$) and time of application (tillering, heading and spike-panicle initiation phases). It revealed the dependence of crop yield enhancement on the dose and time of the application. Thus, the best results were recorded with doses between 10 and 100 $mg.ha^{-1}$ applied at heading time, which increased crop yield between 0.83 and 2.48 $t.ha^{-1}$ compared to control, without any application. Besides, in general, there was a greater crop yield increase during dry season than in the wet. On the other hand, Mariña *et al.* (2002) demonstrated that BIOBRAS-6 or BIOBRAS-16 applied at heading time to rice plants grown in poorly saline soil (4.3 $dS.m^{-1}$) provided a crop yield enhancement by 25-30 %. Diaz *et al.* (2002) using BIOBRAS-16 as yield promoter in this crop. Here, they tested two complete doses (20 and 50 $mg.ha^{-1}$) and made two applications (tillering and spike-panicle initiation or spike-panicle initiation and grain-filling phases) during two seasons. All the treatments with BB-16 gave significantly higher yield than the control (increases between 26.1 and 82.9%) during both seasons. Furthermore, these authors performed experiments in a larger scale and found that 10 $mg.ha^{-1}$ BB-16 sprayed at two times (tillering and spike-panicle initiation phases) significantly increased crop yield under those conditions. (Table 1). Moreover, Almenares *et al.*, (1999) also reported increased corn yield by the foliar application of BB-16.

*Table 1. Effect of BB-16 applications on rice yield at three farms of "Los Palacios" Rice (Enterprise from Pinar del Río province).*

| Farms | Area (ha) | Crop yield ($t.ha^{-1}$) | | Yield increase against control (%) |
|---|---|---|---|---|
| | | Control | BB-16 | |
| Caribe | 155 | 3.38 | 6.72 | 98.8 |
| Cubanacán | 85 | 3.9 | 6.4 | 73.5 |
| Montoto | 114 | 3.7 | 6.2 | 67.6 |

These results show that increase in crop yield was associated with increase in panicle number per square meter and filled grain number per panicle. All this may be the consequence of the increased translocation of assimilates from the vegetative organs to panicles induced by BIOBRAS-6 or BIOBRAS-16, like other authors (Khripach *et al.*, 1999). The results of trials with BIOBRAS-6 or BIOBRAS-16 on corn plants from 1996 to 1998, included in the final report of a research project by

Núñez (2000 a), reported an increase in yield by 10 %; however, the time of foliar spray was important and BIOBRAS-16, was more effective than BIOBRAS-6.

On the other hand, it was demonstrated that soaking the seeds for eight hours in solutions of 0.01, 0.05 or 0.1mg.l$^{-1}$, not only significantly improved plant height at 42 and 70 days after sowing but also increased grain yield. Various authors have found a crop yield enhancement as a result of seed treatment with solutions of very low concentrations of natural brassinosteroids (Khripach *et al.*, 1999) or brassinosteroid analogs (Kamuro and Takatsuto, 1999), which demonstrates that this kind of treatment may be an effective way of increasing crop yield.

*b) Legumes*

In soybean, Corbera and Núñez (1998) observed the effectiveness of BIOBRAS-6 when it was applied at a dose of 20 mg.ha$^{-1}$, before flowering (40-45 days after sowing). Corbera and Núñez (2002) also studied the influence of BB-6 combined with biological fertilizers (arbuscular mycorrhiza, AM, and *Bradyrhizobium japonicum*) on crop yield. BB-6 increased yield of these plants (Table 2).

Results of field trials with the use of these analogs on *Phaseolus vulgaris* are scarce. However, Table 3 shows the results of applying BB-16 on black bean yield. Here, two technologies were tested, on the first one, 10 mg.ha$^{-1}$ of BB-16 was sprayed at flowering time and, on the second one, two applications with 5 and 10 mg.ha$^{-1}$ were performed at 24 and 58 days after sowing, respectively. The best results were obtained by one application of 10 mg.ha$^{-1}$, at flowering. Another experiment, at a larger scale, gave a 39 % more crop yield when 20 mg.ha$^{-1}$ of the product was sprayed, at flowering.

*c) Potato and vegetables*

In potato crop, Núñez *et al.* (1995 a) showed that foliar sprays of BIOBRAS-6 increased fresh weight of total and commercial tubers (tuber fraction with a longer than 35 mm diameter). Later, in 1997, Torres and Núñez applied the same analog on potato var. Desirée that enhanced commercial tuber fresh weight between 9 and 34 %, when doses of 0.5 and 1.0 mg.l$^{-1}$ were sprayed at 30 and 45 days after plantation, respectively. In other investigation, Núñez (2000 b) studied the effect of BIOBRAS-6 and BIOBRAS-16, with a complete dose of 100 mg.ha$^{-1}$, applied in a splitted manner, twice or thrice, during crop life cycle but the results showed no significant change in yield (Table 4).

This author also found an increase in yield between 8 and 13.5 %, although without statistical significance, when treatments with BB-6 or BB-16 at 45 and 55 days after plantation with doses of 10 and 50 mg.ha$^{-1}$, respectively were performed. On the other hand, during 1996-1997 season, field trials, using 100 mg.ha$^{-1}$ of BIOBRAS-16 as total dose with several potato varieties were made in different regions of Cuba. Results revealed that the magnitude of an increase in yield depended on the variety and the region. It confirmed the results described by other authors about brassinosteroids (Kamuro and Takatsuto, 1991).

*Table 2. Effect of BIOBRAS-6 application on soybean yield of two varieties inoculated with biological fertilizers (arbuscular mycorrhiza, AM, or/and Bradyrhizobium japonicum).*

| Treatments | Pod number plant$^{-1}$ | Weight of 100 seeds (g) | Yield (t.ha$^{-1}$) | Yield increase against control (%) |
|---|---|---|---|---|
| CUBASOY-23 | | | | |
| *B.japonicum* (Control) | 10.17 c | 15.67 | 1.15 d | - |
| AM | 12.80 b | 16.67 | 1.34 c | 16.52 |
| *B.japonicum*+ AM | 13.45 ab | 17.00 | 1.54 b | 33.91 |
| AM+ BB-6 | 13.20 ab | 16.75 | 1.48 b | 28.70 |
| *B.jap.*+ AM+ BB-6 | 14.05 a | 17.75 | 1.64 a | 42.61 |
| S.E.x | 0.35 *** | 0.53 NS | 0.03 *** | |
| INCASOY-27 | | | | |
| *B.japonicum* (Control) | 14.60 b | 14.00 | 1.64 d | - |
| AM | 15.60 b | 14.75 | 1.67 d | 1.83 |
| *B. japonicum*+ MA | 16.17 a | 15.25 | 1.92 b | 17.07 |
| AM+ BB-6 | 15.80 b | 15.00 | 1.83 c | 11.59 |
| *B.jap.*+ AM+ BB-6 | 19.12 a | 15.50 | 2.12 a | 29.27 |
| S.E.x | 0.62 *** | 0.65 | 0.03 *** | |

[1]Means followed by the same letters are not different significantly according to Duncan' s test at $P<0.05$

*Table 3. Influence of different doses and application times of BB-16 on black bean yield (Means ± S.E.).*

| Treatments | Pod number/plant | Seed number/pod | Weight of 100 seeds (g) | Crop yield (t.ha$^{-1}$) |
|---|---|---|---|---|
| Control | 12.5±0.86 | 6.20±0.59 | 19.82±0.47 | 3.22±0.20 |
| BB-16 (15 mg.ha$^{-1}$) Two applications | 21.1±1.36 | 5.82±0.44 | 20.04±0.14 | 3.58±0.23 |
| BB-16 (10 mg.ha$^{-1}$) One application | 29.6±3.10 | 5.17±0.20 | 20.24±0.55 | 4.47±0.18 |

*Table 4. Effect of application times of BIOBRAS-6 and BIOBRAS-16 on potato yield var. Desirée*

| Treatments | | Tuber number | | Fresh weight of tubers (g) | | Dry matter (%) |
|---|---|---|---|---|---|---|
| | | Total | >35 mm | Total | > 35 mm | |
| Control | | 7.2 | 5.9 | 536.62 | 503.33 | 19.15 |
| BB-6 | 30 + 45 DAP | 7.4 | 5.7 | 523.25 | 483.83 | 18.40 |
| | 45 + 55 DAP | 7.6 | 6.0 | 562.54 | 528.17 | 18.60 |
| | 30 + 45 + 55 DAP | 6.6 | 5.6 | 499.08 | 478.25 | 19.11 |
| BB-16 | 30 + 45 DAP | 7.4 | 5.9 | 544.29 | 516.58 | 19.92 |
| | 45 + 55 DAP | 7.0 | 5.8 | 555.17 | 528.42 | 18.12 |
| | 30 + 45 + 55 DAP | 7.2 | 5.7 | 542.08 | 506.58 | 18.26 |
| S.E.x | | 0.60 N.S. | 0.46 N.S. | 43.25 N.S. | 40.52 N.S. | 0.49 N.S. |

DAP- Days after planting

The influence of BIOBRAS-6 and BIOBRAS-16 on tomato crop has also been studied. Núñez *et al.* (1995 b) demonstrated that the foliar spray with 1 mg.l$^{-1}$ BIOBRAS-6 at flowering increased crop yield, irrespective of the date of transplantation, although the increase was not always statistically significant. In another experiment with young tomato plants, Núñez *et al.* (1996) showed that the foliar spray of 0.05 or 0.5 mg.l$^{-1}$ BIOBRAS-6 stimulated plant growth and metabolic activity, during this stage. Further studies (Núñez *et al.*, 1998 a) showed that applications of 50 and 100 mg.ha$^{-1}$ BB-6 on tomato crop at flowering gave similar response and this formulation was more effective as yield promoter when the dose of 50 mg.ha$^{-1}$ was splitted, and two foliar sprays (one week after transplanting and at flowering time) were applied (Table 5).The effectiveness of the doses between 40 and 50 mg.ha$^{-1}$ of BB-6, applied at these phases was further confirmed with two varieties at farmers´ field (Table 6).

Later, Núñez (2000 a) showed that BB-6 application improved tomato yield, by increasing fruit size, without affecting acidity, total soluble solids and dry matter. It suggests that BB-6 may also have enhanced assimilate translocation from leaves to fruits. The formulation termed BIOBRAS-16 has also been successfully applied to tomato crop. Generally, two foliar sprays during plant life cycle (one week after transplantation and at flowering) with a complete dose of 40 mg.ha$^{-1}$ are recommended where 10 % more fruits were obtained than the control (Table 7). However, Fernández *et al.* (1998) failed to get beneficial results by the application of BB-16 on tomato plants.

*Table 5. Influence of BIOBRAS-6 application on tomato yield (variety INCA-17)*

| Treatments | Crop yield[1] (t.ha$^{-1}$) | Yield increase against control (%) |
|---|---|---|
| CONTROL | 34.28 d | - |
| 50 mg,ha$^{-1}$ | 38.48 c | 12.25 |
| 100 mg.ha$^{-1}$ | 37.71 c | 10.00 |
| 25 mg.ha$^{-1}$ + 25 mg.ha$^{-1}$ | 46.09 a | 34.45 |
| 50 mg.ha$^{-1}$ + 50 mg.ha$^{-1}$ | 42.86 b | 25.03 |
| S.E.x | 0.916*** | |

[1]Means followed by the same letters are not different significantly according to Duncan's test at P<0.05

*Table 6. Influence of BB-6 on yield of two tomato varieties grown in two localities from Havana province*

| Localities | Varieties | Treatments | Crop yield (t.ha$^{-1}$) | Yield increase against control (%) |
|---|---|---|---|---|
| Batabanó | Campbell-28 | Control | 17.6 | - |
| | | BB-6 | 22.7 | 29.0 |
| Bababanó | Lignon | Control | 14.7 | - |
| | | BB-6 | 21.7 | 47.6 |
| Güines | Lignon | Control | 12.1 | - |
| | | BB-6 | 15.1 | 24.2 |

*Table 7. Results of BIOBRAS-16 applications on various tomato varieties grown in different provinces of Cuba*

| Province | Varieties | Área (ha) | Yield increase against control (%) |
|---|---|---|---|
| *Santiago de Cuba* | HC-3880 | 0.43 | 100.0 |
| | Floradel | 0.45 | 32.3 |
| | Floradel | 0.50 | 43.4 |
| *Las Tunas* | Mamonal | 2.00 | 100.0 |
| *Camagüey* | L 10-3 | 5.00 | 47.1 |
| | Campbell-28 | 3.75 | 16.2 |
| *Ciego de Avila* | Roma | 44.3 | 20.0 |
| *Matanzas* | Rossol | 3.35 | 54.6 |
| *La Habana* | Lignon | 12.00 | 86.7 |
| *Isla de la Juventud* | Campbell-28 | 1.0 | 52.0 |
| | INCA 9-1 | 0.5 | 38.0 |

Recently, a field trial with two new Cuban tomato varieties, Amalia and Mariela, was performed where BIOBRAS-16 was able to stimulate crop yield by 30 and 57 %, respectively without affecting internal fruit quality (Núñez, unpublished). On the other hand, it has been demonstrated that the application of BB-16, at the beginning of flowering, on tomato plants, grown with organic fertilizer tomato plants increased crop yield and allowed the saving of the effect of nitrogen fertilizer (Terry *et al.*, 2001). In onion, Núñez *et al.* (1998b) studied the effect of different doses of both formulations and the times of application. They concluded that BB-16 was better than BB-6 in enhancing plant yield and 0.1 mg-$l^{-1}$, BB-16, sprayed 50 days after plantation (beginning of bulb formation stage) provided best. Núñez *et al.* (1994) studied the influence of BIOBRAS-6 on garlic yield. Crop yield increased by 43 % when 0.1 mg $L^{-1}$ of the formulation was sprayed at 47 days after plantation. Similar results were found by Fernández (1999) who adopted the technique of soaking for 24 hours in combination with foliar spray.

In recent years, urban agriculture has been intensified in Cuba, so many vegetables are being cultivated and, in this production system, BIOBRAS-16 has been successfully applied (Núñez *et al.*, 2001). Some of these results are presented in Table 8. It is important to point out here that crop yield increase of selected species was always noticed, irrespective of the variety and growth conditions. The application of BB-16 on vegetable crops was extended to other provinces and, since 2000, this formulation was officially recommended as a yield-promoting agent (Ministerio de la Agricultura, 2000).

Nearly all the results described above have been obtained using foliar sprays, however, various authors have demonstrated that seed treatment with BB-16 at 0.01-0.05 mg.$L^{-1}$ concentration enhanced lettuce (Alfonso and Núñez, 1996) and tomato (Augustin, 2001) seedling growth, although it is still uncertain if this type of treatment increases crop yield. Seed treatment has not been fully exploited in Cuba and it might be a very useful technique for enhancing plant growth and crop yield and improving disease resistance like that of 24-epibrassinolide (Khripach *et al.*, 1999) and TNZ 303, which is a combination between the brassinosteroid analog TS 303 and a jasmonic acid analog (Kamuro and Takatsuto, 1999).

*d) Miscellaneous*

Preliminary results reported by Diaz *et al.* (1995), found that foliar spray of BIOBRAS-6 at 20 and 50 days after transplantation enhanced leaf growth in tobacco. On the other hand, Pita *et al.* (1996; 1998) demonstrated the effectiveness of DI-31 foliar spray on leaf growth and quality of tobacco. The influence of BB-6 and BB-16 has also been studied on papaw, strawberry and citrus crops. The results showed that papaw seed soaking in 0.1 mg.$L^{-1}$ of BB-6 or BB-16 for 24 hours not only increased seed germination percentage, plant height and shoot diameter but also harvested more fruits (Pozo *et al.*, 1996). In strawberry, the application of these analogs increased crop yield and stolone number per plant (Pozo *et al.*, 1994). A foliar spray of BB-16 on grapefruit trees gave the slight enhanced rate of fruit ripening and favoured peel chlorophyll, degradation, during degreening process, in an ethylene chamber (García *et al*., 1998). In *Coffea arabica*, Soto *et al*. (1997) and Utria *et al*. (2000) reported that seedling growth was enhanced by seed treatment with BB-16. Furthermore, soaking of *Hibiscus sabdariffa* L. seeds for four hours in different concentrations of BB-16 enhanced seedling growth (Núñez, unpublished). On the other hand, in ornamental plants, Plana and Núñez (2001) reported beneficial effect of BIOBRAS-6 on African violet and Fernández and Sotomayor (2000) showed that soaking gladiolus bulbs for 24 hours and a foliar spray at 60 days after plantation with BB-16 enhanced inflorescence emergence and flower opening, and increased bulb number and their diameter. BIOBRAS-16 has been also tested to promote rooting of guava cuttings (Ramírez, 2002).

*Table 8. Main results of BIOBRAS-16 applications on three crops under urban agricultural conditions*

| Province | Variety | Area ($m^2$) | Yield increase against control (%) |
|---|---|---|---|
| LETTUCE (*Lactuca sativa*) | | | |
| *Santiago de Cuba* | B. Simpson | 400 | 55.6 |
| *Granma* | B. Simpson | 70 | 16.7 |
| | B. Simpson | 120 | 33.3 |
| | B. Simpson | 5 200 | 17.4 |
| *Camagüey* | B. Simpson | 5 000 | 23.3 |
| | B. Simpson | 5 000 | 29.5 |
| *Villa Clara* | B. Simpson | 1 800 | 48.4 |
| CUCUMBER (*Cucumis sativus* L.) | | | |
| *Guantánamo* | SS-5 | 22 | 31.6 |
| *Santiago de Cuba* | SS-5 | 5 000 | 25.8 |
| | Poinset | 5 000 | 28.3 |
| *Granma* | SS-5 | 70 | 18.3 |
| | SS-5 | 11 300 | 18.9 |
| *Camagüey* | Poinset | 5 000 | 31.15 |
| *Villa Clara* | SS-5 | 480 | 19.4 |
| GREEN BEANS (*Vigna unguiculata sesquipedalis*) | | | |
| *Santiago de Cuba* | Lina | 5 000 | 47.3 |
| *Camagüey* | Bondadosa | 2 500 | 66.7 |
| *Villa Clara* | Canton-1 | 680 | 33.3 |

Taking into account, the effectiveness of BIOBRAS-16 on different agriculturally important crops, since 1998, this formulation has been officially registered in Cuba (MINAG, 1998) and currently many producers and farmers use it as a growth and yield-promoting agent. Moreover, this formulation has been successfully applied to various crops in other countries (Colombia, Venezuela and Chile). Thus, for example, in Venezuela, where the formulation is termed BIOCRECE, has been sprayed at doses between 10 and 20 mg.ha$^{-1}$ on cotton, corn, rice, coffee, tomato and pepper, to enhance yield between 7 and 40 % (Robaina and Scovino, 1998).

*Applications in Plant Biotechnology*

Several studies have shown the efficiency of brassinosteroids, as plant growth regulators, in plant-cell cultures (Takematsu *et al.*, 1983; Sala and Sala, 1985; Bellincampi and Morpurgo, 1988; Oh and Clouse, 1998). Considering plant biotechnology as a very useful tool for plant breeding and obtaining "high quality seeds" for agriculture, some studies have been made, in Cuba, in order to evaluate the inclusion of spirostanic analogs of brassinosteroids as plant growth regulator in different biotechnological processes. Therefore, BIOBRAS-6 has been used in various *in vitro* propagation phases of bananas and plantains and it has been demonstrated that it may be used as a substitute of 6-bencylaminopurine (6-BAP), used in the culture medium, in the first phase or establishment phase (Rayas *et al.*, 1996) and indoleacetic acid (IAA), used in the third phase or rooting phase (Rodríguez, 1999). Besides, it may be useful in the proliferation phase when added to culture medium in the presence of auxin and cytokinin (Rodríguez *et al.*, 1998). Concerning *in vitro* garlic propagation, BIOBRAS-6 resulted better than 6-BAP and 2-iP in the establishment and proliferation phases, respectively (Izquierdo *et al.*, 1998). Remarkable results in the *Dianthus caryophyllus* micropropagation were also obtained (Montes *et al.*, 1997). Regarding sugarcane, BB-6 stimulated root regeneration from callus when it was used alone or combined with IAA. On the other hand, the use of BB-6 as a single plant growth regulator increased *Citrus aurantium* callus fresh weight more than combined with 2,4-D, in the culture medium (González *et al.*, 1995). Furthermore, BB-16 promoted sugarcane callus growth and embryogenesis under saline and water stresses. A similar effect was observed on the callus under the same growth conditions (González *et al.*, 2001). Favourable results were also obtained in *C. aurantium* plant regeneration from callus, using low concentrations of BB-6 (Diosdado, 1997). Besides BB-6, BB-16 also enhanced rooting during *in vitro* sugarcane propagation (Jiménez *et al.*, 1996).

Prominent results with the use of BB-6 have been reported in other processes such as: *in vitro* potato propagation (Hernández *et al.*, 1999), potato callus differentiation (Hernández, 1994); banana somatic embryo germination (Barranco, 2001), growth and acclimatization of papaw plantlets obtained from somatic embryos (Gómez *et al.*, 1996) and potato plantlet acclimatization (Agramonte *et al.*, 1996).

Using the formulation termed MH-5, which has a trihydroxylated spirostanic analog as active ingredient, compound **8** (Hechavarría, 1998; see Scheme 2), Moré (2000) found that the combination of 0.001 mg.l$^{-1}$ with 2,4-D was much better than 2,4-D and kinetin for potato embryogenic callus formation. On the other hand, García *et al.* (1997) reported that BB-6 or MH-5 combined with 2,4-D promoted embryogenic callus formation of *Coffea canephora* var. Robusta. Later, García *et al.* (1998) showed that 0.01 mg.l$^{-1}$ MH-5 combined with 0.5 mg.l$^{-1}$ 2,4-D or with 0.5 mg.l$^{-1}$ 2,4-D and 2 mg.l$^{-1}$ kinetin in the culture medium gave maximum callus fresh weight. They concluded that MH-5 favoured hormonal balance for increasing callus fresh weight of

*Coffea canephora*, irrespective of the presence or absence of kinetin and the concentration of 2,4-D in the culture medium. The addition of different concentrations of MH-5 (0.01, 0.05 or 0.1 $mg.l^{-1}$) to culture media with variable concentrations of kinetin (2, 1, 0.5 or 0.2 $mg.l^{-1}$) was also studied by García (2000) and the results showed that, after 90 days of culture, 0.05 $mg.l^{-1}$ MH-5 combined with 1 or 0.2 $mg.l^{-1}$ kinetin provided best callus fresh and dry weights, respectively. A similar response was obtained by Coll (1996) in *Citrus aurantium* L., who found the most significant embryogenic callus growth with 0.01 $mg.l^{-1}$ MH-5, as a kinetin substitute in the medium.

These results demonstrate that MH-5, combined with 2,4-D was better than 2,4-D and kinetin for obtaining potato, coffee and sour orange embryogenic calli, cytokinin may be substituted by this brassinosteroid analog in the process. Similar results were reported by Wang *et al.* (1992) using brassinolide in eight cotton cultivars. However, Núñez *et al.* (2002) showed that lettuce callus induction, from intact cotyledons, did not occur after 25 days of culture when BB-6 and MH-5 were employed as single plant growth regulator in the culture media at a concentration range of 0.001 to 0.1 $mg.l^{-1}$. Generally, in this process, the culture medium used, contains only exogenous cytokinin (6-BAP at 0.1 $mg.l^{-1}$) as plant growth regulator, so it implies that endogenous hormonal content of cotyledons and the cytokinin added produce an adequate auxin/cytokinin combination to stimulate cell division; however, 6-BAP, substituted by brassinosteroid analogs, in these concentrations failed to promote cell division, suggesting that these compounds combined with the endogenous hormonal content of these explants were not able to initiate cell division. A distinct response with the addition of brassinosteroid analogs in presence of 6-BAP was observed, because both products hastened callus formation and stimulated plant regeneration in the presence of cytokinin. Similar results were reported by Plana *et al.* (2002), who studied the influence of BB-6 on tomato morphogenesis. These results demonstrate once more, that auxin and cytokinin cell status is critical for determining brassinosteroid effect (Oh and Clouse, 1998). The effect of BB-6 or MH-5 on histodifferentiation and maturation of sugarcane (*Saccharum* spp.) somatic embryos has also been studied by Nieves *et al.* (2002). They evaluated its action as naphtalenacetic acid (NAA) substitute and found that MH-5 at 0.01 $mg.l^{-1}$ gave the best results increasing embryo percentage at advanced maturation stages.

## CONCLUSIONS

The rapid metabolism of natural brassinosteroids in plants is one of the major problems of these growth regulators for their application in agriculture. 23-O-glucosylation seems to be one of the inactivation processes. Therefore, brassinosteroid analogs like TS 303, which have masked hydroxyl groups in the side chain, become very important. Many other brassinosteroid analogs like them may be synthesized from various raw materials through more efficient and economic processes than those used to obtain natural brassinosteroids. Steroidal sapogenins are very useful starting material because they have a spiroketal moiety, which bear two oxygen atoms in the side chain. Therefore, it could be possible that a steroidal sapogenin with typical brassinosteroid functions, in steroidal nucleus, where at least one of the spiroketal moiety oxygen overlaps one of the hydroxyl groups of natural brassinosteroid active configuration. This hypothesis compelled us to synthesize more than a hundred of brassinosteroid spirostanic analogs. Practical applications of the formulations termed BIOBRAS-6 and BIOBRAS-16, which have brassinosteroid spirostanic analogs as active ingredients, to agricultural fields showed that they may be used as growth or yield stimulators for different crops. Thus, BIOBRAS-16 is now recommended as growth and yield promoter for vegetable crops in urban agriculture. It is interesting to point out that, generally, doses between 10 and 50 $mg.ha^{-1}$ BIOBRAS-16 were able to yield stable results for increasing crop yield (Khripach *et al.*, 1997). This characteristic along with a very low toxicity of this formulation (LD50 $5g.kg^{-1}$, oral for mice, 2 $g.kg^{-1}$, dermal for mice and the negative Ames test for mutagenicity) makes BIOBRAS-16 a new product with broad agricultural prospects. Nevertheless, further investigations with this formulation should be performed in order to test some other modes of application, as well as the effects of combining with other plant growth regulators and exploring its potentialities as a plant-protecting agent against biotic and abiotic stresses. On the other hand, it was demonstrated that BIOBRAS-6 and MH-5 might be used as growth regulators in plant biotechnology and like natural brassinosteroids; they can be combined with auxins and/ or cytokinins. However, a deeper investigation about the mechanisms of action of these new brassinosteroid analogs in plants is still required.

## REFERENCES


Abe, H. (1989). Advances in brassinosteroid research and prospects for its agricultural application. Japan Pesticide Information 10-14.

Adam, G., Marquardt, V. (1986). Brassinosteroids. Phytochemistry 25: 1787-1799.

Adam, G., Marquardt, V., Vorbrodt, H., Hörhold, C., Andreas, V., Gartz, J. (1991). Aspects of synthesis and bioactivity of brassinosteroids. In: Brassinosteroids: Chemistry, Bioactivity and Applications, pp 74-85. Eds H G Cutler, T Yokota and G Adam. American Chemical Society, Washington.

Agramonte, D., Jiménez, F., Pérez, M., Gutiérrez, O., Ramírez, D., Núñez, M. (1996). Empleo de sustancias bioestimuladoras (Biobras-6 y Biobras-16) en la fase de adaptación de vitroplantas de papa (*Solanum tuberosum* L.). Programa y Resúmenes X Seminario Científico, Instituto Nacional de Ciencias Agrícolas, La Habana, pp 162.

Alfonso, J., Núñez, M. (1996). BIOBRAS-16, nuevo modo de aplicación en hortalizas. Programa y Resúmenes X Seminario Científico, Instituto Nacional de Ciencias Agrícolas, La Habana, pp 159.

Almenares, J., Cuñarro, R., Ravelo, R., Fitó, E., Moreno, I., Núñez, M. (1999). Influencia de diferentes dosis y momentos de aplicación del BIOBRAS-16 en el cultivo del maíz. Cultivos Tropicales 20 (3): 77-81.

Alonso, E. (1990). Synthesis of spirostanic analogs of brassinosteroids. Ph. D. Thesis. University of Havana, Havana, 105 pp.

Anaya, H., Tacoronte, J. E., Cabrera, M. T., Pérez, C., Enríquez, M. (2000). Síntesis de A-secoespirostanos a partir de la diosgenina en condiciones de catálisis por transferencia de fase. Revista CENIC de Ciencias Químicas 31(2): 119 – 122.

Anónimo. (1988). JRDC-694 (epi-brassinolide). Technical information. Nippon Kayaku Co., Ltd. Japan.

Anónimo. (1998). BIOBRAS-16 (DI-31). Registro Central de Plaguicidas. Centro Nacional de Sanidad Vegetal. Ministerio de la Agricultura. CUBA: Tomo 3, Folio 212.

Anónimo. (2000). Manual técnico de organopónicos y huertos intensivos. Grupo Nacional de Agricultura Urbana. Ministerio de la Agricultura. La Habana, 126 pp.

Augustin, J. M. (2001). Influence of BIOBRAS-16 on tomato plant growth (*Lycopersicon esculentum* Mill) during seedling stage. Ing. Agron. Thesis. Agricultural University of Havana. Havana, 35 pp.

Barranco, L. A. (2001). Somatic embryogenesis of banana (*Musa* AAAB, cv. FHIA-18) using liquid culture

media. Ph D. Thesis. Central University "Marta Abreu" of Las Villas. Santa Clara. 120 pp.
Bellincampi, D., Morpurgo, G. (1988). Stimulation of growth in *Daucus carota* L. cell cultures by brassinosteroid. Plant Science 54: 153-156.
Bobrick, A., Khripach, V., Zhabinskii, V., Zavadskaya, M., Litvinovskaya, R. (1998). A method of production of sanitated seed potato. Pat. Appl. BY 19,981,189.
Brosa, C., Soca, L., Terricabras, E., Ferrer, J., Alsina, A. (1998). New synthetic brassinosteroids: a 5-alpha-hydroxy-6-ketone analog with strong plant growth promoting activity. Tetrahedron 54: 12337-12348.
Brosa, C., Soca, L., Terricabras, E., Zamora, I. (1996). Brassinosteroids: looking for a practical solution. Proceedings of the Plant Growth Regulation Society of America 23: 21-26.
Coll, F., Jomarrón, I., Robaina, C., Alonso, E., Cabrera, M. T. (1995). Polyhydroxyspirostanones as plant growth regulators. PCT Int. Appl. W09713, 780.
Corbera, J., Núñez, M. (1998). Efectividad del empleo de brasinoesteroides o compuestos análogos en el cultivo de la soya (*Glycine max* (L.) Merrill). Programa y Resúmenes XI Seminario Científico, Instituto Nacional de Ciencias Agrícolas, La Habana, pp 131.
Corbera, J., Núñez, M. (2002). Evaluación agronómica del análogo de brasinoesteroides BIOBRAS-6 en soya cultivada sobre suelo Ferralítico Rojo compactado. Cultivos Tropicales. In press.
Dago, A., Pomés, R., Coll, F., Hechavarría, M., Punte, G., Echevarría, G. (1997). (25 R)-5β-hydroxy-spirost-2-en-6-one. Acta Crystallographica 53: 1705 –1706.
Dbaubhadel, S., Chaudhary, S., Dobinson, K., Krishna, P. (1999). Treatment with 24-epibrassinolide, a brassinosteroid increases the basic thermotolerance of *Brassica napus* and tomato seedlings. Plant Molecular Biology 40: 333-342.
Díaz, G., Pérez, N., Núñez, M., Torres, W. (1995). Efecto de un análogo de brasinoesteroide -DAA-6- en el cultivo del tabaco (*Nicotiana tabacum* L.). Cultivos Tropicales 16(3): 53-55.
Díaz, S., Morejón, R., Núñez, M. (2002). BIOBRAS-16, una alternativa para incrementar los rendimientos en el cultivo del arroz. Memorias 2do. Encuentro Internacional de Arroz, La Habana, pp 266-272.
Diosdado, E. (1997). Effect of bio-regulators on somatic embryogenesis and protoplast fusion and culture of sour orange (*Citrus aurantium* L.). Ph D. Thesis. University of Havana, Havana, 105 pp.
Espinoza, L., Coll, D., Coll, F., Preite, M., Cortés, M. (2000). Synthesis and plant growth – activity of three new brassinosteroid analogues. Synthetic Communications 39: 1963 - 1974.
Fernández, A. (1999). Effect of Biobras-6 on garlic (*Allium sativum* L.) growth and development. M. Sc. Thesis. University of Havana. Havana, 60 pp.
Fernández, A., Sotomayor, E. (2000). Influencia del análogo de Brasinoesteroide BB-16 en el gladiolo. Programa y Resúmenes XII Seminario Científico. Instituto Nacional de Ciencias Agrícolas, La Habana, pp. 200.
Fernández, A., Pérez, P., Sotomayor, E., Coll, F., Jomarrón, I., Robaina, C. (1998). Efecto del análogo de brasinoesteroides "Biobras-16" en el rendimiento y calidad del tomate (*Lycopersicon esculentum* Mill). Memorias del Evento Científico "Producción de cultivos en condiciones tropicales". Instituto de Investigaciones Hortícolas "Liliana Dimitrova", La Habana, pp 253-256.
Franco, I., Morales, O., Socorro, M. (2002). Efecto del análogo de brasinoesteroides BIOBRAS 6 sobre el crecimiento y desarrollo en dos variedades de arroz (*Oryza sativa* L.). Memorias 2do. Encuentro Internacional de Arroz, La Habana, pp 273-277.
Fujita, F. (1985). Prospects for brassinolide utilization in agriculture. Chemical Biology 23: 717-725.
García, D. (2000). Acción del análogo de brasinoesteroides MH-5 y la kinetina en la formación de biomasa en callos de *Coffea canephora* var. Robusta. Cultivos Tropicales 21(3): 39-46.
García, D. (2002). Synthesis and characterization of pregnanic and androstanic analogs of brassinosteroids. M.Sc. Thesis, University of Havana, Havana, 67 pp.
García, D., Marrero, M. T., Cuba, M., Núñez, M. (1997). Efecto cualitativo de análogos de brasinoesteroides como sustitutos hormonales en la callogénesis de café (*Coffea canephora* variedad Robusta). Cultivos Tropicales 18(2): 44-46.
García, D., Torres, W., Cuba, M., Núñez, M. (1998). Análisis del crecimiento de callos de *Coffea canephora* var. Robusta en presencia del análogo de brasinoesteroides MH-5. Cultivos Tropicales 19(3): 55-60.
García, M. E., Cáceres, I., Betancourt, M., Pozo, L., Coll, F., Robaina, C., Altuna, B. (1998). Adelanto del inicio de la cosecha de frutos de toronja (*Citrus paradisi* Macf.) mediante aspersiones foliares de fitorreguladores. Programa y Resúmenes XI Seminario Científico. Instituto Nacional de Ciencias Agrícolas, La Habana, pp 123.
Gómez, R., Posada, L., Reyes, M., Núñez, M. (1996). Empleo de sustancias biorreguladoras (Biobras-6) en

la conversión y adaptación de plantas de fruta bomba (*Carica papaya* L.) var. Maradol Rojo obtenidas a partir de embriones somáticos. Resúmenes IV Coloquio Internacional de Biotecnología de las Plantas, Santa Clara, pp 58.

González, S., Gómez, M., Coll, D. (2001). Efecto de análogos de brasinoesteroides en callos de caña de azúcar sometidos a estrés abiótico. Revista Jardin Botánico Nacional 22(2): 247-252.

González, S., Diosdado, E., Rodríguez, J., Xiqués, X., Román, M. I., González, C. (1995). Physiological effects of the synthetic brassinosteroid DAA-6 on propagation from callus culture and on seed germination (Conference abstract). Advances in Modern Biotechnology 3: 4.

Grove, M. D., Spencer, G. F., Rohwedder, W. K., Mandava, N., Worley, J. F., Warthen, J. D., Steffens, G. L., Flippen-Anderson, J. L., Cook, J. C. (1979). Brassinolide, a plant growth-promoting steroid isolated from *Brassica napus* pollen. Nature 281: 216-217.

Hechavarría, M. (1998). Synthesis, characterization and biological activity of steroidal trihydroxyketones. M. Sc. Thesis , University of Havana, Havana, 62 pp.

Hernández, M. (1994). Empleo de brasinoesteroides para la diferenciación de callos de papa (*Solanum tuberosum* L.). Cultivos Tropicales 15(3): 78.

Hernández, M., Moré, O., Núñez, M. (1999). Empleo de análogos de brasinoesteroides en el cultivo *in vitro* de papa (*Solanum tuberosum* L.). Cultivos Tropicales 20(4): 41-44.

Hotta, Y., Tanaka, T., Bingshan, L., Takeuchi, Y., Konai, M. (1998). Improvement of cold resistance in rice seedlings by 5-aminolevulinic acid. Journal of Pesticide Science 23: 29-33.

Iglesias, M. (1996). Syntheses of furostanic and spirostanic analogs of brassinosteroids. Ph. D. Thesis. University of Havana, Havana, 98 pp.

Iglesias, M., Pérez, C., Coll, F. (1996a). Brasinoesteroides naturales y análogos sintéticos. Revista CENIC de Ciencias Químicas 27: 3-12.

Iglesias, M., Pérez, C., Coll, F. (2001 b). Spirostanic analogues of Brassinosteroids. Analogues of castasterone. Revista Cubana de Química XIII (2): 179.

Iglesias, M., Pérez, C., Coll, F. (2002). Spirostanic analogues of castasterone. Steroids 67: 159- 163.

Iglesias, M., Pérez, R., Coll, F., Leliebre, V. (1997a). Dicetonas espirostánicas monohidroxiladas con actividad biológica y su procedimiento de obtención. Cuban Patent 22 492 CO7J73/00.

Iglesias, M., Pérez, R., Coll, F., Leliebre, V. (1997b). Dicetona espirostánica dihidroxilada con actividad promotora del crecimiento vegetal y su procedimiento de obtención. Cuban Patent 22 494 CO7J21/00.

Iglesias, M., Pérez, R., Pérez, C., Coll, F. (1999). Synthesis and characterization of (25R)-2α,3α-epoxy-5α-spirostan-12,23-dione. Synthetic Communications 29: 1811 - 1818.

Iglesias, M., Pérez, R., Pérez, C., Coll, F. (2000). Synthetic steroidal sapogenins. Part III. 23-ketohecogenin and 23-ketoisochiapagenin. Synthetic Communications 30: 163 - 170.

Iglesias, M., Pérez, R., Leliebre, V., Pérez, C., Coll, F. (2001a). Spirostanic analogues of teasterone: synthesis, characterization and biological activity of laxogenin (23S)-hydroxylaxogenin and 23-ketolaxogenin (23-oxalaxogenin). Journal of the Chemical Society, Perkins Transactions I 0: 261–266.

Iglesias, M., Pérez, R., Leliebre, V., Pérez, C., Coll, F. (1996b). Synthesis and biological activity of (22R, 25R)-5α-furostan-2α, 3α, 26-triol. Journal of Chemical Research (S) 11: 504 - 505.

Iglesias, M., Pérez, R., Leliebre, V., Pérez, C., Coll, F. (1998a). Synthesis of (25R)-5α-spirostan-2α,3α,6β-triol –acetate. Synthetic Communications 28: 75 - 81.

Iglesias, M., Pérez, R., Leliebre, V., Pérez, C., Coll, F. (1998b). Synthesis of (25R)-2α,3α-epoxy-5α-spirostan-6,23-dione. Synthetic Communications 28: 4387-4392.

Iglesias, M., Pérez, R., Leliebre, V., Pérez, C., Coll, F. (1998d). Synthesis of (22R,25R)-2α,3α26-trihydroxy-5α-furostan-6-one. Synthetic Communications 28: 1779 - 1784.

Iglesias, M., Pérez, R., Leliebre, V., Pérez, C., Rosado, A., Coll, F. (1998c). Synthesis of (25R)-3β,26-dihydroxy-5α-furostan-6-one. Synthetic Communications 28: 1381-1386.

Ikekawa, N., Zhao, Y. J. (1991). Application of 24-epibrassinolide in agriculture. In: Brassinosteroids. Chemistry, Bioactivity and Applications, pp 280-291. Eds H G Cutler, T Yokota and G Adam. American Chemical Society, Washington.

Izquierdo, H., Quiñones, Y., Fernández, A., Disotuar, R., Pedroso, D., Capote, J., Coll, F., Sotolongo, J., Hernández, M. I., Rodríguez, B. (1998). Biobras-6 y Biobras-16, dos nuevos brasinoesteroides sintéticos cubanos con actividad bio-reguladora en plantas. I. Su influencia en la micropropagación *in vitro* de *Allium sativum* L. Clon "Criollo-9". Memorias del Evento "Producción de cultivos en condiciones tropicales", Instituto de Investigaciones Hortícolas "Liliana Dimitrova", La Habana, pp

110-112.
Jiménez, M., Ortiz, R., De la Fe, C. (1996). Efecto de dos análogos de brasinoesteroides como suplementos o sustitutos de auxinas en la micropropagación de la caña de azúcar. Programa y Resúmenes X Seminario Científico, Instituto Nacional de Ciencias Agrícolas, La Habana, pp 157.
Jomarrón, I. (1995). Synthesis of bioactive spirostanones and spirostanlactones. Ph. D. Thesis. University of Havana, Havana, 120 pp.
Jomarrón, I., Coll, F., Robaina, C. (1997). Espirostanlactonas con funciones oxigenadas en la posición 2 y 3 y su actividad como reguladores del crecimiento vegetal. Cuban Patent 22497, CO7J21/00, AOIN 45/00.
Jomarrón, I., Coll, F., Pérez, C., Robaina, C., Alonso, E. (1996). Synthesis and spectroscopic characterization of (25R)-2β,3β-dihydroxy-5α-spirostan-6-one and their lactone (25R)-2β,3β-dihydroxy-B-homo-7-oxa-5α-spirostan-6-one. Abstracts of XXII Latinoamerican Congress of Chemistry, pp 447.
Kamuro, Y., Takatsuto, S. (1991). Capability for and problems of practical uses for brassinosteroids. In: Brassinosteroids. Chemistry, Bioactivity and Applications, pp 292-297. Eds H G Cutler, T Yokota and G Adam. American Chemical Society, Washington.
Kamuro, Y., Takatsuto, S. (1999). Practical application of brassinosteroids in agricultural fields. In: Brassinosteroids: Steroidal Plant Hormones, pp 223-241. Eds A Sakurai, T Yokota and S D Clouse. Springer-Verlag, Tokyo.
Kamuro, Y., Takatsuto, S., Noguchi, T., Watanabe, T., Fujisawa, H. (1996). Application of a long-lasting brassinosteroid TS 303 in combination with other plant growth regulators. Proceedings of the Plant Growth Regulation Society of America 23: 27-31.
Kamuro, Y., Takatsuto, S., Watanabe, T., Noguchi, T., Kuriyama, H., Suganuma, H. (1997). Practical aspects of brassinosteroid compound [TS 303]. Proceedings of the Plant Growth Regulation Society of America 24: 111-116.
Khripach, V. A., Zhabinskii, V. N., De Groot, A.E. (1999). Brassinosteroids. A new Class of Plant Hormones. Academic Press.
Khripach, V. A., Zhabinskii, V. N., De Groot, A.E. (2000). Twenty years of brassinosteroids: steroidal plant hormones warrant better crops for the XXI century. Annals of Botany 86: 441-447.
Khripach, V, A., Zhabinskii, V, N., Litvinovskaya, R. (1991). Synthesis and some practical aspects of brassinosteroids In: Brassinosteroids: Chemistry, Bioactivity and Applications, pp 43-55. Eds. H G Cutler, T Yokota and G Adam. American Chemical Society, Washington.
Khripach, V, A., Zhabinskii, V, N., Malevannaya, N. (1997). Recent advances in brassinosteroid study and applications. Proceedings of the Plant Growth Regulation Society of America 24: 101-106.
Kohout, L., Kasal, A., Strnad, M. (1996). Pregnane type brassinosteroids with a four carbon ester functionality in position 20. Collection of Czechoslovak Chemical Communication 61: 930-940.
Kohout, L., Slavikova, B., Strnad, M. (1998). 17a-Oxa-17a-homobrassinosteroid analogues. Collection of Czechoslovak Chemical Communication 63: 646-654.
Kohout, L., Strnad, M., Kaminek, M. (1991). Types of brassinosteroids and their bioassays. In: Brassinosteroids: Chemistry, Bioactivity and Applications, pp 56-73. Eds. H G Cutler, T Yokota and G Adam. American Chemical Society, Washington.
Korableva, N., Platonova, T., Dogonadze, M., Bibick, N. (1999). A stability change of potato to premature germination and diseases under the brassinosteroid action. In: Regulators of plant growth and development 5, pp 102-103. Eds V Sheveluch, G Karlov, N Karsunkina, E Salnikova, I Skorovogatov and A Siusheva. Agricultural Academy, Moscow.
Kovganko, N. V., Ananich, S. K. (1991). Synthesis of the 5α-hydroxy analogs of brassinosteroids from $\Delta^5$-sterols. Zhurnal Organicheskoi Khimii, 27(1): 103 - 108.
Kumawat, B., Sharmat, D., Jat, S. (1997). Effect of brassinosteroid on yield attributing characters under water deficit stress condition in mustard (*Brassica juncea* (L.) Czern and Coss.). Annals of Biology. (Ludhiana) 13: 91-93.
Li, L., Van Staden, J. (1998). Effects of plant growth regulators on drought resistance of two maize cultivars. South African Journal of Botany 64: 116-120.
Mandava, N. B. (1988). Plant growth-promoting brassinosteroids. Annual Review of Plant Physiology and Plant Molecular Biology 39: 23 – 52.
Mariña, C., Castillo, P., Morales, O., Alonso, E., Aguilera, R., Rosabal, M., Pérez, B., Licea, L. (2002). Aplicación de reguladores de crecimiento vegetal en plantas de arroz cultivadas en suelos salinizados.

Una alternativa para elevar el rendimiento agrícola. Memorias 2do. Encuentro Internacional de Arroz, La Habana, pp 277.

Marquardt, V., Adam, G. (1991). Recent advances in brassinosteroids research. Chemistry of Plant Protection, 7, 103 -139.

Marquardt, V., Adam, G., Coll, F., Alonso, E. (1989). Preparation of brassinosteroid analogs of spirostanes as plant growth regulators. Ger (East) DD 273, 638.

Maugh, T. H. (1981). New chemicals promise larger crops. Science 212: 33-34.

Mola, J., Magalhães, G. (1998). Síntese de análogo de brassinoesteroíde a partir de vespertilina. Química Nova 21(6): 726 – 730.

Montes, S., Mesa, O., Hernández, M. M., Santana, N., Núñez, M., Varela, M. (1997). Uso del análogo de brasinoesteroide BB-6 en la micropropagación del clavel. Cultivos Tropicales 18(2): 51-55.

Moré, O. (2000). Effect of new bio-regulators on potato (*Solanum tuberosum* L.) embryogenic callus formation. M.Sc. Thesis. University of Havana, Havana.

Nieves, N., Rodríguez, K., Cid, M., Castillo, R., Núñez, M. (2002). Action of brassinoesteroids on histodifferentiation and maturation of sugarcane somatic embryos. In: The importance of Plant Tissue Culture and Biotechnology in Plant Sciences, pp 321-330. Eds. A Taji and R Williams. IAPTC & B. Australian Branch.

Novoa, H., Peter, O., Blanton, N., Súarez, L., Iglesias, M., Coll, F. (2000). 3β-acetoxy-5α,6β-dihydroxybisnorcholanic acid 22→16 lactone. Acta Crystallographica, Sect C: Crystallographic Structure Communications 56(1): 78 - 79.

Núñez, M. (2000a). Análogos de brasinoesteroides cubanos como biorreguladores en la agricultura. Informe final de investigación. Instituto Nacional de Ciencias Agrícolas, La Habana, 88 pp.

Núñez, M. (1999). Aplicaciones prácticas de brasinoesteroides y sus análogos en la agricultura. Cultivos Tropicales 20 (3): 63-72.

Núñez, M. (2000b). Efectividad de la aplicación de dos análogos de brasinoesteroides cubanos en el cultivo de la papa. Memorias XIX Congreso de la Asociación Latinoamericana de Papa, La Habana, pp 135.

Núñez, M., Mazorra, L. M. (2001). Los brasinoesteroides y las respuestas de las plantas al estrés. Cultivos Tropicales 22 (3): 19-26.

Núñez, M., Robaina, C. (2000). Brasinoesteroides, nuevos Reguladores del Crecimiento vegetal con amplias perspectivas para la Agricultura. Documento IAC 68. Instituto Agronômico, Campinas. 83 pp.

Núñez, M., Torres, W., Coll, F. (1995a). Effectiveness of a synthetic brassinosteroid on potato and tomato yields. Cultivos Tropicales 16(1): 26-27.

Núñez, M., Torres, W., Echevarría, I. (1996). Influencia de un análogo de brasinoesteroide en el crecimiento y la actividad metabólica de plantas jóvenes de tomate. Cultivos Tropicales 17(3): 26-30.

Núñez, M., Sosa, J., Alfonso, J. L., Coll, F. (1998b). Influencia de dos nuevos biorreguladores cubanos en el rendimiento de plantas de cebolla (*Allium cepa*) cv. Red Creole. Cultivos Tropicales 19(1): 21-24.

Núñez, M., Alfonso, J. L., Arzuaga, J., Hernández, A., Coll, F. (1998a). Influencia de nuevos biorreguladores cubanos en la producción de hortalizas en condiciones tropicales Proceedings of the Interamerican Society for Tropical Horticulture 42: 335-343.

Núñez, M., Benítez, B., Domingos, J. P., Torres, W., Coll, F., Jomarrón, I. (1994). Influencia de análogos de brasinoesteroides en el rendimiento de diferentes cultivos hortícolas. Cultivos Tropicales 15(3): 87.

Núñez, M., Domingos, J. P., Torres, W., Coll, F., Alonso, E., Benítez, B. (1995b). Influencia del análogo de brasinoesteroide Biobras-6 en el rendimiento de plantas de tomate cultivar INCA-17. Cultivos Tropicales 16(3): 49-52.

Núñez, M., Siqueira, W. J., Hernández, M., Zullo, M. A., Robaina, C., Coll, F. (2002). Influence of two brassinosteroid analogues on callus formation and plant regeneration of lettuce (*Lactuca sativa*). Plant Cell Reports. In press.

Núñez, M., Arzuaga, J., Robaina, C., Alfonso, J. L., Benítez, B., Coll, F. (2001). Utilización de un nuevo biorregulador de origen natural en la agricultura urbana. Libro Resumen IV Encuentro de Agricultura Orgánica, ACTAF, La Habana, pp. 102-103.

Oh, M. H., Clouse, S. D. (1998). Brassinolide affects the rate of cell division in isolated leaf protoplasts of *Petunia hybrida*. Plant Cell Reports 17: 921-924

Pérez, R., Pérez, C., Coll, F. (1998a). Synthesis of analogues of brassinosteroids with 5β-cholanic acid skeleton. Synthetic Communications 28: 3387 - 3396.

Pérez, R., Iglesias, M., Pérez, C., Coll, F., Coll, D., Rosado, A. (1998b). Synthesis of analogues of brassinosteroids from chenodeoxycholic acid. European Journal of Organic Chemistry 11: 2405 -

2407.
Pita, O., Cuéllar, A. Y., Coll, F., Robaina, C. (1998). Efecto de diferentes concentraciones de un análogo de brasinoesteroide DI-31 en el rendimiento y la calidad del tabaco. Programa y Resúmenes XI Seminario Científico, Instituto Nacional de Ciencias Agrícolas, La Habana, pp 123.
Pita, O., Cuéllar, Y., Coll, F., Robaina, C. (1996). Influencia de un análogo de brasinoesteroide, DI-31, en el rendimiento y calidad del tabaco (*Nicotiana tabacum* L.). Programa y Resúmenes X Seminario Científico, Instituto Nacional de Ciencias Agrícolas, La Habana, pp 155.
Plana, D., Alvarez, M., Florido, M., Lara, R. M., Núñez, M. (2002). Efecto del Biobras 6 en la morfogénesis *in vitro* del tomate (*Lycopersicon esculentum*, Mill) var. Amalia. Cultivos Tropicales 23(2): 21-25.
Plana, R., Núñez, M. (2001). Influencia del Biobras-6 en el desarrollo y floración de la violeta africana (*Saintpaulia iontha* Wendl). Proceedings of the Interamerican Society for Tropical Horticulture 43: 151-153.
Pozo, L., Rivera, T., Noriega, C., Iglesias, M., Coll, F., Robaina, C., Velázquez, B., Rodríguez, O. L., Rodríguez, M. E. (1994). Algunos resultados en el cultivo de los frutales mediante la utilización de brasinoesteroides o compuestos análogos. Cultivos Tropicales 15(3): 79.
Pozo, L., Muñoz, S., Noriega, C., Reyes, C., Velázquez, B., Rodríguez, M. E., Ortiz, M., Coll, F., Robaina, C., Iglesias, M., Jomarrón, I., Alonso, E. (1996). Efecto de cuatro brasinoesteroides sintéticos sobre la germinación, el crecimiento y la fructificación de una variedad colombiana de *C. papaya*. Programa y Resúmenes X Seminario Científico, Instituto Nacional de Ciencias Agrícolas, La Habana, pp 154.
Quyen, L., Adam, G., Schreiber, K. (1994a). Partial synthesis of nitrogenous brassinosteroid analogues with 22-26-Epiminocholestane and Spirosolane skeleton. Liebigs Annalen der Chemie 0(11): 1143 - 1147.
Quyen, L., Adam, G., Schreiber, K. (1994b). Partial Synthesis of nitrogenous brassinosteroid analogues with solanidane skeleton. Tetrahedron 50(37): 10923 - 10932.
Ramírez, A. (2002). Uso de bioestimuladores en la reproducción de *Psidium guajava* L. mediante el enraizamiento de esquejes. Cultivos Tropicales. In Press.
Ramírez, J., Centurion, T., Gros, G., Galagovsky, L. (2000a). Synthesis and bioactivity evaluation of brassinosteroid analogs. Steroids 65: 329 - 337.
Ramírez, J., Gros, G., Galagovsky, L. (2000b). Effects on bioactivity due to C-5 heteroatom substituents on synthetic 28-homobrassinosteroid analogs. Tetrahedron 56: 6171 - 6180.
Ramraj, V. M., Vyas, B. N., Godrej, N. B., Mistry, K. B., Swami, B. N., Singh, N. (1997). Effects of 28-homobrassinolide on yields of wheat, rice, groundnut, mustard, potato and cotton. Journal of Agricultural Science 128: 405-413.
Rayas, A., Sánchez, R., Ventura, J., López, J. (1996). Efectos del DAA-6 en el medio de implantación para la propagación *in vitro* del banano (*Musa spp.*). Programa y Resúmenes X Seminario Científico, Instituto Nacional de Ciencias Agrícolas, La Habana, pp 158.
Robaina, C. (1994). Synthesis of spirobrassinosteroids starting from hecogenin. Ph. D. Thesis. University of Havana, Havana, 81 pp.
Robaina, C., Scovino, J. I. (1998). Algunos resultados de la aplicación de BIOCRECE en la agricultura venezolana. Programa y Resúmenes XI Seminario Científico, Instituto Nacional de Ciencias Agrícolas, La Habana, pp 132.
Robaina, C., Coll, F., Jomarrón, I. (1995b). Espirobrasinoesteroides 12-cetofuncionalizados y su procedimiento de obtención. Cuban Patent 22407, CO7J21/00, AOIN 45/00.
Robaina, C., Zullo, M. A., Coll, F. (2001e). Síntesis y caracterización espectroscópica de análogos de brasinoesteroides a partir de ácido cólico. Revista Cubana de Química XIII (2): 396.
Robaina, C., Coll, F., Alonso, E., Enriquez, M. (2001c). (25R)-3β-hidroxi-C-homo-11-oxa-5α-espirostan-12-ona y su epímero 3α como reguladores del crecimiento vegetal y su procedimiento de obtención Cuban Patent 22725, CO7J21/00, AOIN 45/00.
Robaina, C., Coll, F., Pérez, C., Jomarrón, I. (1995a). Reacción de Prevôst - Woodward con el (25R)-5α-2-espirosten-12-ona. Química and Industria 2: 19 - 23.
Robaina, C., Coll, F., Alonso, E., Enriquez, M. (2001b). Espirobrasinoesteroides con funciones oxigenadas en los anillos A y C como reguladores del crecimiento y su procedimiento de obtención. Cuban Patent 22718, CO7J21/00, AOIN 45/00.
Robaina, C., Izquierdo, Y., Alonso, E., Coll, F. (1999a). Síntesis, caracterización y actividad biológica de análogos espirostánicos de brasinoesteroides 5α-hidroxilados. Resúmenes 2do Taller de Química de los Productos Naturales, Centro de Química Farmacéutica, Ciudad de la Habana, pp 3.
Robaina, C., Zullo, M., De Acevedo, M., Coll, F. (2001a). Síntesis de un análogo colestánico de

brasinoesteroides 5α-hidroxilado. Revista Cubana de Química XIII(2): 435.
Robaina, C., Alonso, E., Coll, F., Pérez, C., Reyes, M. E. (1996). Síntesis, caracterización y actividad biológica de la 3-epihecogenina y la 2α-3-epihecogenina. Química and Industria 2: 25 - 27.
Robaina, C., Ramírez, I., Coll, F., Pérez, C., Pascual, A. (2001d). Síntesis de epoxilactonas esteroidales. Revista Cubana de Química XIII (2): 396.
Robaina, C., Ramírez, I., Pascual, A., Coll, F., Pérez, C., Coll, D. (2002). Síntesis y actividad biológica de análogos espirostánicos de brasinoesteroides a partir de la 9(11)-deshidrohecogenina. Revista CENIC de Ciencias Químicas. In press.
Robaina, C., Coll, F., Jomarrón, I., Pérez, C., Alonso, E., Ramírez, I., Anaya, H. (1999b). Síntesis y caracterización espectroscópica de la (25R)-2β,3α-dihidroxi-5α-espirostan-12-ona y su 12α-oxa-lactona. Revista CENIC de Ciencias Químicas 30: 107-110.
Rodríguez, T. (1999). Influence of Cuban bio-regulators on some morphological indicators during *in vitro* proliferation and rooting of banana (*Musa spp*.)M.Sc.. Thesis. University of Havana, Havana, 53 pp.
Rodríguez, T., Núñez, M., Vento, H. (1998). Influencia de un análogo de brasinoesteroides en la fase de multiplicación *in vitro* del banano (*Musa spp*.) variedad Gran Enano. Cultivos Tropicales 19(2): 19-22.
Roth, U., Friebe, A., Schnabl, H. (2000). Resistance induction in plants by a brassinosteroid-containing extract of *Lychnis viscaria* L. Zeitschrift für Naturforschung C – A Journal of Biosciences 55: 552-559.
Sairam, R. (1994a). Effects of homobrassinolide application on plant metabolism and grain yield under irrigated and moisture-stress conditions of two wheat varieties. Plant Growth Regulation 14: 173-181.
Sairam, R. (1994b). Effect of homobrassinolide application on metabolic activity and grain yield of wheat under normal and water-stress condition. Journal of Agronomy and Crop Science 173: 11-16.
Sairam, R., Shukla, D., Deshmukh, P. (1996). Effect of homobrassinolide seed treatment on germination, alpha-amylase activity and yield of wheat under moisture stress conditions. Indian Journal of Plant Physiology 1: 141-144.
Sala, C., Sala, F. (1985). Effect of brassinosteroid on cell division and enlargement in cultured carrot (*Daucus carota* L.) cells. Plant Cell Reports 4: 144-147.
Sasse, J. M. (1997). Recent progress in brassinosteroid research. Physiologia Plantarum 100: 696-701.
Soto, F., Tejeda, T., Núñez, M. (1997). Estudio preliminar sobre el uso de brasinoesteroides en cafetos. Cultivos Tropicales 18(1): 52-54.
Takatsuto, S., Kamuro, Y., Watanabe, T., Noguchi, T., Kuriyama, H. (1996). Synthesis and plant growth promoting effects of brassinosteroid compound TS303. Proceedings of the Plant Growth Regulation Society of America 23: 15-20.
Takematsu, T., Takeuchi, Y., Koguchi, M. (1983). New plant growth regulators. Brassinolide analogues: their biological effects and application to agriculture and biomass production. Chemical Regulation of Plants 18: 12-15.
Takeuchi, Y., Ogasawara, M., Konnai, M., Kamuro, Y. (1996). Promotive effectiveness of brassinosteroid (TS 303) and jasmonoid (PDJ) on emergence and establishment of rice seedlings. Proceedings of the Japanese Society of Chemical Regulation of Plants 31: 100-101.
Terry, E., Núñez, M., Pino, M. A., Medina, N. (2001). Efectividad de la combinación biofertilizantes-análogo de brasinoesteroides en la nutrición del tomate (*Lycopersicon esculentum* Mill.). Cultivos Tropicales 22(2): 59-65.
Tian, W. (1992). Study on the rational use of steroidal sapogenins I. Synthesis of sapogenin with A/B ring structure unit of brassinolide. Acta Chimica Sinica (Huaxuebao) 50: 70 - 77.
Torres, W., Núñez, M. (1997). The application of Biobras-6 and its effect on potato (*Solanum tuberosum* L.) yields. Cultivos Tropicales 18(2): 8-10.
Utria, E., Rodríguez, V., Suárez, A., Calderón, J. O. (2000). Influencia del Biobras-16 aplicado en diferentes dosis y momentos en el crecimiento y desarrollo de posturas de *Coffea arabica* L. Programa y Resúmenes XII Seminario Científico, Instituto Nacional de Ciencias Agrícolas, La Habana, pp 191.
Vardhini, S., Rao, S. (1997). Effect of brassinosteroids on salinity induced growth inhibition of groundnut seedlings. Indian Journal of Plant Physiology 2: 156-157.
Wang, W., Zhang, X. L., Liu, J. L (1992). Effect of brassinolide on somatic embryogenesis of *Gossypium hirsutum*. Plant Physiology Communications 28: 15-18.
Xu, H., Shida, A., Futatsuya, F., Kumura, A. (1994). Effects of epibrassinolide and abscisic acid on sorghum plants growing under soil water deficit. I. Effects on growth and survival. Japanese Journal of Crop Science 63: 671-675.

CHAPTER 5

GERHARD LEUBNER-METZGER

# BRASSINOSTEROIDS PROMOTE SEED GERMINATION

Seed germination of *Arabidopsis thaliana, Nicotiana tabacum,* and of parasitic angiosperms (*Orobranche* and *Striga* species) is determined by the balance of forces between the growth potential of the embryo and the mechanical restraint of the micropylar testa and/or endosperm tissues. Brassinosteroids (BR) and gibberellins (GA) promote seed germination of these species and counteract the germination-inhibition by abscisic acid (ABA). Severe mutations in GA biosynthetic genes in *Arabidopsis*, such as *ga1-3*, result in a requirement for GA application to germinate, but germination in this phenotype can also be rescued by BR. Germination of both the BR biosynthetic mutant *det2-1* and the BR-insensitive mutant *bri1-1* is more strongly inhibited by ABA than is germination of wild type. In contrast to GA, BR does not release tobacco photodormancy; i.e. seed germination in darkness remains blocked. BR promotes germination of non-photodormant tobacco seeds, but did not appreciably affect the induction of class I ß-1,3-glucanase (ßGlu I) in the micropylar endosperm. BR and GA promote tobacco seed germination by distinct signal transduction pathways and distinct mechanisms. Xyloglucan endo-transglycosylase (XET) enzyme activity accumulates in the embryo and the endosperm of germinating tobacco seeds and this appears to be partially controlled of BR. GA and light seem to act in a common pathway to release photodormancy, whereas BR does not release photodormancy. Induction of ßGlu I in the micropylar endosperm and promotion of release of 'coat-imposed' dormancy seem to be associated with the GA-dependent pathway, but not with BR signaling. It is proposed that BR promote seed germination by directly enhancing the growth potential of the emerging embryo in a GA-independent manner.

## HORMONAL REGULATION OF SEED DORMANCY AND GERMINATION

Brassinosteroids (BR) and gibberellins (GA) both regulate elongation growth of shoots and photomorphogenesis of seedlings and seem to antagonize the growth-inhibiting actions of abscisic acid (ABA) (Altmann, 1999; Neff *et al*., 2000; Bishop and Koncz, 2002). Only little is known about the interconnected molecular key processes regulating seed dormancy and germination in response to plant hormones and environmental causes. Seed germination of species with 'coat-imposed' dormancy is determined by the balance of forces between the growth potential of the embryo and the constrain, exerted by the covering layers, e.g. testa (seed coat) and endosperm. Seed dormancy can be coat-imposed and/or determined by the embryo itself and is a temporary failure or block of a viable seed to complete germination under physical conditions that normally favor the process (Hilhorst, 1995; Bewley, 1997b; Koornneef

*S.Hayat and A.Ahmad (eds.), Brassinosteroids,* 119-128.

*et al.*, 2002). The process of germination commences with theuptake of water by imbibition of the dry seed, followed by embryo expansion growth. This usually culminates in rupture of the covering layers and emergence of the radicle is generally considered as the completion of germination. The testa is no hindrance during the germination of *Pisum sativum* (Petruzzelli *et al.*, 2000) and *Brassica napus* (Schopfer and Plachy, 1984). *Arabidopsis thaliana* seeds have only a remainder of the endosperm and the testa characteristics are responsible for the degree of coat-imposed dormancy (Debeaujon and Koornneef, 2000; Debeaujon *et al.*, 2000). In tomato seeds the micropylar testa and endosperm tissues, also termed the micropylar cap, confer the primary control of germination timing (Liptay and Schopfer, 1983; Toorop *et al.*, 2000; Wu *et al.*, 2000). Radicle protrusion, during seed germination depends on embryo expansion, which is a growth process driven by water uptake. Treatment with ABA of non-endospermic, non-dormant *B. napus* seeds has no effect on the kinetics of testa rupture, but it inhibits the post-germinational extension growth of the radicle (Schopfer and Plachy, 1984). Thus, ABA does not inhibit initial imbibition of water (water uptake phases 1 and 2) needed for initial embryo extension growth. ABA inhibits the transition to the seedling growth phase (water uptake phase 3) after radicle emergence. In contrast, ABA inhibits the germination of seeds with testa- and/or endosperm-imposed dormancy including *A. thaliana* (Beaudoin *et al.*, 2000; Debeaujon and Koornneef, 2000; Steber and McCourt, 2001), and tobacco (Leubner-Metzger *et al.*, 1995; Leubner-Metzger and Meins, 2000). In the Solanaceae this is achieved by an inhibitory action of ABA on the final step of radicle emergence through the micropylar endosperm (Liptay and Schopfer, 1983; Toorop *et al.*, 2000). ABA treatment does not inhibit the germination of tomato scored as initial radicle extension growth of detipped (surgical removal of the micropylar layers covering the radicle tip) seeds (Liptay and Schopfer, 1983; Groot and Karssen, 1992; Bewley, 1997a). Even 1000 μM ABA does not inhibit the germination of detipped, whereas 100 μM ABA results in a substantial inhibition of the germination of intact tomato seeds (Liptay and Schopfer, 1983).

Rupture of the testa and the endosperm are distinct and temporally separate events during the germination of tobacco seeds (Arcila and Mohapatra, 1983; Leubner-Metzger, 2003). ABA treatment of tobacco seeds does not appreciably affect the kinetics of testa rupture, but it delays endosperm rupture and results in the formation of a novel structure, consisting of the enlarging radicle with a sheath of greatly elongated endosperm tissue. Class I ß-1,3-glucanase (ßGlu I) is induced after testa rupture and just prior to endosperm rupture. This induction is exclusively localized in the micropylar endosperm at the site where the radicle will emerge. ABA inhibits the induction of the ßGlu I genes during tobacco seed germination and specifically delays endosperm rupture (Leubner-Metzger *et al.*, 1995). The close correlation between ßGlu I induction and the onset of endosperm rupture under a variety of physiological conditions support the hypothesis that ßGlu I contributes to endosperm rupture. Direct evidence for a causal role of ßGlu I during endosperm rupture comes from sense-transformation with a chimeric ABA-inducible ßGlu I transgene (Leubner-Metzger and Meins, 2000; Leubner-Metzger, 2003). This has been achieved by transformation of tobacco with a sense-ßGlu I construct consisting of the genomic DNA fragment of the tobacco *ßGlu I B* gene regulated by the castor bean *Cat1* gene promoter, which is known to confer ABA-inducible, endosperm-specific transgene expression in germinating tobacco seeds. ABA down-regulates the ßGlu I host genes in wild-type seeds, but due to the ABA-inducible ßGlu I-transgene it causes high-level ßGlu I expression in sense-ßGlu I- seeds. ABA treatment delays endosperm rupture of after-ripened seeds, but due to ßGlu I over-expression this delay is significantly reduced in sense-ßGlu I- seeds. ßGlu I over-expression reduces the ABA-mediated delay in endosperm rupture of seeds, but ABA treatment does not effect the kinetics of testa rupture. Taken together, these results support the view that a threshold ßGlu I content is necessary, but not sufficient, for endosperm rupture (Leubner-Metzger *et al.*, 1995; Leubner-Metzger, 2003). In the presence of ABA ßGlu I becomes a limiting factor for endosperm rupture, and removal of this block due to expression of the ABA-inducible ßGlu I-transgene in sense-ßGlu I- seeds promotes endosperm rupture until other ABA-sensitive processes become limiting (Leubner-Metzger and Meins, 2000). While these results do not exactly show how ßGlu I promote endosperm rupture, they directly show that ßGlu I is causally involved and that it substantially contributes to endosperm rupture. Possible mechanisms of ßGlu I-action have been discussed in a recent review (Leubner-Metzger, 2003).

In contrast to the inhibition by ABA, seed dormancy release and germination are promoted by GA (Hilhorst and Karssen, 1992; Hilhorst, 1995; Bewley, 1997b; Koornneef *et al.*, 2002; Leubner-Metzger, 2003). Seeds of GA-deficient mutants do not germinate without exogenous treatment with GA. This is also the case for the GA-deficient *gib1* mutant of tomato (Groot and Karssen, 1992; Wu *et al.*, 2000), but detipping can replace the requirement for the GA-treatment and induce germination of *gib1* seeds. Photodormant tobacco seeds do not germinate in darkness, but treatment with a red-light pulse or with GA is sufficient to release photodormancy, induce testa rupture and subsequent endosperm rupture in the dark (Leubner-Metzger et al., 1996; Leubner-Metzger, 2001, 2002). Endosperm rupture during dark-germination of tobacco is accompanied by GA-enhanced expression of ßGlu I in the micropylar endosperm. Thus, GA can release coat-imposed dormancy, induce testa rupture, enhance ßGlu I expression in the endosperm, promote endosperm rupture and counteract the inhibitory action of ABA on seed germination. Red light has been shown to up-regulate the biosynthesis of bioactive $GA_1$ and $GA_4$ by inducing GA biosynthetic genes in germinating seeds of lettuce and Arabidopsis (Toyomasu *et al.*, 1993, 1998; Yamaguchi *et al.*, 1998, 2001). Recent publications demonstrate that BR also interact with light and ABA in regulating seed germination of Arabidopsis and tobacco (Leubner-Metzger, 2001; Steber and McCourt, 2001). The role of BR in seed germination and the interactions of BR with other plant hormones during this process are the focus of this review.

## PROMOTION OF SEED GERMINATION BY BRASSINOSTEROIDS

Brassinosteroids (BR) and GA interact with light in regulating elongation growth of shoots and photomorphogenesis of seedlings by what appear to be independent pathways (Altmann, 1999; Neff *et al.*, 2000; Bishop and Koncz, 2002). Endogenous

BRs have been identified in the seeds of several species, including pea (Yokota *et al.*, 1996), *A. thaliana* (Schmidt *et al.*, 1997) and *Lychnis viscaria* (Friebe *et al.*, 1999). BR application has been reported to enhance germination of certain parasitic angiosperms (Takeuchi *et al.*, 1991, 1995), cereals (Gregory, 1981; Yamaguchi *et al.*, 1987), *Arabidopsis* (Steber and McCourt, 2001), and tobacco (Leubner-Metzger, 2001). Pretreatment with brassinolide stimulates the germination and seedling emergence of aged rice seeds (Yamaguchi *et al.*, 1987) and seed treatment of barley accelerated subsequent seedling growth (Gregory, 1981). It is, however, not known, whether the promoting effect of BR on cereal grains is manifested only on the level of seedling growth or also on the level of germination *per se*. While BR treatment was found to affect the early seedling growth of cress, in the same publication was stated (based on data not shown) that BR do not affect the germination of non-photodormant, non-endospermic cress seeds imbibed in the dark (Jones-Held *et al.*, 1996). In contrast to cereals and cress, the effects of BR on germination *per se* were studied at length in experiments in seeds of parasitic angiosperms, *Arabidopsis* and tobacco.

Germination of the endospermic seeds of parasitic *Orobranche* and *Striga* species is, in contrast to *Arabidopsis* and tobacco, inhibited by light (Takeuchi *et al.*, 1991, 1995; Babiker *et al.*, 2000). Neither BR, ethylene nor GA can substitute for the conditioning treatment with strigol, which is needed for inducing germination of unconditioned (i.e. dormant) seed. Conditioning removes the restriction on the ethylene biosynthetic pathway and increases the capacity to produce ethylene (Babiker *et al.*, 2000). Treatment with BR promotes the germination of conditioned (i.e. non-dormant) *Orobranche* and *Striga* seeds, imbibed in the light and in the dark. These results demonstrate that BR alone is not able to replace strigol and release the dormancy of these seeds. BR promotes the seed germination of parasitic angiosperms after dormancy has been released by counteracting the inhibitory effects of light, acting independently of GA and possibly by promoting ethylene action.

In *A. thaliana* BR promotes the germination of pre-chilled (i.e. non-dormant) seeds of the BR-deficient biosynthesis mutant *det2-1* and the BR-insensitive response mutant *bri1-1* imbibed in the light (Steber and McCourt, 2001). Seed germination of *det2-1* and *bri1-1* is more strongly inhibited by ABA than is germination of the wild type and BR is, therefore, able to partially overcome the inhibition of germination by ABA. BR treatment rescues the germination phenotype of the severe GA-deficient biosynthesis mutant *ga1-3*, which normally requires GA treatment for dormancy release and germination. BR treatment also partially rescues the germination phenotype of the severe GA-insensitive response mutant *sly1* (*sleepy1*), which cannot be rescued by treatment with GA. Interestingly, a new allele for *sly1* was identified in a screen for BR-dependent germination and suggests interactions between BR and GA signaling in seeds (Steber *et al.*, 1998; Steber and McCourt, 2001). This is further supported by the germination phenotype of the *gpa1* mutant of Arabidopsis (Ullah *et al.*, 2002). The *GPA1* gene encodes the alpha subunit of a heterotrimeric G protein. Seeds with the *gpa1* null mutation are 100-fold less responsive to GA and GPA1 overexpressing seeds are hypersensitive for GA. The *gpa1* mutant seeds are also completely insensitive to BR rescue of germination when the level of GA in seeds is reduced. These results point to a role for BR in stimulating germination of *Arabidopsis* seeds via embryo expansion and this effect is likely to be specific to germination. The interactions between hormonal signaling pathways appear to be of utmost importance for the regulation of germination and the inhibitory effects of ABA are counteracted by BR, GA and ethylene (Beaudoin *et al.*, 2000; Debeaujon and Koornneef, 2000; Steber and McCourt, 2001). The *Arabidopsis sax1* (*hypersensitive to abscisic acid and auxin*) dwarf mutant is impaired in BR biosynthesis and exhibits pleiotropic seedlings effects with respect to ABA, auxin, GA, ethylene and BR, but seed germination of *sax1* and wild type is not differentially inhibited by ABA (Ephritikhine *et al.*, 1999).

BR promotes seedling elongation and germination of non-photodormant tobacco seeds, but do not appreciably affect testa rupture and the subsequent induction of ßGlu I in the micropylar endosperm (Leubner-Metzger, 2001). Treatment with BR, but not GA, accelerates endosperm rupture of tobacco seeds imbibed in the light. BR and GA promote endosperm rupture of dark-imbibed non-photodormant seeds, but only GA enhances ßGlu I induction. Promotion of endosperm rupture by BR is dose-dependent and 0.01 μM brassinolide is most effective. BR and GA promote ABA-inhibited dark-germination of non-photodormant seeds, but only GA replaces light in inducing ßGlu I. These results indicate that BR and GA promote tobacco seed germination by distinct signal transduction pathways and distinct mechanisms. GA and light act in a common pathway to release photodormancy, whereas BR does not release photodormancy. ßGlu I induction in the micropylar endosperm and release of coat-imposed dormancy seem to be associated with the GA/light pathway, but not with BR signaling. These findings suggest as a model for the endosperm-limited germination of tobacco: (1) Photodormancy is released exclusively by the GA/light-pathway; (2) Promotion of subsequent endosperm rupture by the BR and the GA/light signal transduction pathways is achieved by independent and distinct mechanisms; (3) ABA inhibits endosperm rupture by interfering with both pathways; (4) The GA/light pathway regulates ßGlu I induction in the micropylar endosperm and seems to control endosperm weakening; (5) It is proposed that the BR pathway promotes endosperm rupture of non-dormant seeds by enhancing the growth potential of the embryo (Leubner-Metzger, 2001, 2003).

## POSSIBLE MECHANISMS OF BR-PROMOTED SEED GERMINATION

Taken together, the findings in parasitic angiosperms, *Arabidopsis* and tobacco suggest that GA and BR act in parallel to promote cell elongation and germination and to counteract the inhibitory action of ABA on seeds. Since BR stimulates germination of the GA-insensitive mutant *sly1*, it is unlikely that BR acts by increasing GA sensitivity. It is possible that BR acts by stimulating GA biosynthesis in *Arabidopsis* seeds imbibed in the light (Steber and McCourt, 2001). BR action via stimulation of GA biosynthesis is however unlikely for tobacco, because BR does not promote the expression of ßGlu I, which is induced by GA in the dark (Leubner-Metzger, 2001). It is known that BR can stimulate ethylene production and ethylene treatment can rescue the germination phenotype of the GA-deficient *Arabidopsis ga1-1* mutant (Karssen *et al.*, 1989; Koornneef and Karssen, 1994; Steber and McCourt, 2001). However, there are several arguments against the hypothesis that BR acts via ethylene: (1) Ethylene levels are not increased in cress seedlings following BR treatment of seeds (Jones-Held *et al.*, 1996). (2) Endogenous ethylene promotes ßGlu I accumulation in the micropylar endosperm of tobacco, but BR treatment promotes endosperm rupture without enhancing ßGlu I accumulation (Leubner-Metzger *et al.*, 1998; Leubner-Metzger, 2001). (3) Ethylene rescue of *ga1-1* seed germination results

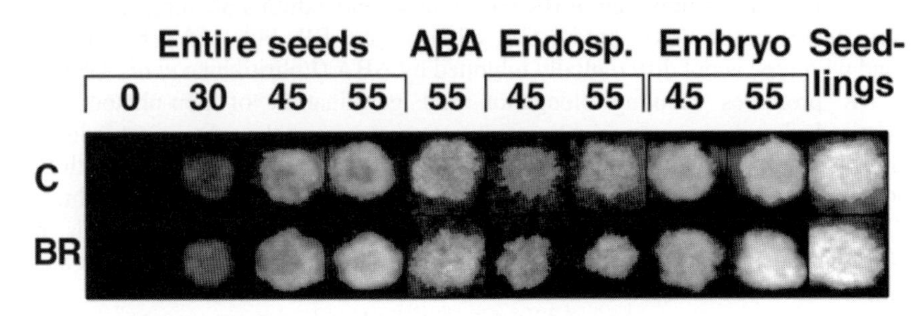


*Figure 1. The accumulation of XET enzyme activity in germinating seeds of Nicotiana tabacum cv. Havana 245. Tobacco seeds were imbibed without (Control; C) and with 10 nM brassinolide (BR) in the medium and incubated for the times indicated (hours) in continuous light; in addition 10 µM ABA was added to the medium of one series (ABA) (Leubner-Metzger, 2001). Protein extracts from entire seeds, seed tissues (Endosperm, Embryo), or seedlings were used. Testa rupture in the populations of ca. 150 seeds was 0 % at 30 h and 100 % at 45 h. Endosperm rupture was 0 % (45 h), ca. 0 % (Control, 55 h) and ca. 30 % (BR, 55 h); only seeds without endosperm rupture were used for the extracts. Tissues were homogenized and XET enzyme activities were determined by the XET 'dot blot' assay as described by Fry (1997). For the semiquantitative XET assay ca. 3 µl (60 µg protein) were applied onto the XET 'dot blot' test paper and incubated for 7 h at 25 °C. Elevated fluorescence under the UV lamp is indicative for accumulating XET enzyme activity.*

in seedlings exhibiting triple response, but BR rescue of *ga1-3* seed germination results in seedlings that do not exhibit triple response (Steber and McCourt, 2001). Another possibility would be BR-action via auxin. Auxin also stimulates cell elongation but it does not rescue germination of *ga1-3* (Koornneef and Karssen, 1994). Thus, if BR stimulates germination via embryo expansion, this effect is likely to be specific to seed germination.

Xyloglucan is a major structural polysaccharide in the cell walls of higher plants and modification of xyloglucan bondages is proposed to be involved in cell expansion growth (Fry, 1995; Campbell and Braam, 1999). Xyloglucan endo-transglycosylases (XET) are enzymes with potential wall-modifying functions and are thought to be involved in the regulation of cell wall loosening necessary for cell expansion growth. XETs cleave a xyloglucan chain and then conserve the energy of the cut bond by synthesizing a new bond on another xyloglucan chain. XETs are encoded by a multigene family and their expression is highly regulated by plant hormones and environmental factors. Gene expression in vegetative tissues of several

XETs is known to be induced by BR, auxin and/or GA (Fry, 1997; Campbell and Braam, 1999). In seeds XETs are involved in the post-germinational mobilization of cell wall xyloglucan reserves (Edwards *et al*., 1985; De Silva *et al*., 1993; Fanutti *et al*., 1993; Tine *et al*., 2000). The transcripts of the GA-regulated XET gene LeXET4 is expressed in the endosperm cap during tomato seed germination (Chen *et al*., 2002). LeXET4 mRNA was strongly expressed in germinating seeds, was much less abundant in stems, and was not detected in roots, leaves or flower tissues. During germination, LeXET4 mRNA was induced prior to endosperm rupture and was localized exclusively to the endosperm cap region. Expression of LeXET4 was dependent on exogenous GA in GA-deficient *gib-1* mutant seeds. ABA had no effect on LeXET4 mRNA expression in wild-type tomato seeds. The temporal, spatial and hormonal regulation pattern of LeXET4 gene expression suggests that LeXET4 has a role in endosperm cap weakening, a key process regulating tomato seed germination (Chen *et al*., 2002). In tomato seeds it was not tested whether XET activity accumulates during germination and whether the LeXET4 mRNA or any other tomato seed XET is induced by BR. Steven Fry introduced a specific XET 'dot-blot' assay for the semiquantitative determination of XET enzyme activity (Fry, 1997). This 'XET test paper' was used to detect XET-catalyzed transglycosylation in protein extracts of germinating tobacco seeds (Figure 1). XET enzyme activity in dry tobacco seeds was very low and accumulated during imbibition. The onset of this accumulation was already visible at the onset of testa rupture (30 h) and further accumulation to high specific XET enzyme activities was detected prior to the onset of endosperm rupture (45 h). XET enzyme activity accumulation was detected in the endosperm and in the embryo and ABA did not appreciably affect the accumulation. Treatment of tobacco seeds with BR enhanced the accumulation of XET enzyme activity especially in the embryo (Fig. 1) and BR is known to promote endosperm rupture of tobacco (Leubner-Metzger, 2001). It is, therefore, possible that XETs in the endosperm and in the embryo play roles in mediating the promotion of seed germination by BR via cell wall loosening. However, this can only be a part of the story, since the counteracting effects of BR on ABA-mediated inhibition of endosperm rupture are not correlated with inhibited XET enzyme activity accumulation. After cloning of the corresponding cDNAs the regulation of specific XETs in the endosperm and the embryo will be investigated in future experiments.

## CONCLUSION

The results obtained with *Arabidopsis*, tobacco and parasitic angiosperms suggest that the BR pathway promotes germination of non-dormant seeds by directly enhancing the growth potential of the embryo (Takeuchi *et al*., 1991, 1995; Leubner-Metzger, 2001; Steber and McCourt, 2001). Exploring the molecular mechanisms of the BR-mediated promotion of cell extension growth and seed germination is a challenging field for future experiments.

## ACKNOWLEDGEMENTS

I thank Stephen C. Fry for kindly providing XET 'test paper'. My research is supported by a grant from the Deutsche Forschungsgemeinschaft (LE 720/3), which is gratefully acknowledged.

## REFERENCES

Altmann, T. (1999). Molecular physiology of brassinosteroids revealed by the analysis of mutants. Planta 208: 1-11.

Arcila, J.,Mohapatra, S C. (1983). Development of tobacco seedling. 2. Morphogenesis during radicle protrusion. Tobacco Science 27: 35-40.

Babiker, A. G. T., Ma, Y. Q., Sugimoto, Y., Inanaga, S. (2000). Conditioning period, $CO_2$ and GR24 influence ethylene biosynthesis and germination of *Striga hermonthica*. Physiologia Plantarum 109: 75-80.

Beaudoin, N., Serizet, C., Gosti, F., Giraudat, J. (2000). Interactions between abscisic acid and ethylene signaling cascades. The Plant Cell 12: 1103-1115.

Bewley, J. D. (1997a). Breaking down the walls - a role for endo-ß-mannanase in release from seed dormancy? Trends in Plant Science 2: 464-469.

Bewley, J D. (1997b). Seed germination and dormancy. The Plant Cell 9: 1055-1066.

Bishop, G. J., Koncz, C. (2002). Brassinosteroids and plant steroid hormone signaling. The Plant Cell 14: 97-110.

Campbell, P., Braam, J. (1999). Xyloglucan endotransglycosylases: diversity of genes, enzymes and potential wall-modifying functions. Trends in Plant Science 4: 361-366.

Chen, F., Nonogaki, H., Bradford, K. J. (2002). A gibberellin-regulated xyloglucan endotransglycosylase gene is expressed in the endosperm cap during tomato seed germination. Journal of Experimental Botany 53: 215-223.

De Silva, J., Jarman, C. D., Arrowsmith, D., Stronach, M. S., Chengappa, S., Sidebottom, C., Reid, J. S. G. (1993). Molecular characterization of a xyloglucan-specific endo-1,4-ß-D-glucanase (xyloglucan endotransglycosylase) from nasturtium seeds. The Plant Journal 3: 701-711.

Debeaujon, I., Koornneef, M. (2000). Gibberellin requirement for *Arabidopsis* seed germination is determined both by testa characteristics and embryonic abscisic acid. Plant Physiology 122: 415-424.

Debeaujon, I., Léon-Kloosterziel, K. M., Koornneef, M. (2000). Influence of the testa on seed dormancy, germination, and longevity in *Arabidopsis*. Plant Physiology 122: 403-413.

Edwards, M., Dea, I. C. M., Bulpin, P. V., Reid, J. S. G. (1985). Xyloglucan amyloid mobilization in the cotyledons of *Tropaeolum majus* seeds following germination. Planta 163: 133-140.

Ephritikhine, G., Pagant, S., Fujioka, S., Takatsuto, S., Lapous, D., Caboche, M., Kendrick, R. E., Barbier-Brygoo, H. (1999). The *sax1* mutation defines a new locus involved in the brassinosteroid biosynthesis pathway of *Arabidopsis thaliana*. The Plant Journal 18: 315-320.

Fanutti, C., Gidley, M. J., Reid, J. S. G. (1993). Action of a pure xyloglucan endo-transglycosylase (formaly called xyloglucan-specific endo-1,4-ß-D-glucanase) from the cotyledons of germinated nasturtium seeds. The Plant Journal 3: 691-700.

Friebe, A., Volz, A., Schmidt, J., Voigt, B., Adam, G., Schnabl, H. (1999). 24-*Epi*-secasterone and 24-*epi*-castasterone from *Lychnis viscaria* seeds. Phytochemistry 52: 1607-1610.

Fry, S C. (1995). Polysaccharide-modifying enzymes in the plant cell wall. Annual Review of Plant Physiology and Plant Molecular Biology 46: 497-520.

Fry, S C. (1997). Novel 'dot-blot' assays for glycosyltransferases and glycosylhydrolases: Optimization for xyloglucan endotransglycosylase (XET) activity. The Plant Journal 11: 1141-1150.

Gregory, L. E. (1981). Acceleration of plant growth through seed treatment with brassins. American Journal of Botany 68: 586-588.

Groot, S. P. C., Karssen, C. M. (1992). Dormancy and germination of abscisic acid-deficient tomato seeds. Plant Physiology 99: 952-958.

Hilhorst, H. W. M. (1995). A critical update on seed dormancy. I. Primary dormancy. Seed Science Research 5: 61-73.

Hilhorst, H. W. M., Karssen, C. M. (1992). Seed dormancy and germination: the role of abscisic acid and gibberellins and the importance of hormone mutants. Plant Growth Regulation 11: 225-238.
Jones-Held, S., Vandoren, M., Lockwood, T. (1996). Brassinolide application to *Lepidium sativum* seeds and the effects on seedling growth. Journal of Plant Growth Regulation 15: 63-67.
Karssen, C. M., Zagórsky, S., Kepczynski, J., Groot, S. P. C. (1989). Key role for endogenous gibberellins in the control of seed germination. Annals of Botany 63: 71-80.
Koornneef, M., Bentsink, L., Hilhorst, H. (2002). Seed dormancy and germination. Current Opinion in Plant Biology 5: 33-36.
Koornneef, M., Karssen, C. M. (1994). Seed dormancy and germination. In Arabidopsis, pp 313-334. Eds E M Meyerowitz and C R Somerville, Cold Spring Harbor Laboratory Press, New York
Leubner-Metzger, G. (2001). Brassinosteroids and gibberellins promote tobacco seed germination by distinct pathways. Planta 213: 758-763.
Leubner-Metzger, G. (2003). Functions and regulation of ß-1,3-glucanase during seed germination, dormancy release and after-ripening. Seed Science Research 13: 17-34.
Leubner-Metzger, G. (2002). Seed after-ripening and over-expression of class I ß-1,3-glucanase confer maternal effects on tobacco testa rupture and dormancy release. Planta 215: 659-698.
Leubner-Metzger, G., Fründt, C., Meins, F. Jr. (1996). Effects of gibberellins, darkness and osmotica on endosperm rupture and class I ß-1,3-glucanase induction in tobacco seed germination. Planta 199: 282-288.
Leubner-Metzger, G., Fründt, C., Vögeli-Lange, R., Meins, F. Jr. (1995). Class I ß-1,3-glucanase in the endosperm of tobacco during germination. Plant Physiology 109: 751-759.
Leubner-Metzger, G., Meins, F. Jr. (2000). Sense transformation reveals a novel role for class I ß-1,3-glucanase in tobacco seed germination. The Plant Journal 23: 215-221.
Leubner-Metzger, G., Petruzzelli, L., Waldvogel, R., Vögeli-Lange, R., Meins, F. Jr. (1998). Ethylene-responsive element binding protein (EREBP) expression and the transcriptional regulation of class I ß-1,3-glucanase during tobacco seed germination. Plant Molecular Biology 38: 785-795.
Liptay, A., Schopfer, P. (1983). Effect of water stress, seed coat restraint, and abscisic acid upon different germination capabilities of two tomato lines at low temperature. Plant Physiology 73: 935-938.
Neff, M. M., Fankhauser, C., Chory, J. (2000). Light: an indicator of time and place. Genes and Development 14: 257-271.
Petruzzelli, L., Coraggio, I., Leubner-Metzger, G. (2000). Ethylene promotes ethylene biosynthesis during pea seed germination by positive feedback regulation of 1-aminocyclopropane-1-carboxylic acid oxidase. Planta 211: 144-149.
Schmidt, J., Altmann, T., Adam, G. (1997). Brassinosteroids from seeds of *Arabidopsis thaliana*. Phytochemistry 45: 1325-1327.
Schopfer, P., Plachy, C. (1984). Control of seed germination by abscisic acid. II. Effect on embryo water uptake in *Brassica napus* L. Plant Physiology 76: 155-160.
Steber, C. M., Cooney, S. E., McCourt, P. (1998). Isolation of the GA-response mutant *sly1* as a suppressor of *ABI1-1* in *Arabidopsis thaliana*. Genetics 149: 509-521.
Steber, C. M., McCourt, P. (2001). A role for brassinosteroids in germination in *Arabidopsis*. Plant Physiology 125: 763-769.
Takeuchi, Y., Omigawa, Y., Ogasawara, M., Yoneyama, K., Konnai, M., Worsham, A. D. (1995). Effects of brassinosteroids on conditioning and germination of clover broomrape (*Orobanche minor*) seeds. Plant Growth Regulation 16: 153-160.
Takeuchi, Y., Worsham, A. D., Awad, A. E. (1991). Effects of brassinolide on conditioning and germination of witchweed (*Striga asiatica*) seeds. In Brassinosteroids: chemistry, bioactivity and applications, pp 298-305. Eds H G Cuttler, T Yokota and G Adam, American Chemical Society, Washington.
Tine, M. A. S., Cortelazzo, A. L., Buckeridge, M. S. (2000). Xyloglucan mobilization in cotyledons of developing plantlets of *Hymenaeae courbaril* L. (Leguminoseae-Caesalpinoideae). Plant Science 154: 117-126.
Toorop, P. E., van Aelst, A. C., Hilhorst, H. W. M. (2000). The second step of the biphasic endosperm cap weakening that mediates tomato (*Lycopersicon esculentum*) seed germination is under control of ABA. Journal of Experimental Botany 51: 1371-1379.
Toyomasu, T., Kawaide, H., Mitsuhashi, W., Inoue, Y., Kamiya, Y. (1998). Phytochrome regulates gibberellin biosynthesis during germination of photoblastic lettuce seeds. Plant Physiology 118: 1517-1523.

Toyomasu, T., Tsuji, H., Yamane, H., Nakayama, M., Yamaguchi, I., Murofushi, N., Takahashi, N., Inoue, Y. (1993). Light effects on endogenous levels of gibberellins in photoblastic lettuce seeds. Journal of Plant Growth Regulation 12: 85-90.

Ullah, H., Chen, J. G., Wang, S. C., Jones, A. M. (2002). Role of a heterotrimeric G protein in regulation of Arabidopsis seed germination. Plant Physiology 129: 897-907.

Wu, C.T., Leubner-Metzger, G., Meins, F. Jr., Bradford, K. J. (2000). Class I ß-1, 3-glucanase and chitinase are expressed in the micropylar endosperm of tomato seeds prior to radicle emergence. Plant Physiology 126: 1299-1313.

Yamaguchi, S., Kamiya, Y., Sun, T. P. (2001). Distinct cell-specific expression patterns of early and late gibberellin biosynthetic genes during *Arabidopsis* seed germination. The Plant Journal 28: 443-453.

Yamaguchi, S., Smith, M. W., Brown, R. G. S., Kamiya, Y., Sun, T. P. (1998). Phytochrome regulation and differential expression of gibberellin 3ß-hydroxylase genes in germinating *Arabidopsis* seeds. The Plant Cell 10: 2115-2126.

Yamaguchi, T., Wakizuka, T., Hirai, K., Fujii, S., Fujita, A. (1987). Stimulation of germination in aged rice seeds by pretreatment with brassinolide. Proceeding of Plant Growth Regulation Society of America 14: 26-27.

Yokota, T., Matsuoka, T., Koarai, T., Nakayama, M. (1996). 2-Deoxybrassinolide, a brassinosteroid from *Pisum sativum* seed. Phytochemistry 42: 509-511.

CHAPTER 6

A. B. PEREIRA-NETTO, S. SCHAEFER, L. R. GALAGOVSKY AND J. A. RAMIREZ

# BRASSINOSTEROID-DRIVEN MODULATION OF STEM ELONGATION AND APICAL DOMINANCE: APPLICATIONS IN MICROPROPAGATION

In horticulture, inhibition of shoot branching during orchards establishment is highly desired to avoid stem break by wind while increased branching during crop production usually leads to increased yields. In forestry, reduced branching significantly improves wood quality, especially for pulp production. Thus, branching control is an important component on the establishment of productivity rates for crop and forestry species. Brassinosteroids (BRs) are natural polyhydroxy steroidal lactones and ketones with phytohormone action. These plant steroids are known to stimulate stem elongation and to control apical dominance as well. Our research efforts have been focused on the use of BRs to control stem elongation and apical dominance aiming to improve crop and forestry species productivity, and yield in *in vitro*-based plant propagation techniques. In this chapter we describe how BRs can be used to significantly improve microprogation techniques, more especifically for the marubakaido apple rootstock.

## INTRODUCTION

In all multicellular organisms growth and morphogenesis must be coordinated, but in the case of higher plants, this is of particular importance because the timing of organogenesis is not fixed but occurs in response to environmental constraints (Schumacher *et al.*, 1999). Coordinate control of plant growth is regulated by both external stimuli and internal mechanisms. The internal components of plant signaling are generally mediated by chemical growth regulators (Azpiroz *et al.*, 1998). The most recently discovered class of plant growth substances, the brassinosteroids, has been the least studied up to date; however, rapid progress toward understanding BR biosynthesis (Fujioka and Sakurai, 1997; Choe *et al.*, 1999a; Noguchi *et al.*, 2000), metabolism (Fujioka and Sakurai, 1997; Yokota, 1997) and signal transduction pathway is now being made (Friedrichsen *et al.*, 2000; Li and Chory, 1999; Li *et al.*, 2001; Schumacher *et al.*, 1999; Wang *et al.*, 2001, 2002; Yin *et al.*, 2002). Plant growth is accomplished by orderly cell division and tightly regulated cell expansion. For plants, the contribution of cell expansion to growth is of much greater significance than for most of the other organisms and the final size reached by all plant organs

*S.Hayat and A.Ahmad (eds.), Brassinosteroids*, 129-157

depends upon a period of significant cell elongation, which usually follows an active cell division period (Azpiroz *et al.*, 1998).

## BRASSINOSTEROIDS

Since the isolation and identification of brassinolide (**1**) (see Figure 1) by Grove *et al.* (1979) more than 40 natural brassinosteroids have been detected in different organs of plant species belonging to several families. Most of them show a 5 a-cholestane carbon skeleton with the main following structural characteristics:
i) Ring A: presents one to three oxygenated groups (mostly hydroxyl groups), always oxygenated at carbon 3.
ii) Ring B: presents a 6-oxo-7-oxalactone, a 6-keto group or the ring can be completely saturated.
iii) Rings A to D fusion: all of them *trans*.
iv) Side chain: 22,23-dihydroxy groups; 24-methyl or 24-ethyl groups (the last one called 28-homo series); sometimes a methyl group at carbon 25 and sometimes an unsaturated bond between carbons 24 and 28.

Brassinolide (**1**) and castasterone (**2**) are thought to be the main natural brassinosteroids because of their wide distribution and their potent biological activity (Arteca, 1995).

Brassinosteroid analogs are those synthetic compounds that show structural similarity with natural brassinosteroids and/or brassinolide-like activity (Zullo *et al.*, 2002b). The main synthetic brassinosteroid are 28-homobrassinolide (**3**), 28-homocastasterone (**4**) (also named homocastasterone), 24-epibrassinolide (**5**), 24-epicastasterone (**6**) and 22, 23-diepi brassinolide (**7**), because of their high bioactivity and accessible synthetic production from stigmasterol (compounds **3** and **4**) or ergosterol (compounds **5**-**7**). These compounds, formerly thought to be exclusively synthetic, have now been demonstrated to be of natural occurrence in plants (Abe *et al.*, 1983; Fujioka, 1999).

Brassinosteroids, the only steroids known to influence plant development (Kim *et al.*, 1998), induce a broad spectrum of responses such as cell division, bending, reproductive and vascular development, membrane polarization and proton pumping, source/sink relationships, and modulation of stress (Clouse and Sasse, 1998). However, stimulation of longitudinal growth of young tissues via cell elongation and cell division is a major biological effect of brassinosteroids (Clouse, 1996). Besides its effects on cell elongation and division, brassinosteroids have also been shown to stimulate vascular differentiation, a developmental process critical for stem elongation. After 24 hours of the application of brassinolide **(1)**, a 10-fold increase in differentiation of tracheary elements was observed in explants of *Helianthus tuberosus* (Clouse and Zurek, 1991). For untreated explants, tracheary

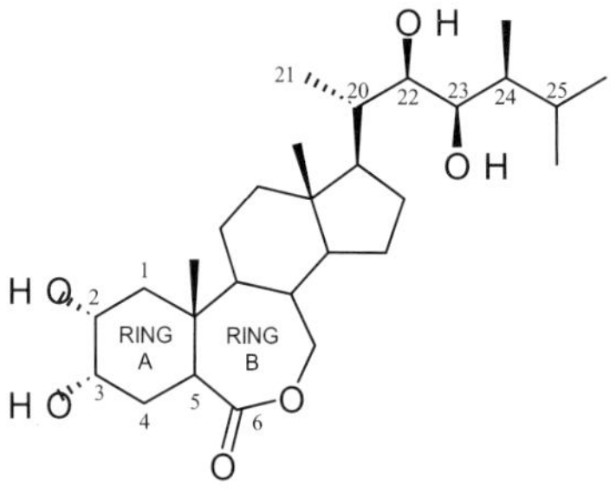

**1** brassinolide

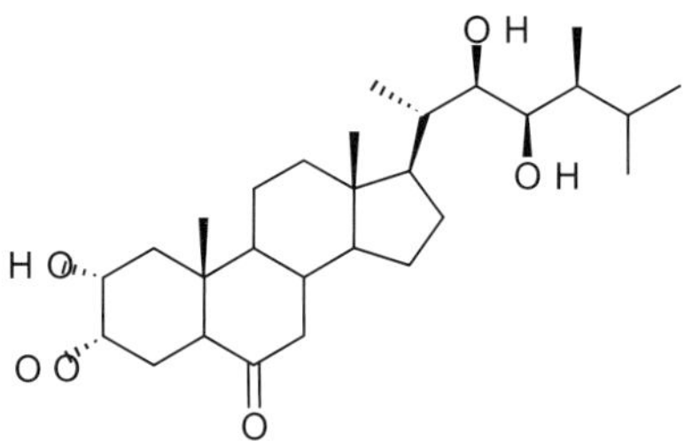

**2** castasterone

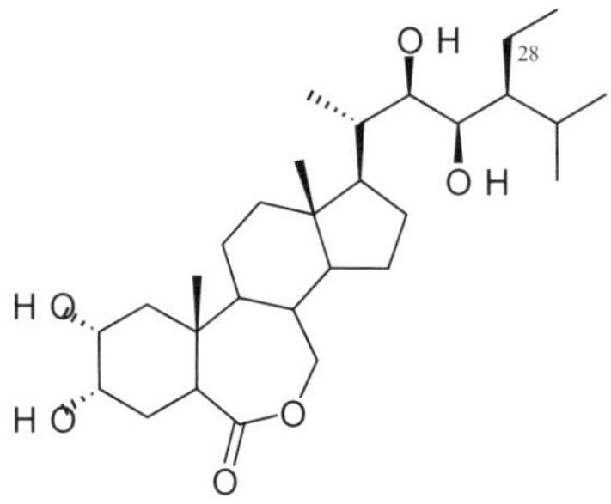

**3** 28-homobrassinolide

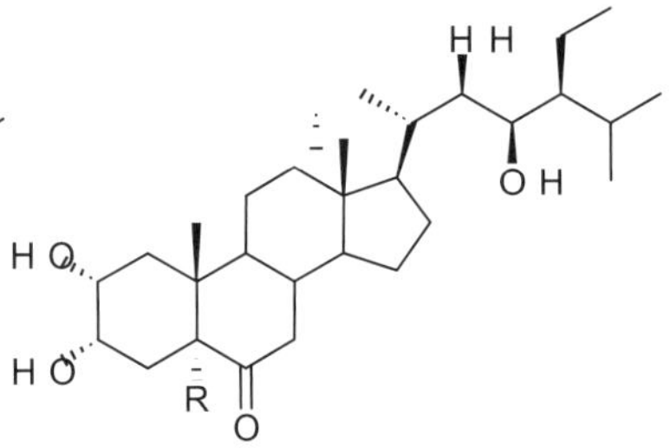

| | | |
|---|---|---|
| **4** | R= H | 28-homocastasterone |
| **8** | R= F | 5α-fluoro-28-homocastasterone |
| **9** | R= OH | 5α-hydroxy-28-homocastasterone |

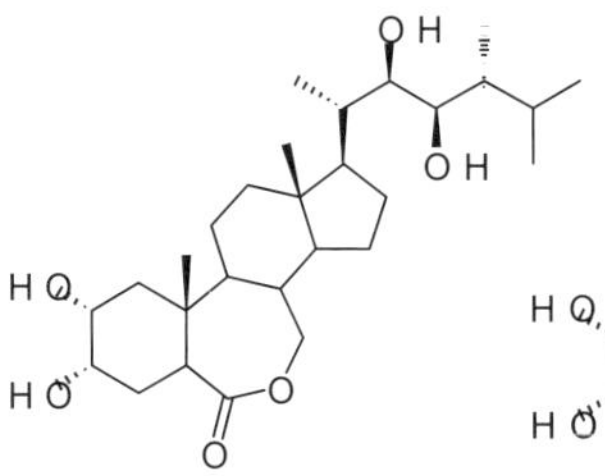

5 24-epibrassinolide

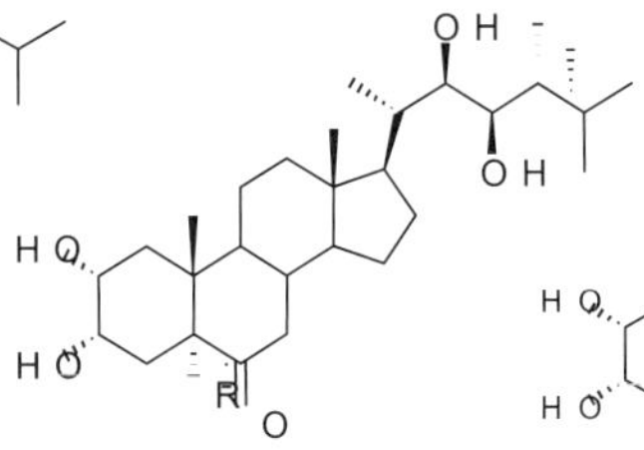

**6** 24-epicastasterone

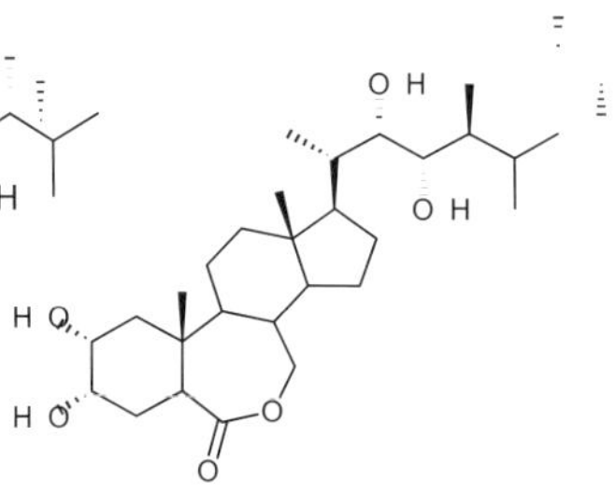

**7** 22,23-diepibrassinolide

*Figure 1. Natural brassinosteroids (structures 1-7) and synthetic analogs*

elements differentiation requires 72 hours (Clouse and Sasse, 1998). Furthermore, abnormal vascular bundle organization has been consistently observed in several BR mutants (Szekeres *et al.* 1996; Choe *et al.*, 1999b). For example, in *dwf7-1*, an *Arabidopsis* mutant defective in the $\Delta^7$ sterol C-5 desaturation step leading to brassinosteroid biosynthesis, within a single vascular bundle, the size and number of xylem cells in generally are reduced, whereas the number of phloem cells is similar to or even greater than that in the wild type (Choe *et al.*, 1999b).

*Brassinosteroids action on cell elongation*

Before their maturation, plant cells usually enlarge 10- to 1000-fold in volume by a process that entails massive vacuolation and irreversible expansion of the cell wall (Cosgrove, 1997). Cell elongation is a developmental process that is regulated by light and phytohormones and is of critical importance for plant growth (Azpiroz *et al.*, 1998).

The plant cell wall forms a highly cross-linked, rigid matrix, which makes cell expansion and differentiation difficult (Clouse and Sasse, 1998). Cell wall modification by wall relaxation or loosening, and by incorporation of new polymers into the expanding wall to maintain wall integrity are requirements for cell elongation and other morphogenetic preocesses take place (Clouse and Sasse, 1998). Various proteins with possible roles in cell wall modification processes have been reported, including glucanases, xyloglucan endotransglycosylases (XETs), and expansins (Cosgrove, 1997). It has been pointed out that expansins might be primarily responsible for wall relaxation, but glucanases and XETs would affect the extent of expansin activity by changing the viscosity of the hemi-cellulose matrix (Cosgrove, 1997). Thus, by altering the viscosity of the matrix, glucanases, other wall hydrolases, and XET affect the amount of wall enlargement that results from expansin activity (Cosgrove, 1997).

Brassinosteroids are able to stimulate elongation of vegetative tissue in a wide variety of plants at very low concentrations. BR application at nanomolar to micromolar levels causes pronouced elongation of hypocotyls and epicotyls of dicots, as well as coleoptiles and mesocotyls of monocots (Clouse and Sasse, 1998).

It has been well demonstrated that brassinosteroids change the biophysical properties of plant cell walls (Zurek *et al.*, 1994), and young vegetative tissues are particularly responsive to brassinosteroids (Clouse and Sasse, 1998). Analysis of the distribution of exogenously supplied $^{125}$I-brassinosteroid demonstrated that the radiolabelled brassinosteroid accumulated in the elongating zone of mung bean epicotyls and the apex of cucumber seedlings (Xu *et al.*, 1994). Treatment of elongating soybean epicotyls with brassinosteroids results in increased plastic extensibility of the walls within two hours with a concomitant enhancement in the BRU1's mRNA (Zurek and Clouse, 1994), a gene which encodes for a XET (Oh *et al.*, 1998). It has also been demonstrated that increasing concentrations of applied brassinosteroid during early stages of elongation resulted in a linear enhancement of extractable XET activity in the soybean epicotyls (Oh *et al.*, 1998).

Cell expansion requires not only wall relaxation but also an adequated supply of wall components (Sasse, 1997). In epicotyls of azuki beans, it has been demonstrated that inhibitors of cellulose biosynthesis or microtubule re-orientation inhibit stem elongation induced by brassinosteroid, and also that brassinolide (BL), isolated or in combination with auxins, increase the percentage of cortical microtubules transversally oriented (Mayumi and Shibaoka, 1995), which contributes to the establishement of the growth direction once the orientation of cortical microtubules usually correlates with orientation of microfibrils (Clouse and Sasse, 1998).

Auxins stimulate cell elongation through a process postulated to require cell wall acidification caused by proton extrusion, being this proton extrusion driven by a $K^+$-dependent ATPase located at the plasma membrane (Cleland, 1995). Brassinosteroid-induced cell expansion is also accompanied by proton extrusion, and hyperpolarization of cell membranes (Clouse and Sasse, 1998). The treatment of a vacuolar ATPase with BL was able to activate the enzyme indicating that the brassinosteroid-induced elongation might be due to a reduction on the water potential of the vacuole via ion and/or sugar uptake (Sasse, 1997). In addition to that, 22S, 23S-homobrassinolide and 24-epibrassinolide have been shown to be able to cause, at least temporarily, hyperpolarization of the plasma membrane. However, no correlation between the ability to hyperpolarize the plasmalemma and known growth effects was found (Dahse *et al.*, 1991). Furthermore, in systems such as the plasma membrane vesicles of potato (*Solanum tuberosum* L.) dormant tubers, epibrassinolide has been shown to inhibit ATP-dependent accumulation of $H^+$ (Ladyzhenskaya and Korableva, 2001). When seen together, these data suggests that although cell elongation induced by both, auxins and brassinosteroids, seem to involve ATPase activation, the enzymes, located in different subcellular compartments, would play different roles on cell wall elongation.

Although both, auxins and brassinosteroids, are known to stimulate cell elongation, their kinetics is considerably distinct. Upon application, auxins usually present a very short lag time of at least eight minutes before the elongation rate starts to increase, with maximum rates of elongation being achieved after 30 to 60 minutes (Evans, 1985; Cleland, 1995). For brassinosteroids, the lag time is at least 45 minutes with elongation rates keep increasing for several hours (Zurek *et al.*, 1994; Mayumi and Shibaoka, 1996; Clouse and Sasse, 1998). This difference in kinetics can also be found at the gene expression level once auxins induce the TCH4 gene much more rapidly compared to brassinosteroid (Xu *et al.*, 1995; Clouse and Sasse, 1998). Additive effects of brassinosteroids have been found on elongation with gibberellins, and on lateral enlargement with cytokinins, while inhibitory effects of cytokinins, abscisic acid and ethylene on brassinosteroid-stimulated stem elongation have also been reported (Clouse and Sasse, 1998).

### *Brassinosteroids action on cell division*

Although microscopy analysis of *Arabidopsis* mutants deficient and insensitive to brassinosteroids has shown that dwarfism is caused by reduced cell size and not cell number (Kauschmann *et al*., 1996), brassinosteroids have been demonstrated to stimulate cell divison in systems such as cultured parenchyma cells of *Helianthus tuberosus*. In this system, an increase of at least 50% on the rate of cell division was found after treatment with nanomolar concentrations of brassinosteroid, in the presence of cytokinin and auxin (Clouse and Zurek, 1991).

Hu *et al*. (2000) identified brassinosteroid-responsive genes in det2, an *Arabidopsis* mutant deficient on brassinosteroids biosynthesis, suspension culture and found that epi-brassinolide upregulated transcription of the CycD3, a D-type plant cyclin gene through which cytokinin activates cell division. Epi-brassinolide was also shown to be able to replace cytokinin in culturing of *Arabidopsis* callus and suspension cells. The epi-brassinolide-driven induction of CycD3 was further demonstrated to involve de novo protein synthesis, but no protein phosphorylation or dephosphorylation (Hu *et al*., 2000).

## STRUCTURE-BIOACTIVITY RELATIONSHIPS

Brassinosteroid bioactivity is evaluated by different bioassays whose results are not systematically consistent (Khripach *et al*., 1999a). Rice lamina inclination test (RLIT) is the most commonly used bioassay among research groups; however, different groups perform particular modifications on the basic protocol of RLIT and use different rice cultivars as well, which makes quantitative data crossing invalid for comparison purposes.

The RLIT, originally developed for auxins (Maeda, 1965), was modified by Wada and co-workers (1981, 1984) for brassinosteroid detection. While the RLIT assay has a detection limit of 50 ppm for indolacetic acid, its detection limit for brassinolide (**1**) and 28-homobrassinolide (**3**) is 0.5 and 5 ppb, respectively. This bioassay can be performed either on explants or on whole seedlings (Takeno and Pharis, 1982; Kim *et al*., 1990; Khripach *et al*, 1999a; Ramírez *et al*. 2000a). However, since there is not a single system to assess bioactivity, each different research team determines quantitative structure-bioactivity relationships according to their own data, obtained through the use of specific standards, cultivars and bioassay protocols (Khripach *et al*., 1999a).

Besides the lack of standardization for bioassays, cross data analysis is also complicated by the combination of substituents since bioactivity values are affected by the nature of the substituents; *i.e.* moving from lactone to 6-ketone it is observed that the activity decreases from 100% to 50% in the pair brassinolide (**1**)/castasterone (**2**) (Takatsuto *et al*., 1983a) in the RLIT, while 24-epicastasterone (**6**) is about 3 times more active than castasterone (**2**) in the bean second inter-node assay (Thompson *et al*., 1982). Notwithstanding these limitations, the following qualitative view of structure-bioactivity relationships has been accepted:

i) Ring A: $2\alpha,3\alpha$ dihydroxy > $3\alpha$ hydroxy $\cong$ 3 dehydro > $3\beta$ hydroxy

Figure 2 shows the chemical structures of compounds **10-28,** involved in the analysis of structure-activity relationships of the most frequently found types of substitution for ring A, ring B and side chain, and configuration of the A/B ring fusion as well.

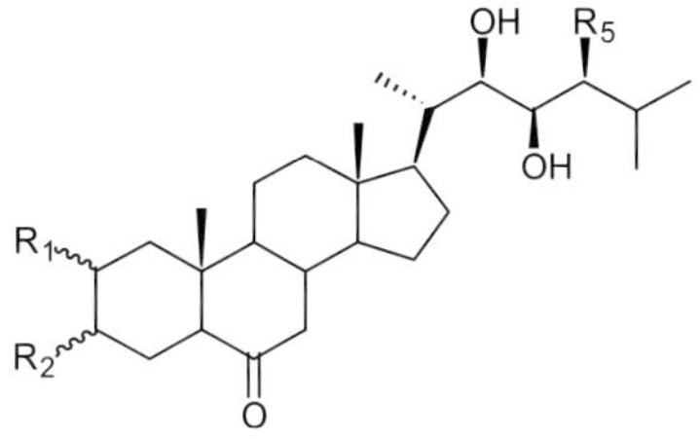

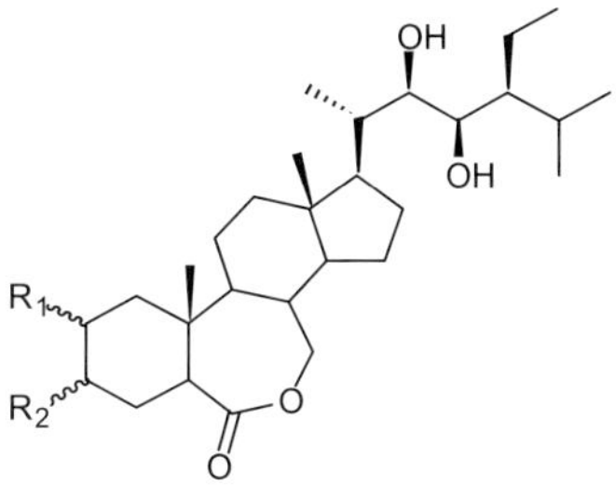

| | $R_1$ | $R_2$ | $R_5$ | |
|---|---|---|---|---|
| **12** | H | αOH | Et | 28-homotyphasterol |
| **13** | H | βOH | Et | 28-homoteasterone |
| **16** | H | O | Et | 3-dehydro-28-homoteasterone |
| **17** | H | H | Et | 2,3-dideoxy-28-homocastasterone |
| **18** | H | O | Me | 3-dehydroteasterone |
| **19** |  O | | Me | 2β,3β-epoxycastasterone (secasterone) |
| **20** |  O | | Me | 2α,3α-epoxycastasterone |

| | $R_1$ | $R_2$ | |
|---|---|---|---|
| **10** | H | αOH | 2-deoxy-28-homobrassinolide |
| **11** | H | βOH | 2-deoxy-3-epi-28-homobrassinolide |
| **14** | H | O | 3-dehydro-2-desoxy homobrassinolide -28- |
| **15** | H | H | 2,3-dideoxy-28-homobrassinolide |

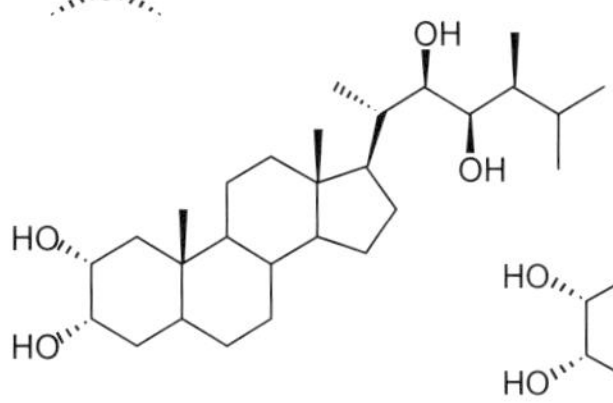

**21** 6-deoxocastasterone

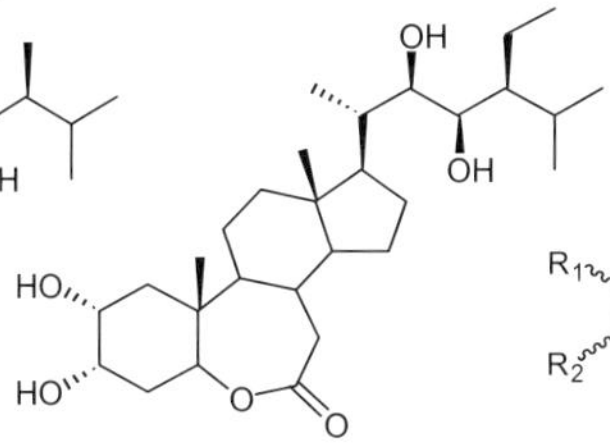

**22** 6-oxa-7-oxo-28-homobrassinolide

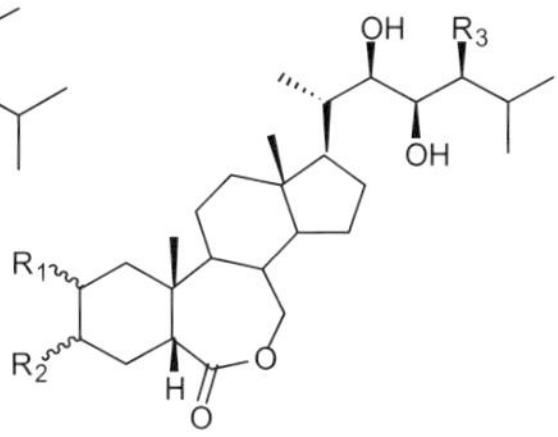

| | $R_1$ | $R_2$ | $R_3$ | |
|---|---|---|---|---|
| **23** | αOH | αOH | Me | 5-epibrassinolide |
| **24** | βOH | βOH | Me | 2,3,5-triepibrassinolide |
| **25** | H | O | Et | 2,3,5-triepi-28-homobrassinolide |

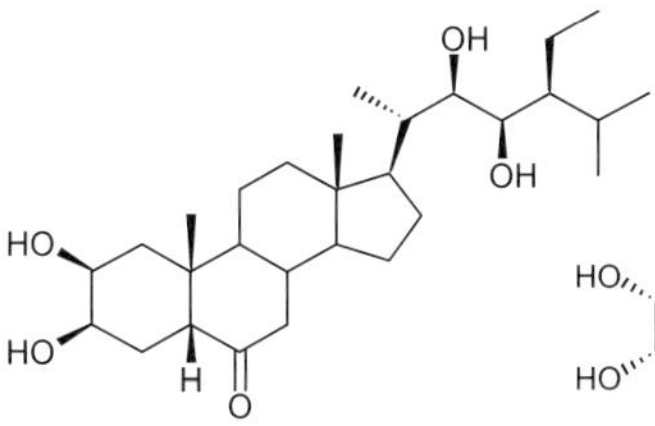

**26** 2,3,5-triepi-28-homocastasterone

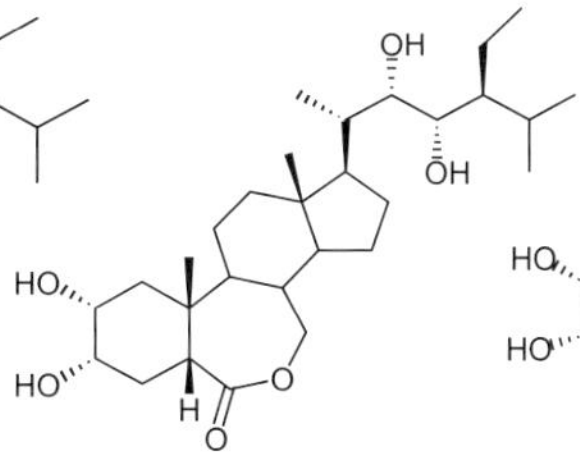

**27** 22, 23 diepi-28-homobrassinolide

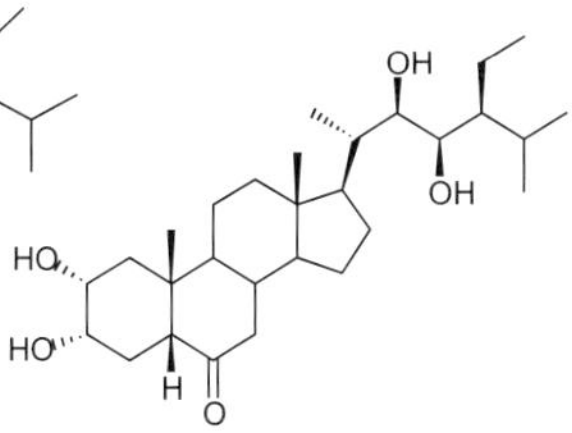

**28** 22, 23 diepi-28-homocastasterone

*Figure 2. Chemical structures of compounds **10** to **28***

The effect of ring A substituents on the brassinolide activity has been studied with some detail within the 28-homo series.

Takatsuto *et al.* (1987) found that the change in the hydroxyls positions from 2α,3 α to 3 α,4 α in 28-homobrassinolide (**3**) slightly reduce activity on RLIT (using *Oryza sativa* L, cultivar *Arborio* Jl). Those authors also reported that 2-deoxy-28-homobrassinolide (**10**) and its 3β epimer (**11**) are about 10 times less active than 28-homobrassinolide (**3**), while 28-homotyphasterol (**12**) is about ten times less active than 28-homocastasterone (**4**) in the same type of assay, indicating that the hydroxyl group at 2-α position is not indispensable to elicit the biological activity. Takatsuto and co-workers (1987) also reported that 3-dehydro-2-deoxy-28-homobrassinolide (**14**) and 2,3-dideoxy-28-homobrassinolide (**15**) possess *ca.* 5% of the activity of 28-homobrassinolide **(3),** while 3-dehydro-28-homoteasterone (**16**) and 2,3-dideoxy-28-homoteasterone (**17**) are, respectively, ten and one hundred times less active than 2,3-dideoxy-28-homobrassinolide (**15**). These results suggest that the 7-oxa-lactone group play a more important role in the plant growth promoting activity than the 6-oxo group.

Galagovsky *et al.* (2001) reported smaller decreases in bioactivity through the series 28-homobrassinolide (**3**), 28-homocastasterone (**4**), 28-homotyphasterol (**12**) and 28-homoteasterone (**13**), in RLIT, using *Oryza sativa* L, cultivar *Chuy*. These results were consistent with the downstream biosynthetic C-6 early oxidation pathway (Sakurai, 1999; Noguchi *et al.*, 2000) (Figure 5).

The 3β-isomers, like 2-deoxy-3-epi-28-homobrassinolide (**11**) and 28-homoteasterone (**13**) are also ten times less active than 28-homobrassinolide (**3**) and 28-homocastasterone (**4**), respectively (Takatsuto *et al.*, 1987).

3-Dehydroteasterone (**18**), secasterone (**19**) and 2,3-diepisecasterone (**20**) show, respectively, 74%, 59% and 89% of the bioactivity of 24-epicastasterone (**6**) in the rice lamina inclination assay (Voigt *et al.*, 1995).

ii) Ring B: 6-oxo-7-oxalactone > 6-keto >> 7-oxo-6-oxalactone ≅ 6-deoxo

Most bioactive brassinosteroids bear the 6-oxo-7-oxalactone moiety, followed by the 6-keto brassinosteroids. The absence of an oxygen function at ring B decreases significantly the brassinosteroid activity, as in the case of 6-deoxocastasterone (**21**) that shows only 1% of the bioactivity of castasterone (**2**) (Yokota *et al.*, 1983). Mandava (1988) confirmed that 6-deoxo brassinosteroid analogs are almost inactive. Transforming the 6-oxo-7-oxalactone to 6-oxa-7-oxolactone (see compound **22**) or to ether, thialactone, lactam, 6-aza-7-oxalactone and 6-aza-7-thiolactone (Okada and Mori, 1983; Kishi *et al.*, 1986; Takatsuto *et al.*, 1987) dramatically reduces its activity.

*iii)* Junctions of rings: A/B *trans* > A/B *cis*

Figure 3 shows different conformations of a rigid steroidal nucleus with either A/B *trans* or A/B *cis* fusion.

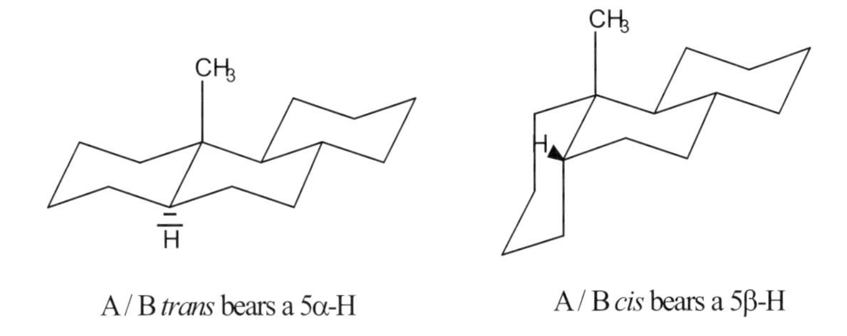


*Figure 3. Different conformation of the rigid steroidal nucleus with A/B trans and A/B cis ring fusions.*

The almost rigid structure of the steroidal nucleus of brassinosteroids has been confirmed by molecular orbital calculations, nuclear magnetic resonance experiments and X-ray diffraction studies, which also revealed that in the 5α-series, the A and C rings assume a chair conformation. This type of conformation has been also confirmed for the ring B of 6-ketobrassinosteroids, while in the 6-oxo-7-oxalactones the 7-membered B ring tends to lie in the same plane as rings C and D (Grove *et al.*, 1979; Thompson *et al.* 1979). In brassinosteroids of the 5β-series, the ring A also assumes a chair conformation, but it is set almost perpendicularly to the plane formed by the rings B, C and D.

Although brassinosteroids with *cis* A/B ring fusion (5 β configuration) have not yet been isolated from natural sources, some analogs have been synthesized. Evaluation of synthetic 5-epibrassinolide (**23**) (Seto *et al.* 1998) as well as 2,3,5-triepibrassinolide (**24**) showed a nearly complete loss of bioactivity in the RLIT indicating that *trans*-fusion of rings A/B is essential for bioactivity (Seto *et al.*, 1999). However, results found by Brosa *et al.* (1994) when testing analogs of the 28 homo series by RLIT, employing the *Bahia* rice cultivar and a high brassinosteroid dose (1μg/segment) are not in agreement with the data that Seto and colleagues (1999) have found. She reported that either 28-homobrassinolide (**3**) or 2,3,5-triepi-28-homobrassinolide (**25**) showed 87% of the bioactivity of brassinolide (**1**). She also reported that in the 6-keto series, 28-homocastasterone (**4**) presents 97% of the brassinolide (**1**) activity while 2,3,5-triepi-28-homocastasterone (**26**) shows only 51% of the activity of brassinolide (Brosa *et al.*, 1996). Ring A/B *cis* fusion, together with a 2β, 3β diol function are characteristic groups in ecdysteroids; Richter and Koolman (1991) have reported that 22,23-diepi-28-homobrassinolide (**27**) and 22,23-diepi-28-homocastasterone (**28**) slightly elicit both, ecdysteroid agonist and antagonist activities.

*iv)* Side chain:

22*R*,23*R* dihydroxy >> 22*S*,23*S* dihydroxy > mono hydroxy

and

24*S* methyl > 24*R* methyl ≅ 28*S* ethyl

Thompson *et al.* (1981) were the first to report high bioactivity of 22*S*, 23*S* unnatural brassinosteroids epimers. 28-Homo and 24-epi brassinosteroids bearing a 22*S*, 23S-diol moiety have been synthesized from stigmasterol or ergosterol, respectively. Mori (1980) was the first to report the synthesis and the high bioactivity of 22,23-diepi-28-homobrassinolide (**27**) on RLIT using *Oryza sativa* L, cultivar *Kinmaze*. 28-homobrassinolide (**3**) and 28-homocastasterone (**4**) are reported to exhibit 100 and 87% of the brassinolide (**1**) and castasterone (**2**) bioactivity, respectively, in RLIT (Khripach *et al.*, 1999a).

22*S*, 23S analogs are obtained as by products either during the last synthetic step involving a chiral asymmetric dihydroxylation (Sharpless *et al.*, 1992) on the insaturated side chain of the precursors (Mc Morris *et. al.*, 1996), or as the main products when osmium tetroxide is used without any quiral ligand (Thompson *et al.*, 1979).

In addition to the structure-bioactivity relationships already mentioned, useful informations can also be obtained from computer analysis of energy-minimized structures of several brassinosteroids. These analysis have suggested that two independent factors may explain the activity, an entropic one, that is related to the flexibility of the side chain, and an enthalpic one, that is related to the oxygen atoms's spatial situation which is involved with the interaction brassinosteroid-receptor (McMorris *et al.*, 1994, Brosa, 1999; Brosa *et al.*, 1996).

### *The presence of atypical functional groups*

The introduction of a hydroxyl group at 5α induces 1,000 times decrease on the brassinolide activity when moving from 7-dehydro-24-epicastasterone (**29**) (Figure 4) to 7-dehydro-5α-hydroxy-24-epicastasterone (**30**), and about 100 times in the pair 7-dehydro-22, 23,24-triepicastasterone (**31**)/ 7-dehydro-5α-hydroxy-22, 23,24-triepicastasterone (**32**) (Takatsuto *et al.*, 1987).

A decrease in bioactivity of 5α-hydroxy-28-homocastasterone (**9**) on the RLIT, using 28-homocastasterone (**4**) as the reference compound, has also been reported (Brosa *et al.*, 1998). Replacement of the 5α-hydrogen by a 5α-fluoro group on 28-homocastasterona (**4**), 28-homotyphasterol (**12**) and 28-homoteasterone (**13**) standards yielded compounds **8**, **33** and **34**, respectively, which showed similar bioactivities, compared to the standards, in RLIT at the highest dose used in the study, 1μg/plant (Ramírez, 2000). Replacement of 3-hydroxy by a 3-fluoro group on 28-homotyphasterol (**12**) and 28-homoteasterone (**13**) also yielded the active analogs **35** and **36**, respectively (Galagovsky et al., 2001).

Figure 4 below shows the chemical structures of compounds **29-38** involved in the analysis of structure- activity relationships of synthetic analogs bearing unnatural 5α-hydroxy and 5α-fluoro substituents.

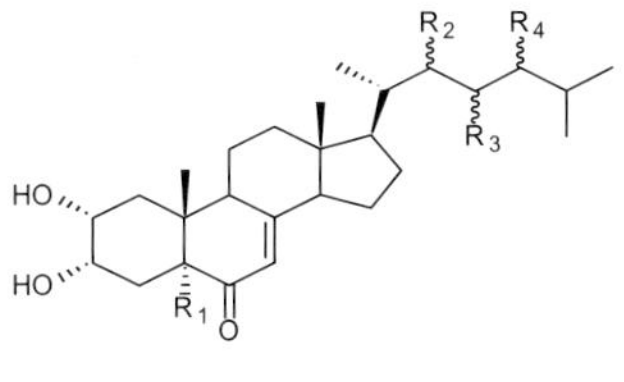

| | $R_1$ | $R_2$ | $R_3$ | $R_4$ | |
|---|---|---|---|---|---|
| **29** | H | *R*-OH | *R*-OH | *R*-Me | 7-dehydro-24-epicastasterone |
| **30** | OH | *R*-OH | *R*-OH | *R*-Me | 7-dehydro-5α-hydroxy-24-epicastasterone |
| **31** | H | *S*-OH | *S*-OH | *R*-Me | 7-dehydro-22,23,24-triepicastasterone |
| **32** | OH | *S*-OH | *S*-OH | *R*-Me | 7-dehydro-5α-hydroxy-22,23,24-triepicastasterone |

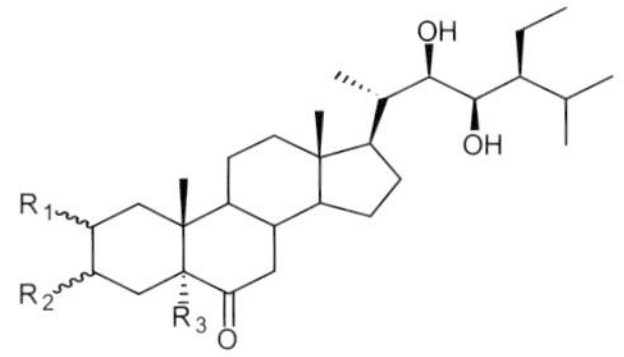

| | $R_1$ | $R_2$ | $R_3$ | |
|---|---|---|---|---|
| **4** | αOH | αOH | H | 28-homocastasterone |
| **8** | αOH | αOH | F | 5-fluoro-28-homocastasterone |
| **9** | αOH | αOH | αOH | 5-hydroxy-28-homocastasterone |
| **12** | H | αOH | H | 28-homotyphasterol |
| **13** | H | βOH | H | 28-homoteasterone |
| **33** | H | αOH | αF | 5α-fluoro-28-homotyphasterol |
| **34** | H | βOH | αF | 5α-fluoro-28-homoteasterone |
| **35** | H | αF | H | 3α-fluoro-28-homotyphasterol |
| **36** | H | βF | H | 3β-fluoro-28-homoteasterone |
| **37** | H | αOH | αOH | 5-hydroxy-28-homotyphasterol |
| **38** | H | βOH | αOH | 5-hydroxy-28-homoteasterone |

*Figure 4. Chemical structures of compounds* ***29*** *to* ***38***

*The brassinosteroid-receptor theory*

Progresses in molecular biology have allowed the identification of the protein encoded by the BRI1 gene as a putative brassinosteroid-receptor in *Arabidopsis*. BRI1 encodes a receptor kinase which presents an extracellular domain containing 25 leucine-rich repeats (LRRs), which are interrupted by a 70-amino-acid island, a transmembrane domain, and a cytoplasmic kinase domain with serine/threonine specificity (Li and Chory, 1997; Friedrichsen *et al.*, 2000).

Experiments combining induction of plasmolysis and confocal fluorescence microscopy demonstrated that the BRI1-*GFP* (Green Fluorescent Protein) fusion protein is located at the plasma membrane (Friedrichsen *et al.*, 2000). The BRI1-*GFP* fusion protein was expressed in all tissues in the seedling and in adult organs as well, being this expression pattern of the fusion protein consistent with previous.
BRI1 mRNA expression data (Li and Chory, 1997; Friedrichsen et al., 2000). Although there is no tissue-specific expression of BRI1, there is temporal regulation. Fully expanded leaves and elongated root or inflorescence cells express BRI1::GFP at low levels, being this expression pattern also consistent with physiological data that showed that exogenously applied brassinosteroids promote growth only in younger tissues (Mandava, 1988; Friedrichsen et al., 2000).

It has been suggested that BL induction of BRI1 phosphorylation requires the kinase activity of BRI1, indicating that BL-binding induces autophosphorylation of BRI1 (Wang et al., 2001). Thus, having in sight that: The BRI1-GFP fusion protein is located at the plasma membrane (Friedrichsen et al., 2000); The receptor activation usually involves auto-phosphorylation; 3. The near identical phenotypes of bri1 to brassinosteroid-biosynthetic mutants (Wang et al., 2001); it might be concluded that plants perceive steroids at the cell surface and that BRI1 is likely to be the primary brassinosteroid receptor in Arabidopsis (Wang et al., 2001) or a protein of the receptor-complex in Arabidopsis (Friedrichsen et al., 2000).

Friedrichsen *et al.*, (2000) proposed two models for the function of BRI1 ibrassinolide signaling. In the first, similarly to animal RPKs, the binding of ligand to the LRR or the island domain may result in the dimerization of BRI1 with itself or another receptor kinase. This dimerization would result in transphosphorylation and activation of the kinase domain. The activated kinase would then send a phosphorylation signal to alter gene expression and induce cell expansion, among other effects. Conversely, BRI1 may not be the receptor itself but may be a protein in the brassinolide receptor complex whose extracellular domain is involved in interactions with other receptor complex proteins. The formation of an active complex results in activation of the kinase phosphorylation signal.

Other brassinosteroid-insensitive mutants such as *cbb2*, 18 *bin* and 3 alleles of *dwf2* are allelic to *bri1* (Clouse and Sasse, 1998; Li and Chory, 1999). The fact that all brassinosteroid-insensitive mutants are allelic to a single gene suggests that BRI1 is the only unique and specific component of the brassinosteroids signal transduction pathway and that the other components of the pathway areeither redundant or shared with other signalling cascade (Li and Chory, 1999). It has been recently pointed out that the signal transduction pathway from BRI1 has two branches. The first branch controls rapid changes in the rate of cell elongation through regulated assembly of the V-ATPase, while the second branch induce changes in gene expression which control cell expansion and other processes (Chory, 2001).

It is usually assumed that brassinosteroids bind to three points in its receptor: the 2α, 3 α -hydroxyls (Wada *et al.*, 1981), the B ring lactone and the 22 α, 23 α -hydroxyls (Kishi *et al.*, 1986). It was formerly considered that the receptor affinity to the 2 α, 3 α -hydroxyls would be greater than to the 22 α, 23 α -hydroxyls, as variations in side chain structure are less important for brassinolide activity than structure variations in ring A (Takatsuto *et al.*, 1983b). A study on quantitative structure-activity relationships indicated, however, that the contributions of the ring A and of the side chain hydroxyls configurations accounted for 25% and 35% of the total activity of brassinolide, and that study also indicated that the activity of a brassinosteroid or analog would be larger the greater was the similarity between the compound and the brassinolide (**1**) itself (Brosa *et al.*, 1996). In a later study, Brosa *et al.* (1999) reported a decrease in bioactivity when using 5α-hydroxylated analogs on RLIT. Those authors suggested that an H-bonding between the 3α and the 5α-hydroxy groups could be responsible for the decreased ability of the molecule to bind to the active site of the receptor through its C-3 hydroxyl group. However, Ramírez *et al.* (2000a) reported that 5α-fluorinated analogs of 28-homocastasterone (**4**), 28-homotyphasterol (**12**) and 18-homoteasterone (**13**), compounds **8**, **33** and **34**, respectively, showed high bioactivity (RLIT), even at low doses, when compared with the related natural 5α-H 28-homo series, while a decreased bioactivity was confirmed when their corresponding 5α-hydroxylated analogs were tested (see compounds **9, 37** and **38**).

The high activity of the 5α-fluorinated analogs was unexpected since this electronegative group is able to form a hydrogen bond with the C-3 hydroxy group, and this fact was predicted to decrease bioactivity. Molecular modeling calculations confirmed that a very close contact of 1.97 Å was established between the hydrogen of the 3α-hydroxyl group and the fluorine atom, forming a favorable and stable conformation of a six membered ring. These results were consistent with a typical hydrogen bond involving fluorine (Howard, 1996; O'Hagan, 1997; Dunitz, 1997).

Collectively, these results, together with a look at the reactions involved in the metabolism, and also the multiple bioactivities displayed by brassinosteroids in plant cells (Khripach *et al.*, 1999b) suggest that there is likely to be more than a single receptor site for brassinosteroids (Clouse, 2002). Each receptor site must display different structural requirements for exhibiting the maximum activity; this may be the reason why there are different structure-activity relationships depending on the bioassay employed (Zullo *et al.*, 2002a). According to these results, further improvements on the methodology used to predict the activity of both, natural brassinosteroids and analogs are required.

*Fluorinated analogs*

Organofluorine compounds have recently attracted considerable attention in the fields of agrochemistry, pharmaceutical and material science (Liebman *et al.*, 1988; Welch and Eswarakrishnan, 1991; Filler, 1997). Fluorinated analogs have been recognized as useful tools for pharmacological and physiological studies of natural products since

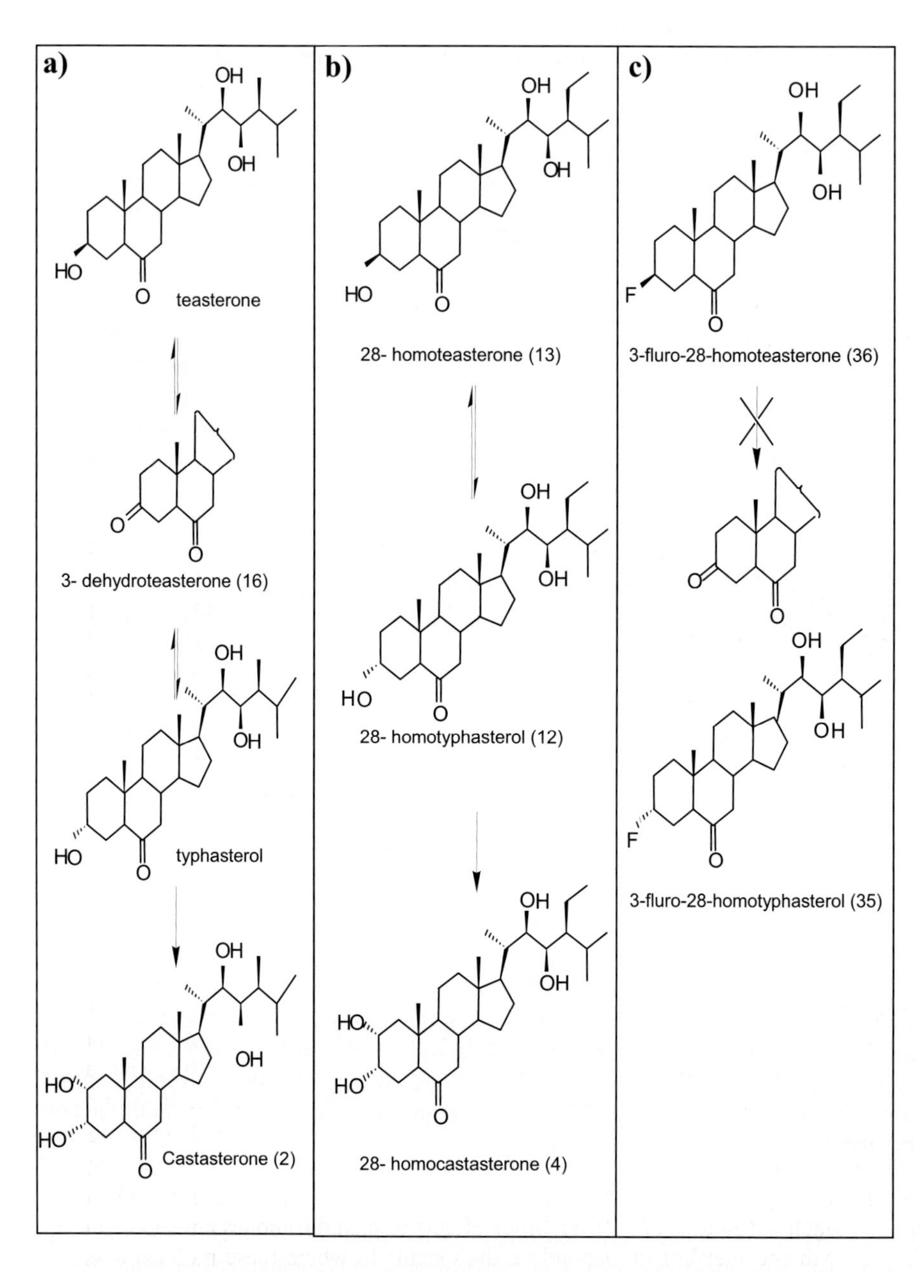


a) The "early C-6 oxidation pathway" proposed for 24-methyl natural brassinosteroids (Noguchi *et al.* 2000)
b) The 28-homo series shows increasing activity consistent with its downstream biosynthetic pathway.
c) Compound **36** is not expected to undergo downstream biosynthesis, however it shows activity in the RLIT.

*Figure 5: Biosynthetic relationships among C-6 oxidized brassinosteroids of the C-24 methyl and the C-24 ethyl series, and two C-3 fluorinated analogs.*

the introduction of a fluorine atom into a molecule often leads to a significant change in its physical and biological properties (Martin *et al*, 1992). In view of their unique biological properties, fluorinated steroids have been widely studied (Kobayashi *et al.* 2000) and several fluoro-substituted compounds are now considered as analogs of plant hormones (Saito *et al.*, 1998).

Although substantial effort has been made to develop methods for introducing fluorine groups in bioactive brassinosteroids (Jin *et al.*, 1993; Back *et al.*, 1999; Jiang *et al.*, 2000; Ramírez *et al.*, 2000; Galagovsky *et al.*, 2001), only a few studies on their properties have been reported. Back *et al.* (1999) found that 25-fluor analogs of brassinolide and castasterone showed almost no activity with or without the presence of IAA (indol acetic acid) in the RLIT, while the presence of a 25-hydroxy group yields a molecule with potent biological activity. Ramírez *et al.* (2000) reported that 5α-fluoro-28-homocastasterone (**8**), 5α-fluoro-28-homotyphasterol (**33**) and 5α-fluoro-28-homoteasterone (**34**) show very high *in vitro* bioactivity instead of a predicted low bioactivity, based on the assumed formation of an intramolecular H-bonding between substituents at C-3 and 5 (Brosa, 1997; 1999).

In a recent paper, Galagovsky *et al.* (2001) reported that activities of 28-homobrassinosteroids are higher as the compound tested is closer to the 28-homobrassinolide (**3**) in the biosynthetic pathway (Figure 5). This fact is consistent with the biosynthetic route described as "the early C-6 oxidation pathway" proposed by Noguchi *et al.* (2000) for the 24-methyl series. In a complementary trial, synthetic 3-fluoro-28-homotyphasterol (**35**) and 3-fluoro-28-homoteasterone (**36**) were tested. The 3-fluoro-28-homoteasterone (**36**) can not yield down stream biosynthetic metabolites because it can not afford the 3-dehydro biosynthetic intermediate (**16**) proposed by Yokota (1994) (Figure 5). Notwithstanding, compound **36** showed 60% of the activity of 28-homocastasterone (**4**). Those authors proposed that this high activity is due to a *per se* activity. These three unexpected results obtained while testing fluorinated brassinosteroids evidence the importance of assessing the bioactivities of analogs bearing fluoro substituents.

Gibberellins (GAs) are diterpenoids known to stimulate stem elongation. However, GAs bearing no 3β-hydroxyl group are not able to stimulate elongation by itself, requiring a 3β-hydroxylation in order to become active (Spray *et al.*, 1984; Nakayama *et al.*, 1991). Fluorination is an interesting modification once fluorine presents smallest van der Waar's radius second to hydrogen and the largest electronegativity, and consequently acts as a hydrogen mimic regarding size and a hydroxyl mimic regarding electronegativity (Saito *et al.*, 1998). Indeed, 3β-Fluor-$GA_9$ and 3β-Fluor-$GA_{20}$ have been shown to be active *per se* in promoting shoot elongation in plants such as rice (*Oryza sativa*) and cucumber (*Cucumis sativus*), which suggests that 3β-fluorine actually acts as a mimic of 3β-hydroxyl group towards stimulation of shoot elongation (Saito *et al.*, 1998). Since effects of plant hormones and analogues on plant growth and metabolism depend on the extents to which these molecules satisfy the structural requirements of the receptors and enzymes, the differential effects we describe for 28-HCTS and two of their analogs in this chapter may provide important clues to probe into the signal transduction pathways and metabolism of brassinosteroids.

### *The marubakaido apple rootstock*

Today, marubakaido (*Malus prunifolia*, Willd, Borkh) is one of the most widely used apple rootstock in several countries. This rootstock, also known as maruba, chinese apple or plumleaf crabapple, is vigorous and show good compatibility with all comercial (scion) apple cultivars in use (Flores *et al.*, 1999). Although marubakaido shows resistance against several pathogens such as *Phytophthora* (Zanol *et al.*, 1998), it is susceptible to several viruses, which make *in vitro* propagation one of the most promising techniques for its multiplication (Flores *et al.*, 1999). However, i*n vitro* multiplication rates for woody plants are typically low, which make the micropropagation techniques available for several species barely feasible for commercial purposes and the marubakaido apple rootstock does not consist an exception to this fact.

We have used the marubakaido rootstock as a model to study plant growth regulators metabolism and signal transduction pathways aiming to overcome constrictions in several biotechnological applications, such as the low *in vitro* multiplication rates found for woody plants. Shoot proliferation is a powerful tool to increase *in vitro* multiplication rates for tree species (Shekhawat *et al.*, 1993; Shaefer *et al.*, 2002). Since brassinosteroids are known to stimulate stem elongation and to change apical dominance patterns in several plant species, we investigated the hypothesis that brassinosteroids might be able to increase *in vitro* multiplication rates, via shoot proliferation, for the marubakaido apple rootstock. In this chapter, we introduce our experience with the performance of 28-homocastasterone (**4**) and two synthetic 5α substituted analogs: 5α-fluorohomocastasterone (**8**) and 5α-hydroxyhomocastasterone (**9**) (Figure 1) [(22R,23R)-5α-fluoro-2α,3α,22,23-tetrahydroxystigmastan-6-one and (22R,23R)-2α,3α,5α,22,23-pentahydroxystigmastan-6-one, respectively] on the *in vitro* multiplication of marubakaido and we also discuss potential uses of brassinosteroids for the improvement of desired features for horticultural and forestry species, and for biotechnological processes as well, more especifically for *in vitro* plant propagation.

## EXPERIMENTAL

### *Plant material and culture conditions*

Shoot apices measuring between 10 and 20 mm in length were taken from 30 day-old aseptically-grown shoots of a clone of *Malus prunifolia* (Willd.) Borkh var. Marubakaido and used as explant sources in the experiments. Explants were grown on 40 mL of MS (Murashige and Skoog 1962) basal medium supplemented with (μM): 555 myo-inositol, 4.06 nicotinic acid, 2.43 pyridoxine. HCl, 26.64 glycine, 6.25

thiamine. HCl, 2.2 $N^6$-Benzyladenine, $30g.L^{-1}$ sucrose and $6g.L^{-1}$ agar. The pH was adjusted to 5.7 prior to autoclaving.

Cultures were maintained in a culture room, using a completely randomized design. Photoperiod (16/8 hours - light/dark) was provided by cool-white fluorescent tubes giving a photosynthetic photon flux density (400-700 nm) of 40 $\mu mol.m^{-2}.s^{-1}$ at the culture level. Relative humidity was kept at 70 ± 5%. Air temperature around the cultures was 27.0 ± 1.0°C.

*Synthesis of 5F-HCTS (8) [(22R,23R)-5-fluoro-2α,3α,22,23-tetrahydroxy-5α-stigmastan-6-one]*

The synthesis of compound **8** is summarized in Scheme 1 (Ramirez *et al*., 2000). Stigmasterol was treated with mesyl chloride to afford the corresponding 3β-mesylate **39**. Stereo and regioselective epoxidation of **39** with a mixture of potassium permanganate / ferric nitrate / t-butanol afforded the epoxide **40**. Proper stereochemistry at C-5 was achieved by *trans*-diaxial acid catalyzed nucleophilic opening of the 5β,6β-epoxide **40** with boron trifluoride etherate, to give the key compound **41**. Subsequent oxidation with PCC (pyridinium chlorochromate), and elimination of the 3β-mesiloxy moiety by refluxing **42** with lithium bromide gave the dienone **43**. Osmium-catalyzed asymmetric dihydroxylation (CAD) (Sharpless *et al*., 1992), of double bonds of **43** using $K_3Fe(CN)_6$ as cooxidant and hydroquinidine-1,4-phthalazinediyl diether [$(DHQD)_2$-PHAL] as chiral ligand gave, after purification, 19 % yield of compound **8**. In all cases CAD yielded diastereomeric 22S, 23S isomers in minor proportion (ratio 3:1) and more than 30% of the starting material, which could be recycled.

The position and configuration of the 5α-fluoro moiety was deduced mainly from its effect on the $^{13}C$-NMR spectra. The acidic condition involved in the 5β, 6β-epoxide opening favored the attack of the fluoride at the most substituted position (C-5) with inversion of configuration at this center. A DEPT (Distortionless Enhancement by Polarization Transfer) experiment on compound 8 revealed that the resonance at 98.6-ppm (the carbon bearing the fluorine) corresponded to a trisubstituted carbon. Other easily assigned carbons, such as C-6, C-10 and C-19, appeared as doublets with $^{19}F$-$^{13}C$ coupling constants those were consistent with the proposed structure. Furthermore, the coupling pattern of the H-3 and H-6 multiplets in the $^1H$-NMR spectra agrees with an A/B *trans* junction. The fluorine chemical shifts are consistent with those expected for structurally related tertiary alkyl fluorides (Joseph-Nathan *et al*., 1984). Spectral resolution allowed signals to appear as doublets ($J \approx 40$ - 45 Hz.) due to a coupling with the vicinal H-4β.

*Synthesis of 5OH-HCTS (9) [(22R,23R)- 2α,3α,5α,22,23-pentahydroxystigmastan-6-one]*

Analog bearing a 5α-hydroxyl group (9) was obtained by the synthetic pathway diagrammed in Scheme 2 (Ramírez *et al*., 2000). Compound 45 was obtained as single product by hydrolytic opening of the 5β, 6β epoxide 40. Oxidation of 45 with PCC yielded the 5α-hydroxy-6-oxo steroid 46. Compound 46 was subjected to elimination to give (22*E*)-5-hydroxy-5α-stigmasta-2, 22-dien-6-one(47), which was tetrahydroxylated to obtain (22*R*, 23R)-2α, 3α, 5α, 22,23-pentahydroxystigmastan-6-one (9).

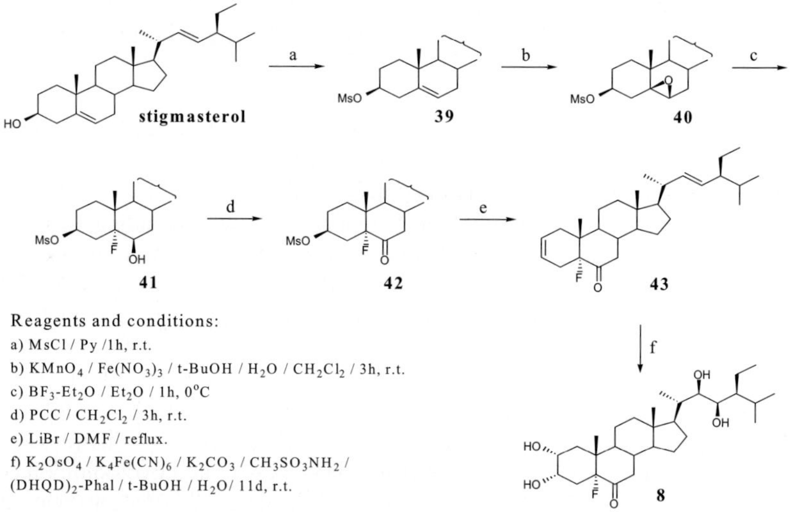

Reagents and conditions:
a) MsCl / Py /1h, r.t.
b) $KMnO_4$ / $Fe(NO_3)_3$ / t-BuOH / $H_2O$ / $CH_2Cl_2$ / 3h, r.t.
c) $BF_3$-$Et_2O$ / $Et_2O$ / 1h, 0°C
d) PCC / $CH_2Cl_2$ / 3h, r.t.
e) LiBr / DMF / reflux.
f) $K_2OsO_4$ / $K_4Fe(CN)_6$ / $K_2CO_3$ / $CH_3SO_3NH_2$ / $(DHQD)_2$-Phal / t-BuOH / $H_2O$/ 11d, r.t.

*Scheme 1. Synthesis of 5α-fluor-28-homocastasterone (**8**) (5-F-HCTS)*

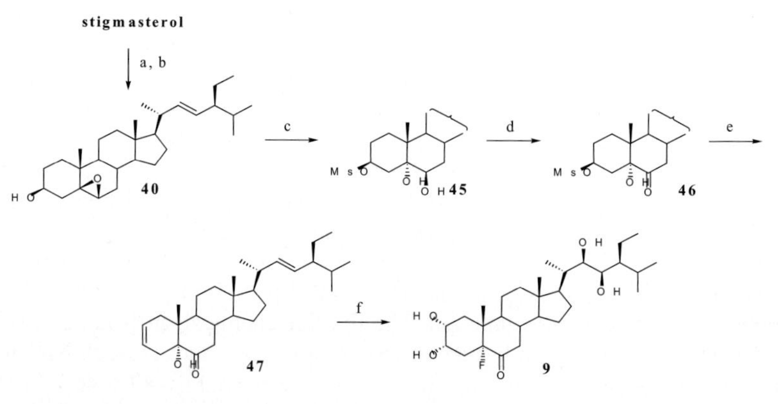

Reagents and conditions:
a) MsCl / Py /1h, r.t.
b) $KMnO_4$ / $Fe(NO_3)_3$ / t-BuOH / $H_2O$ / $CH_2Cl_2$ / 3h, r.t.
c) $HClO_4$ / THF / $H_2O$, r.t
d) PCC / $CH_2Cl_2$ / 3h, r.t.
e) LiBr / DMF / reflux.
f) $K_2OsO_4$ / $K_4Fe(CN)_6$ / $K_2CO_3$ / $CH_3SO_3NH_2$ / $(DHQD)_2$-Phal / t-BuOH / $H_2O$/ 9d, r.t.

*Scheme 2. Synthesis of 5α- hydroxy – 28- homocastasterone (9)*

*Nuclear Magnetic Resonance data*

*General*

$^1$H-NMR spectra were recorded on a Bruker AM-500 at 500 MHz; $^{13}$C NMR spectra were recorded on a Bruker AC-200 at 50.3 MHz. Chemical shifts (δ) are given in ppm downfield from TMS as the internal standard. $^{19}$F-NMR spectra were recorded on a Bruker AM-500 at 470.4 MHz., chemical shifts (δ) are given in ppm upfield from $CFCl_3$ as the internal standard. Coupling constant ($J$) values are in Hz.

*Spectroscopic characterization of compound 8.*

**$^1$H-NMR**: 0.68 (18-$H_3$, 3H, s), 0.91 (21-$H_3$, 3H, d, $J$ = 6.8 Hz), 0.92 - 0.99 (26-$H_3$, 27-$H_3$ and 29-$H_3$, 9H, m), 0.96 (19-$H_3$, 3H, s), 2.61 (7α-H, 1H, dd, $J$ = 12.5 Hz, 12.5 Hz ), 3.58 (22-H, 1H, dd, $J$ = 8.4 Hz, 1.3 Hz), 3.71 (23-H, 1H, dd, $J$ = 8.4 Hz, 1.3 Hz), 3.76 (2β-H, 1H, m), 4.05 (3β-H, 1H, dd, $J$ = 6.0 Hz, 3.0 Hz).

**$^{13}$C-NMR**: 11.6 (C21), 11.7 (C18), 13.4 (C29), 14.2 (C19, $J_{CF}$ = 5.2 Hz), 18.7 (C28), 19.2 and 20.9 (C26 and C27), 20.9 (C11), 23.6 (C15), 27.3 (C16), 28.8 (C25), 29.9 (C4, $J_{CF}$ = 19.3 Hz), 34.3 (C1), 36.8 (C20), 37.4 (C8), 39.9 (C12), 42.0 (C13), 42.8 (C7), 42.8 (C10, $J_{CF}$ = 24.8 Hz), 45.2 (C9, $J_{CF}$ = 3.9 Hz), 46.4 (C24), 52.3 and 55.9 (C14 and C17), 66.8 (C3), 67.6 (C2), 74.1 (C22), 72.2 (C23), 98.2 (C5, $J_{CF}$ = 176.9 Hz), 207.7 (C6, $J_{CF}$ = 27.0 Hz). $^{19}$F-NMR ($CDCl_3$): -155.2 ($J$= 45.4 Hz).

*Spectroscopic characterization of compound 9.*

**$^1$H-NMR**: 0.67 (18-$H_3$, 3H, s), 0.77 (19-$H_3$, 3H, s), 0.91 (21-$H_3$, 3H, d, $J$ = 6.8 Hz), 0.92 - 0.99 (26-$H_3$, 27-$H_3$ and 29-$H_3$, 9H, m), 2.60 (7α-H, 1H, dd, $J$ = 12.5 Hz, 12.5 Hz ), 3.57 (22-H, 1H, dd, $J$ = 8.4 Hz, 1.3 Hz), 3.70 (23-H, 1H, dd, $J$ = 8.4 Hz, 1.3 Hz), 3.80 (2β-H, 1H, m), 4.15 (3β-H, 1H, dd, $J$ = 6.0 Hz, 3.0 Hz).

**$^{13}$C-NMR**: 11.6 (C21), 11.7 (C18), 13.4 (C29), 14.2 (C19), 18.7 (C28), 19.2 and 20.9 (C26 and C27), 21.0 (C11), 23.6 (C15), 27.3 (C16), 28.8 (C25), 30.3 (C1), 34.0 (C4), 36.8 (C20), 37.2 (C8), 39.5 (C12), 42.0 (C13), 42.8 (C7), 44.4 (C10), 45.3 (C9), 46.4 (C24), 52.3 and 55.9 (C14 and C17), 67.4 (C3), 69.5 (C2), 72.2 (C23), 74.1 (C22), 79.5 (C5), 207.7 (C6).

*Application of 28-HCTS, 5F-HCTS and 5OH-HCTS*

Ninety five per cent (v/v) ethanol microdrops (5 μL) containing known amounts of compounds **4**, **8** or **9** were pipetted onto the main vein of the uppermost neo-formed leaf, measuring at least three mm wide, of 15 days-old shoots originated from shoot apices, as above described. Only single applications were used, and control shoots were treated with 5 μL 95% (v/v) ethanol microdrops. Single microdrops were used for each leaf. Each treatment consisted of eight replicates (one replication = one culture vessel) with four explants per replication. Data analysis was carried out with the help of the software JMP (Statistical Analysis System, SAS Institute Inc., USA, 1989-2000). Each experiment was repeated at least twice. The entire data set obtained in the experiments was used for data analysis. For the purpose of this book chapter, "multiplication rate" is defined as the number of neoformed branches ≥15-mm in length, the minimum length suitable for propagation purposes, 30 days after the treatment.

## RESULTS

Increase on the multiplication rate (MR) for *in vitro*-grown *Malus prunifolia* shoots was associated with leaf application of 5F-HCTS (compound **8**) in the 100 to 10,000 ng per shoot range (Figure 6), being the effect statistically significant at p=0.05% for the 500 and 1000 ng per shoot doses. However, the 500 ng per shoot was the most effective dose for the enhancement of MR, resulting in a 112% increase on MR, compared to shoots treated with five microliters 95% ethanol, grown in culture medium enriched with 2.2 μM $N^6$-Benzyladenine. Virtually no change on multiplication rates was found for shoots treated with either 28-HCTS or 5OH-HCTS (Figure 6).

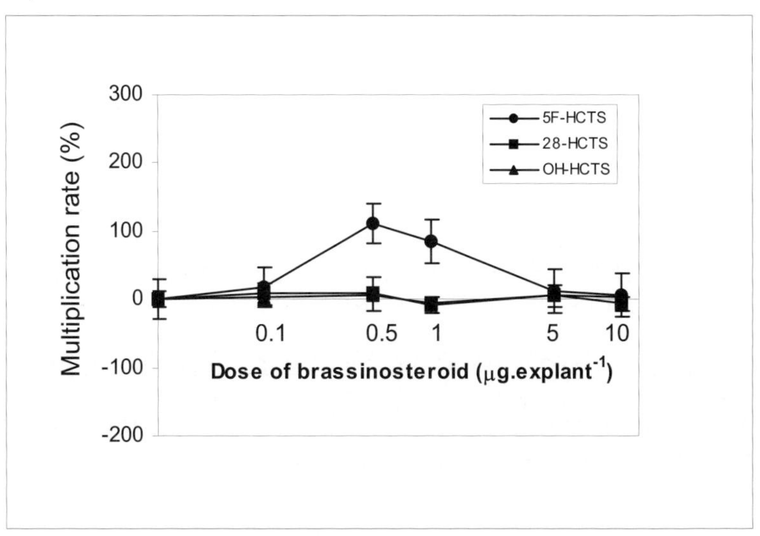


*Figure 6 – Effect of 5F-HCTS, 28-HCTS and 5OH-HCTS on the average in vitro multiplication rate of Malus prunifolia. Values plotted in the graph are relative to the average multiplication rate of 3.4 found for the control (95% v/v ethanol). Vertical bars indicate standard error.*

Shoots treated with 1000 ng of 5F-HCTS showed a significant (p=0.05) 146% increase on the number of main branches (branches originated directly from the initial explant, Figure 7) formed during the culture cycle, while shoots treated with 500 ng of 5F-HCTS presented an also significant (p=0.05) 238% and 250 % increase, respectively, for the number of primary lateral branches (branches originated from the main branches, Figure 8) and for the number of secondary lateral branches (branches originated from the primary lateral branches, Figure 9), measuring at least 15 mm in length. Thus, it can be concluded from figures 6 to 9 that the increase on the muliplication rate found for shoots treated with 500 ng per shoot of 5F-HCTS was mainly due to an increase in both, number of primary and secondary lateral branches while the increase on the multiplication rate found for shoots treated with 1000 ng per shoot of 5F-HCTS was due essentially to an increase on the number of main branches.

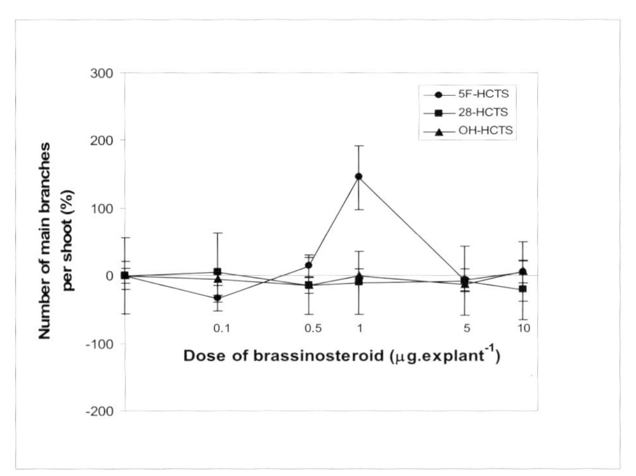


*Figure 7 – Effect of 5F-HCTS, 28-HCTS and 5OH-HCTS on the average number of main branches of Malus prunifolia measuring at least 15 mm in length. Values plotted in the graph are relative to the average number of main branches of 1.9 found for the control (95% v/v ethanol). Vertical bars indicate standard error.*

Differently from 5F-HCTS, which induced remarkable changes in the architecture of the *in vitro*-grown marubakaido shoots, 28-HCTS and 5OH-HCTS applications resulted in no significant (p=0.05) change in any of the features evaluated.

## DISCUSSION

28-homocastasterone has been widely employed in field trials because of its greater synthetic accessibility compared to the brassinolide **(1)**. When preliminarily tested in *in vitro*-grown plant systems, 28-homocastasterone and a 3β-acetoxy derivative of 28-homoteasterone, showed promising results towards the improvement of micropropagation techniques for tropical plants such as cassava [*Manihot esculenta*,

Crantz], yam [*Dioscorea alata* L.] and pineapple (*Ananas comosus* L. Merril] (Bieberach, 2000).

When tested in a tree species micropropagation system (this study), compounds **4, 8** and **9** presented contrasting effects on the architecture of *in vitro*-

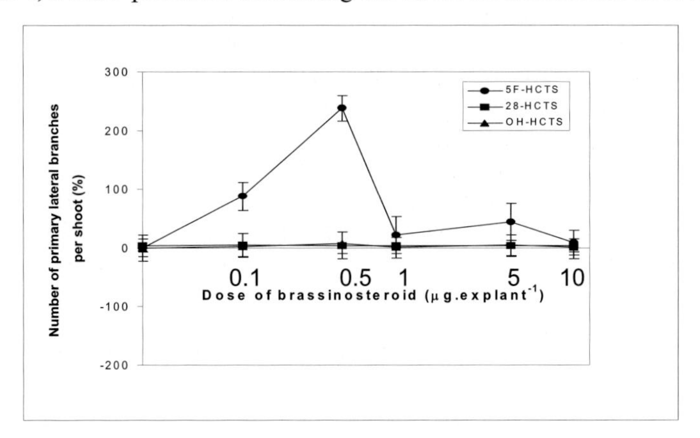


*Figure 8 – Effect of 5F-HCTS, 28-HCTS and 5OH-HCTS on the average number of primary lateral branches of Malus prunifolia measuring at least 15 mm in length. Values plotted in the graph are relative to the average number of primary lateral branches of 1.4 found for the control (95% v/v ethanol). Vertical bars indicate standard error.*

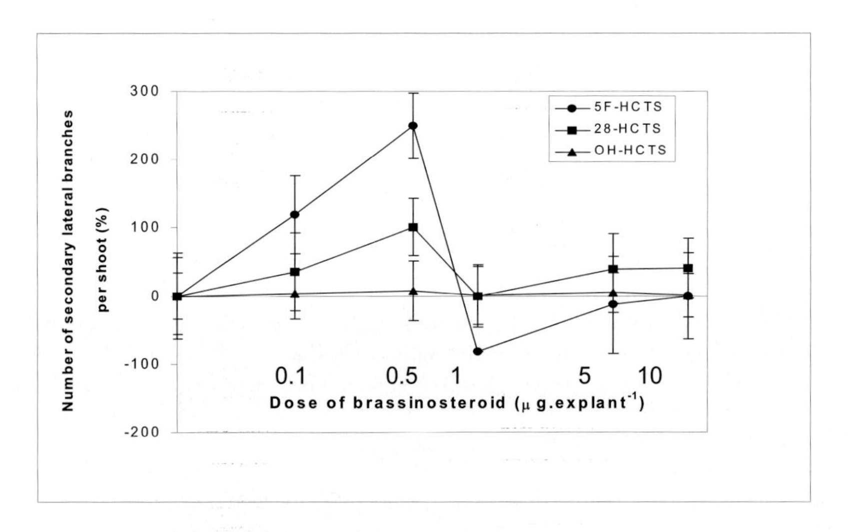


*Figure 9 – Effect of 5F-HCTS, 28-HCTS and 5OH-HCTS on the number of secondary lateral branches of Malus prunifolia measuring at least 15 mm in length. Values plotted in the graph are relative to the average number of secondary lateral branches of 0.2 found for the control (95% v/v ethanol). Vertical bars indicate standard error.*

grown shoots. Compound **8** stimulated shoot proliferation, through stem elongation, but especially through an increase on lateral branching, which resulted in enhanced multiplication rate for the marubakaido apple rootstock. Conversely, compounds **4** and **9** induced no significant change on shoot proliferation.

Since mutants defectives on the biosynthesis (Chory *et al.*, 1989; Chory *et al.*, 1991; Li *et al.*, 1996; Szekeres *et al.*, 1996; Li and Chory, 1999) or on the signal transduction pathway (Clouse *et al.*, 1996; Kauschmann *et al.*, 1996; Li and Chory 1997) of brassinosteroids display reduced apical dominance (Schumacher *et al.*, 1999), it was surprising to find that 5F-HCTS at 1000 ng per shoot and especially at 500 ng per shoot significantly stimulated lateral branch elongation. It was also surprising to find that doses as high as 10 μg per shoot did not change the branching pattern for the marubakaido shoots. Although brassinosteroids are capable of eliciting strong growth responses and a variety of physiological changes through exogenous application to plants (Altmann, 1998), still little is known about the mechanism of action of these plant growth regulators. Having in sight that all of the mutants defectives on the biosynthesis or on the signal transduction pathway of brassinosteroids described until now are herbaceous, our results might indicate that in tree species brassinosteroids might act differently regarding apical dominance control, when compared to herbaceous species.

*Possible involvement of other plant growth regulators on the 5F-HCTS induced branching stimulation*

Cytokinins are known to stimulate lateral branching in several plant species (Schwartzenberg *et al.*, 1994). Like cytokinins, brassinosteroids have also been reported to be involved on branching responses. Application of compounds **1** and **2** to the dumpy (dpy) mutant of tomato, a mutant presenting reduced axillary branching, rescued the dpy phenotype, as did C-23-hydroxylated, 6-deoxo intermediates of brassinolide biosynthesis. The brassinolide precursors campesterol, campestanol, and 6-deoxocathasterone failed to rescue, suggesting that dpy may be affected in the conversion of 6-deoxocathasterone to 6-deoxoteasterone (Koka *et al.*, 2000). Brassinosteroids have also been demonstrated to change endogenous cytokinin levels in various plant species. When added to a culture medium containing growth-limiting amounts of auxin, 24-epibrassinolide (24-epiBR) increased the endogenous predominant cytokinins N-6- (Δ-2-isopentenyl) adenine (iP) and trans-zeatin (Z) on tobacco (*Nicotiana tabacum*) callus tissue (Gaudinova *et al.*, 1995). Thus, the 5F-HCTS-driven branching stimulation observed in our system, might be due to: 1. A stimulation of lateral branching by the 5F-HCTS itself; 2. An eventual 5F-HCTS-driven stimulation of cytokinins biosynthesis; 3. A synergistic effect of 5F-HCTS and the $N^6$-Benzyladenine added to the culture medium; 4. A combination of any of the above.

When used at doses over one microgram per shoot, in the case of the main branches, or over 500 ng per shoot, in the case of the primary and secondary branches, 5F-HCTS inhibited stem elongation, compared to the 500 ng per shoot treatment.

Brassinosteroids such as BL and 24-epiBL have previously been reported to inhibit stem elongation in species such as rice (*Oryza sativa*) (Chon *et al.*, 2000) and pea (*Pisum sativum*) (Kohout *et al.*, 1991), respectively. Brassinosteroids are known to stimulate the biosynthesis of aminocyclopropane-1-carboxylic acid (ACC), the immediate precursor of ethylene and the ethylene biosynthesis itself, in various systems (Arteca *et al.*, 1991). Furthermore, ethylene is known to inhibit stem elongation in different plant species. So, a possible way 5F-HCTS might inhibit stem elongation in our system, when used at higher doses, would be through a stimulation of ethylene production. In addition to a possible inhibitory effect of the ethylene itself on stem elongation, cyanide, a by-product of the ethylene biosynthesis, when ethylene is produced from ACC (Chon *et al.*, 2000), might be, at least in part, responsible for the observed brassinosteroid-induced inhibition of stem elongation.

## CONCLUSIONS

Besides the broad spectrum of physiological responses induced by brassinosteroids, several agricultural applications have been found for this group of plant growth regulators such as increasing yield and improving stress resistance of various major crop species (Cutler *et al.*, 1991). The morphogenetic responses such as increased number of leaves, leaf area, fresh weight and dry weight of foliage and roots, and the number and growth of productive branches and tillers observed after treatment with brassinosteroids are thought to be responsible for the enhancement on the number of ears in gramineous crops, pods in leguminous crops, fruits, and tubers, which results in yield increase of these crops (Kamuro and Takatsuto, 1999).

The results we presented in this chapter show a new application for brasinosteroids in horticulture. The C-5 fluoro derivative of 28-homocastasterone-induced shoot proliferation is an effective method to enhance the *in vitro* multiplication rate for *M. prunifolia*, which significantly contributes to make the micropropagation technique for this apple rootstock and commercially feasible and consequently increase the availability of certified, virus-free propagules.

## ACKNOWLEDGEMENTS

The Brazilian authors are thankful to CNPQ-Brazil for the fellowship granted to S. Schaefer and to Dr. E. L. Pedrotti (University of Santa Catarina, Brazil), for providing the starting shoot cultures used in this study. Argentine authors are greatful to the University of Buenos Aires for UBACyT grants and to UMYMFOR (UBA-CONICET) for spectroscopic analysis.

## REFERENCES

Abe, H., Morishita, T., Uchiyama, M., Takatsuto, S., Ikekawa, N., Ikeda, M., Sassa, T., Kitsuwa, T., Marumo, S. (1983). Occurrence of three new brassinosteroids: brassinone, (24S)-24-ethylbrassinone and 28-norbrassinolide in higher plants. Experientia 39:351-353

Altmann, T. (1998). A tale of dwarfs and drugs: brassinosteroids to the rescue. Trends in Genetics 14:490-495.

Arteca, R. N. (1995). Brassinosteroids. In Plant hormones: Physiology, biochemistry and molecular biology, pp. 206-213. Eds P J Davies. Kluwer Academic Publishers, Netherlands.

Arteca, R. N., Tsai, D. S., Mandava, N. B. (1991). The inhibition of brassinosteroid-induced ethylene biosynthesis in etiolated mung bean hypocotyl segments by 2,3,5-triiodobenzoic acid and 2-(*p*-chlorophenoxy)-2-methylpropionic acid. J of Plant Physiology 139:52-56

Azpiroz, R., Wu, Y., LoCascio, J. C., Feldmann, K. A. (1998). An *Arabidopsis* brassinosteroid-dependent mutant is blocked in cell elongation. Plant Cell 10:219–230.

Back, T., Janzen, L., Nakajima, S., Pharis, R. (1999). Synthesis and biological activity of 25-methoxy-, 25-fluoro-, and 25 azabrassinolide and 25-fluorocastasterone: surprising effects of heteroatm substituent at C-25. Journal of Organic Chemistry 64:5494-5498.

Bieberach, C., de León, B., Teme Centurión, O., Ramírez, J., Gros, E., Galagovsky, L. (2000). Estudios preliminares sobre el efecto de dos brassinosteroides sintéticos sobre el crecimiento in vitro de yuca, ñame y piña. Anales de la Asociación Quimica Argentina, 88:N°1 / 2, 1-7.

Brosa, C. (1999). Structure-activity relationship. In Brassinosteroids: Steroidal Plant Hormones, pp. 191-222. Eds A Sakurai, T Yokota and S D Clouse. Springer Verlag, Tokyo.

Brosa, C., Capdevila, J. M., Zamora, I. (1996). Brassinosteroids: a new way to define the structural requirements. Tetrahedron 52:2435-2448

Brosa, C., Nusimovich, S., Peracaula, R. (1994). Synthesis of new brassinosteroids with potential activity as antiecdysteroids. Steroids 59:463-467

Brosa, C., Soca, L., Terricabras, E., Ferrer, J., Alsina, A. (1998). New synthetic brassinosteroids: a 5α-hydroxy-6-ketone analog with strong plant growth promoting activity. Tetrahedron 54:12337-48.

Brosa, C., Zamora, I., Terricabras, E., Soca, L., Peracaula, R., Rodriguez-Santamarta, C. (1997). Synthesis and molecular modelling: related approaches to the progress in brassinosteroid research. Lipids 32: 1341-1347.

Choe, S., Dilkes, B.P., Gregory, B.D., Ross, A.S., Yuan, H., Noguchi, T., Fujioka, S., Takatsuto, S., Tanaka, A., Yoshida, S., Tax, F. E., Feldmann, K.A. (1999a). The *Arabidopsis* dwarf1 mutant is defective in the conversion of 24-methylenecholesterol to campesterol in brassinosteroid biosynthesis. Plant Physiology 119:897-907.

Choe, S., Noguchi, T., Fujioka, S., Takatsuto, S., Tissier, C.P., Gregory, B.D., Ross, A.S., Tanaka, A., Yoshida, S., Tax, F. E., Feldmann, K.A. (1999b). The *Arabidopsis* dwf7/ste1 mutant is defective in the Delta (7) sterol C-5 desaturation step leading to brassinosteroid biosynthesis. Plant Cell 11:207-221.

Chon, N. M., Nishikawa-Koseki, N., Hirata, Y., Saka, H., Abe, H. (2000). Effects of brassinolide on mesocotyl, coleoptile and leaf growth in rice seedlings. Plant Production Science 3:360-365.

Chory, J. (2001). Light, brassinosteroids, and *Arabidopsis* development. Proceedings of the Symposium: Plant Physiology 2000 and Beyond: Breaking the Mold, Plant Biology 2001-ASPP, Providence, Rhode Island, Abstract 30005.

Chory, J., Nagpal, P., Peto, C. A. (1991). Phenotypic and genetic analysis of det2, a new mutant that affects light-regulated seedling development in *Arabidopsis*. Plant Cell 3:445-459.

Chory, J., Peto, C., Feinbaum, R., Pratt, L., Ausubel, F. (1989). *Arabidopsis thaliana* mutant that develops as a light-grown plant in the absence of light. Cell 58:991-999.

Cleland, R. E (1995). In Plant hormones: Physiology, biochemistry and molecular biology, pp. 214-227. Eds P J Davies. Kluwer Academic Publishers, Netherlands.

Clouse, S.D. (1996). Molecular genetic studies confirm the role of brassinosteroids in plant growth and development. Plant Journal 10:1-8.

Clouse, S.D. (2002). Brassinosteroid signal transduction: Clarifying the pathway from ligand perception to gene expression. Molecular Cell 10: 973-982.

Clouse, S. D., Sasse, J. M. (1998). Brassinosteroids: Essential regulators of plant growth and development. Annual Review of Plant Physiology and Plant Molecular Biology 49: 427-451.

Clouse, S. D., Zurek, D. (1991). Molecular analysis of brassinolide action in plant growth and development. In Brassinosteroids: Chemistry, Bioactivity, and Applications, pp. 122-140. Eds H G Cutler, T Yokota and G Adam. American Chemical Society, Washington.

Cosgrove, D. (1997). Relaxation in a high-stress environment: the molecular basis of extensible cell walls and enlargement. Plant Cell 9:1031-1041.

Cutler, H. G., Yokota, T., Adam, G. (1991). Brassinosteroids: chemistry, bioactivity, and applications.pp. 358. American Chemical Society, Washington.

Dahse, I., Petzold, U., Willmer, C. M., Grimm, E. (1991). Brassinosteroid-induced changes of plasmalemma energization and transport and of assimilate uptake by plant-tissues. In Brassinosteroids: Chemistry, bioactivity, and applications, pp. 167-175. Eds H G Cutler, T Yokota and G Adam. American Chemical Society, Washington

Dunitz, J., Taylor, R. (1997). Organic fluorine hardly ever accepts hydrogen bonds. Chemistry- A European Journal. 3: 89-98.

Evans, M. L. (1985). The action of auxin on plant cell elongation. Critical Review of Plant Sciences. 2: 317-365.

Filler, R., Kobayashi, Y., Yagupolskii, L. (1993). Organofluorine Compounds in Medicinal Chemistry and Biomedical Applications. Elsevier, Amsterdam.

Flores, R., Lessa, A. O., Peters, J. A., Fortes, G. R. L. (1999). Efeito da sacarose e do benomyl na multiplicação *in vitro* da macieira. Pesquisa Agropecuaria Brasilleira., 34:2363-2368.

Friedrichsen, D.M., Joazeiro, C. A. P., Li, J., Hunter, T., Chory, J. (2000). Brassinosteroid-insensitive-1 is a ubiquitously expressed leucine-rich repeat receptor serine/threonine kinase. Plant Physiology 123:1247-1256.

Fujioka, S. (1999). Natural occurrence of brassinosteroids in the plant kingdom. In Brassinosteroids: Steroidal Plant Hormones, pp. 21-45. Eds A Sakurai, T Yokota and S D Clouse. Springer Verlag, Tokyo.

Fujioka, S., Sakurai, A. (1997). Biosynthesis and metabolism of brassinosteroids. Physiology Plantarum 100: 710-715.

Galagovsky, L., Gros, E., Ramírez, A. (2001). Synthesis and bioactivity of natural and C-3 fluorinated biosynthetic precursors of 28-homobrassinolide. Phytochemistry 58:973-980.

Gaudinova, A., Sussenbekova, H., Vojtechova, M., Kaminek, M., Eder, J., Kohout, L. (1995). Different effects of 2 brassinosteroids on growth, auxin and cytokinin content in tobacco callus-tissue. Plant Growth Regulation 17:121-126

Grove, M. D., Spencer, G. F., Rohwedder, W. K., Mandava, N., Worley, J. F., Warthen, J. D. Jr., Steffens, G. L., Flippen-Anderson, J. L., Cook, J. C. Jr. (1979). Brassinolide, a plant growth-promoting steroid isolated from Brassica napus pollen. Nature 281: 216-217.

Howard, J., Hoy, V., O'Hagan, D., Smith, G. (1996). How good is fluorine as a hydrogen bond acceptor? Tetrahedron 38:12613-12622.

Hu, Y., Bao, F., Li, J. (2000). Promotive effect of brassinosteroids on cell division involves a distinct CycD3-induction pathway in *Arabidopsis*. Plant Journal 24:693-701.

Jin, F., Xu, Y., Huang, W. (1993). 2,2 Difluoro enol silyl ethers: convenient preparation and application to the synthesis of a novel fluorinated brassinosteroid. Journal of the Chemical Society, Perkin Transactions I: 795-799.

Jiang, B., Ying, L., Zhou, W-S. (2000). Stereocontrolled synthesis of the 22*E*,24β (*S*) –trifluoromethyl steroidal side chain and its application to the synthesis of fluorinated analogues of naturally occurring sterols. Journal of Organic Chemistry 65:2631-6236.

Joseph-Nathan, P., Espiñeira, J., Santillan, R. (1984). 19F-NMR study of fluorinated corticosteroids. Spectrochimica Acta 40A: 347-349.

Kamuro, Y., Takatsuto, S. (1999). Practical applications of brassinosteroids in agricultural fields. In: Brassinosteroids: Steroidal Plant Hormones. pp. 223-241. Eds A Sakurai, T Yokota and S D Clouse. Springer Verlag, Tokyo.

Kauschmann, A., Jessop, A., Koncz, C., Szekeres, M., Willmitzer, L., Altmann, T. (1996). Genetic evidence for an essential role of brassinosteroids in plant development. Plant Journal 9:701–713.

Khripach, V. A., Zhabinskii, V. N., de Groot, A. E. (1999a). Bioassays and structure-activity relationships of BS. In Brassinosteroids: A New Class of Plant Hormones, pp 301-324. Eds V A Khripach, V N Zhabinskii and A E de Groot. Academic Press, San Diego.

Khripach, V. A., Zhabinskii, V. N., de Groot, A. E. (1999b). Physiological mode of action of BS. In Brassinosteroids: a New Class of Plant Hormones, pp 219-300. Eds V A Khripach, V N Zhabinskii, A E de Groot. Academic Press, San Diego.

Kim, G –T., Tsukaya, H., Uchimiya, H. (1998). The Rotundifolia 3 gene of *Arabidopsis thaliana* encodes a new member of the cytochrome P-450 family that is required for the regulated polar elongation of leaf cells. Genes and Development, 12:2381–2391.

Kim, S., Abe, H., Little, C., Pharis, R. (1990). Identification of two brassinosteroids from the cambial region of Scots pine (*Pinus silvestris*) by gas-chromatography-mass spectrometry, after detection using a dwarf lamina inclination bioassay. Plant Physiology 94:1709-1713.
Kishi, T., Wada, K., Marumo, S., Mori, K. (1986). Synthesis of brassinolide analogs with a modified ring B and their plant growth-promoting activity. Agricultural and Biological Chemistry 50:1821-1830
Kobayashi, Y., Taguchi, T. (2000). Studies on organofluorine compounds: an overview of our 30 years. Yakugaky Zasshi 120: 951-958.
Kohout, L., Strand, M., Kaminek, M. (1991). Types of brassinosteroids and their bioassay. In Brassinosteroids: Chemistry, Bioactivity, and Applications, pp. 56-73. Eds H G Cutler, T Yokota and G Adam. American Chemical Society, Washington.
Koka, C. V., Cerny, R. E., Gardner, R. G., Noguchi, T., Fujioka, S., Takatsuto, S., Yoshida, S., Clouse, S. D. (2000). A putative role for the tomato genes DUMPY and CURL-3 in brassinosteroid biosynthesis and response. Plant Physiology 122:85-98.
Ladyzhenskaya, E. P., Korableva, N. P. (2001). Effects of growth regulators on $H^+$ translocation across the membranes of plasma membrane vesicles from potato tuber cells. Applied Biochemistry and Microbiology 37:521-523.
Li, J., Chory, J. (1997). A putative leucine-rich receptor kinase involved in brassinosteroid signal transduction. Cell 90:929-938.
Li, J., Chory, J. (1999). Brassinosteroid actions in plants. Journal of Experimental Botany 50:275-282.
Li, J., Nagpal, P., Vitart, V., McMorris, T. C., Chory, J. (1996). A role for brassinosteroids in light-dependent development of *Arabidopsis*. Science 272:398-401.
Li, J., Nam, K. H., Vafeados, D., Chory, D. (2001). BIN2, a new brassinosteroid-insensitive locus in *Arabidopsis*. Plant Physiology 127:14-22.
Liebman, J., Greenberg, A., Dolbier, W. Jr., Eswarakrishnan, S. (1988). Fluorine-Containing Molecules: Structure, Reactivity, Synthesis. VCH Publisher, New York.
Maeda, E. (1965). Rate of lamina inclination in excised rice leaves. Physiology Plantarum 18:813-827.
MacMorris, T., Chávez, R., Patil, P. (1996). Improved synthesis of brassinolide. Journal of the Chemical Society Perkin Transactions I, 295-302.
Mandava, N. B. (1988). Plant growth-promoting brassinosteroids. Annual Review of Plant Physiology and Plant Molecular Biology 39:23-52.
Martin, C., Galdwell, J., Graham, M., Grierson, J., Kroll, K., Cowan, M., Lwellen, T., Rasey, J. Casciari, J., Krohn, K. (1992). Non invasive detection of hypoxic myocardium using fluorine-18-fluoromisonidazole and positron emission tomography. Journal of Nuclear Medicine 22:2202.
Mayumi, K., Shibaoka, H. (1995). A possible double role for brassinolide in the reorientation of cortical microtubules in the epidermal cells of Azuki bean epicotyls. Plant Cell Physiology 36:173-181.
McMorris, T. C., Patil, P. A., Chavez, R. G., Baker, M. E., Clouse, S. D. (1994). Synthesis and biological activity of 28-homobrassinolide and analogues. Phytochemistry 36:585-589.
Mori, K. (1980). Synthesis of a brassinolide analog with high plant growth promoting activity. Agricultural and Biological Chemistry 44:1211-1212.
Murashige, T., Skoog, F. (1962). A revised medium for rapid growth and bioassay with tobacco tissue cultures. Physiologia Plantarum 15:473-497.
Nakayama, M., Yamane, H., Murofushi, N., Takahashi, N., Mander, L. N., Seto, H. (1991). Gibberellin biosynthetic pathway and the physiologically active gibberellin in the shoot of *Cucumis sativus* L. Journal of Plant Growth Regulation 10: 115-119.
Noguchi, T., Fujioka, S., Choe, S., Takatsuto, S., Tax, F. E., Yoshida, S., Feldmann, K. A. (2000). Biosynthetic pathways of brassinolide in *Arabidopsis*. Plant Physiology 124:201-209.
Oh, M. H., Romanow, W., Smith, R., Zamski, E., Sasse, J., Clouse, S. (1998). Soybean BRU1 encodes a functional xyloglucan endo-transglycosylase that is highly expressed in inner epicotyl tissues during brassinosteroid-promoted elongation. Plant Cell Physiology 39:124-130.
O'Hagan, D., Rzepa, H. (1997). Some influences of fluorine in bioorganic chemistry. Chemical Communications 645-652.
Okada, K., Mori, K. (1983). Stereoselective synthesis of dolicholide, a plant growth-promoting steroid. Agricultural Biological Chemistry 47:925-926.
Ramírez, A., Gros, E., Galagovsky, L. (2000). Effect on bioactivity due to C-5 heteroatom substituents on synthetic 28-Homobrassinosteroids analogs. Tetrahedron 56:6171-6181.

Richter, K., Koolman, J. (1991). Antiecdysteroid effects of brassinosteroids. In Brassinosteroids - Chemistry, Bioactivity and Applications, pp. 265-278. Eds H G Cutler, T Yokota and G Adam. American Chemical Society, Washington.

Saito, T., Kamiya, Y., Yamane, H., Murofushi, N., Sakurai, A., Takahashi, N. (1998). Effects of fluorogibberellins on plant growth and gibberellin 3β-hydroxylases. Plant Cell Physiology 39: 574-580.

Sakurai, A. (1999). Biosynthesis. In Brassinosteroids: Steroidal Plant Hormones, pp. 91-111. Eds A Sakurai, T Yokota and S D Clouse. Springer Verlag, Tokyo.

Sasse, J. M. (1997). Recent progress in brassinosteroid research. Physiologia Plantarum 100:696-701.

Schaefer, S., Medeiro, A. S., Ramirez, J. A., Galagovsky, L. R., Pereira-Netto, A. B. (2002). Brassinosteroid-driven enhancement of the *in vitro* multiplication rate for the marubakaido apple rootstock [*Malus prunifolia* (Willd.) Borkh]. Plant Cell Reports 20:1093-1097.

Schumacher, K., Vafeados, D., McCarthy, M., Sze, H., Wilkins, T., Chory, J. (1999). The *Arabidopsis* det3 mutant reveals a central role for the vacuolar H+-ATPase in plant growth and development. Genes and Development 13:3259-3270.

Schwartzenberg, K., Doumas, P., Jouanin, L., Pilate, G. (1994). Enhancement of the endogenous cytokinin concentration in poplar by transformation with *Agrobacterium* T-DNA gene ipt. Tree Physiology 14:27-35.

Seto, H., Fujioka, S., Koshino, H., Suenaga, T., Yoshida, S., Watanabe, T., Takatsuto, S. (1998).Epimerization at C-5 of brassinolide with sodium methoxide and the biological activity of 5-epi-brassinolide in the rice lamina inclination assay. Journal of the Chemical Society, Perkin Transactions 1: 3355-3358.

Seto, H., Fujioka, S., Koshino, H., Suenaga, T., Yoshida, S., Watanabe, T., Takatsuto, S. (1999). 2,3,5-Tri-epi-brassinolide: preparation and biological activity in rice lamina inclination test. Phytochemistry 52:815-818.

Sharpless, K. B., Amberg, W., Bennani, Y. L. (1992). The osmium-catalyzed asymmetric dihydroxylation: a new ligand class and a process improvement. Journal of Organic Chemistry 57:2768-2771.

Shekhawat, N. S., Rathore, T. S., Singh, R. P., Deora, N. S., Rao, S. R. (1993). Factors affecting *in vitro* clonal propagation of *Prosopis cineraria*. Plant Growth Regulation 12:273-280.

Spray, C., Phinney, B. O., Gaskin, P., Gilmour, S. J., MacMillan, J. (1984). Internode length in *Zea mays* L. Planta1 60:464-468

Szekeres, M., Nemeth, K., Koncz-Kalman, Z., Mathur, J., Kauschmann, A., Altmann, T., Redei, G. P., Nagy, F., Schell, J., Koncz, C. (1996). Brassinosteroids rescue the deficiency of CYP90, a cytochrome P450, controlling cell elongation and de-etiolation in *Arabidopsis*. Cell 85: 171–182.

Takatsuto, S., Ikekawa, N., Morishita, T., Abe, H. (1987). Structure-activity relationship of brassinosteroids with respect to the A/B-ring functional groups. Chemical Pharmaceutical Bulletin. 35:211-216.

Takatsuto, S., Yazawa, N., Ikekawa, N., Morishita, T., Abe, H. (1983a). Synthesis of (24R)-28-homobrassinolide and structure-activity relationships of brassinosteroids in the rice lamina inclination test. Phytochemistry 22:1393-1397.

Takatsuto, S., Yazawa, N., Ikekawa, N., Takematsu, T., Takeuchi, Y., Koguchi, M. (1983b). Structure-activity relationship of brassinosteroids. Phytochemistry 22:2437-2441.

Takeno, K., Pharis, R. (1982). Brassinosteroid-induced bending of the leaf lamina of dwarf rice seedlings: an auxin-mediated phenomenon. Plant Cell Physiology 23:1275-1281.

Thompson, M. J., Mandava, N., Flippen-Anderson, J. L., Worley, J. F., Dutky, S. R., Robbins, W. E., Lusby, W. (1979). Synthesis of brassinosteroids: new plant-growth promoting steroids. Journal of Organic Chemistry 44:5002-5004.

Thompson, M. J., Mandava, N. B., Meudt, W.J., Lusby, W. R., Spaulding, D. W. (1981). Synthesis and biological activity of brassinolide and its 22β, 23 β -isomer: novel plant growth promoting steroids. Steroids 38:567-580.

Thompson, M. J., Meudt, W. J., Mandava, N. B., Dutky, S. R., Lusby, W. R., Spaulding, D. W. (1982). Synthesis of brassinosteroids and relationship of structure to plant growth-promoting effect. Steroids 39:89-105.

Voigt, B., Takatsuto, S., Yokota, T., Adam, G. (1995). Synthesis of secasterone and further epimeric 2,3-epoxybrassinosteroids. Journal of the Chemical Society, Perkin Transactions I: 1495-1498.

Wada, K., Marumo, S., Abe, H., Morishita, T., Nakamura, K., Uchiyama, M., Mori, K. (1984). A rice lamina inclination test – a micro-quantitative bioassay for brassinosteroids. Agricultural and Biological Chemistry 48:719-726.

Wada, K., Marumo, S., Ikekawa, N., Morisaki, M., Mori, K. (1981). Brassinolide and homobrassinolide promotion of lamina inclination of rice seedlings. Plant and Cell Physiology 22:323-326.
Wang, Z. Y., Nakano, T., Gendron, J., He, J. X., Chen, M., Vafeados, D., Yang, Y. L., Fujioka, S., Yoshida, S., Asami, T., Chory, J. (2002). Nuclear-localized BZR1 mediates brassinosteroid-induced growth and feedback suppression of brassinosteroid biosynthesis. Developmental Cell 2:505-513
Wang, Z. Y., Seto, H., Fujioka, S., Yoshida, S., Chory, J. (2001). BRI1 is a critical component of a plasma-membrane receptor for plant steroids. Nature 410:380-383.
Welch, J., Eswarakrishnan, S. (1991). Fluorine in Bioorganic Chemistry. John Wiley & Sons, New York.
Yokota, T., Baba, J., Arima, M., Morita, M., Takahashi, N. (1983). Isolation and structures of new brassinolide-related compounds in higher plants. Tennen Yuki Kagob. Toronkai Koen Yoshishu 26: 70-77 [C. A. 100:48616].
Yokota, T., Nakayama, N., Wakisaka, T. (1994). 3-Dehydroteasterone, a 3,6 diketobrassinosteroid as a possible biosynthetic intermediate of brassinolide from wheat grain. Bioscience Biotechnology and Biochemistry 58:1183-1185.
Xu, R., He, Y-J., Wang, Y-Q., Zhao, Y-J. (1994). Preliminary study of brassinosterone binding sites from mung bean epicotyls. Acta Phytophysiologica Sinica 20:298-302.
Xu, W., Prugganan, M. M., Polisensky, D. H., Antosiewicz, D. M., Fry, S. C., Braam, J. (1995). *Arabidopsis* TCH4, regulated by hormones and the environment, encodes a xyloglucan endotransglycosylase. Plant Cell 7:1555-1567.
Yin, Y., Wang, Z. Y., Mora-Garcia, S., Li, J., Yoshida, S., Asami, T., Chory, J. (2002). BES1 accumulates in the nucleus in response to brassinosteroids to regulate gene expression and promote stem elongation. Cell 109:181-191.
Yokota, T. (1997). The structure, biosynthesis and function of brassinosteroids. Trends in Plant Science 2:137–143.
Zanol, G. C., Fortes, G. R. L., Silva, J. B., Faria, J. T. C., Gottinari, R. A., Centellas, A. Q. (1998). Uso do ácido indolbutírico e do escuro no enraizamento *in vitro* do porta-enxerto de macieira Marubakaido. Ciência Rural, 28:387-391.
Zullo, M. A. T., Adam, G. (2002). Brassinosteroid phytohormones – structure, bioactivity and applications. Brazilian Journal of Plant Physiology 14: 143-181.
Zullo, M. A. T., Kohout, L., De Azevedo, M. B. M. (2003). Some notes on the terminology of brassinosteroids. Plant Growth Regulation 39: 1-11.
Zurek, D. M., Clouse, S. D. (1994). Molecular cloning and characterization of a brassinosteroid-regulated gene from elongating soybean (*Glycine max* L.) epicotyls. Plant Physiology 104: 161-170.
Zurek, D. M., Rayle, D. L., McMorris, T. C., Clouse, S. D. (1994). Investigation of gene expression, growth kinetics, and wall extensibility during brassinosteroid-regulated stem elongation. Plant Physiology 104:505-513.

CHAPTER 7

ZHAO YU JU AND CHEN JI-CHU

# STUDIES ON PHYSIOLOGICAL ACTION AND APPLICATION OF 24-EPIBRASSINOLIDE IN AGRICULTURE

EpiBL showed strong activity of stimulating the growth of root explants from tobacco seedlings at low concentration as compared with other known plant hormones. Rootlet number of explants incubated on MS medium containing 0.01-0.05 ppm EpiBL was apparently increased. *Arabidopsis thaliana* has been used to investigate the role of EpiBL in cell differentiation and regeneration *in vitro.* The result showed that calli cultured on MS medium supplemented with 0.05, 0.5 or 5.0 mg /l EpiBL formed much green buds and shoots. Supplementing the culture medium with EpiBL and 0.1mg /l KT induced the greening of callus and bud formation, but KT alone did not induce differentiation. Moreover, electron microscopic examination showed that normal chloroplasts are contained in the cells of green callus, cultured on medium supplemented with EpiBL.Using detached cucumber cotyledons, we found that EpiBL accelerated destruction of chloroplasts in cotyledon. The results also showed that EpiBL promoted senescence in mung bean seedling, accompanied by enhanced peroxidase activity and malondialdehyde and decreased the activity of superoxide dismutase and catalase. The comparison of the ultrastructure of cells, in elongating region of treated hypocotyl segments, with those of control indicated that EpiBL exhibited retarding action in the maintenance of various organelles, against deterioration. The morphological examination demonstrated that the promoting effect of EpiBL on stem growth was mainly due to the stimulation of cell elongation. Using $^{3}H$ labeled $H_2O$, the promoting effect of EpiBL on water absorption was found. EpiBL also has been shown to affect the fatty acid composition of membrane lipid of tissues in mung bean hypocotyls. BRs resemble with cytokinins in regulating de-etiolation as positive regulators, and that the inhibition of hypocotyl elongation and the development of leaves and epicotyls in de-etiolation are independent processes. In China, large-scale field trials, over 10 years, have demonstrated that significant effect of EpiBL on the production of crops is caused partially by the improved tolerance against environmental stress. In Henan Province, treatment with EpiBL during booting stage or flower stage resulted in increase in wheat yield due to promoted flower development and reduced the abortion of grains. Spike weight and thousand weight were increased as compared to the control. Sprayed on wheat leaves also increased resistance against leaf wilt, one of the most harmful diseases induced by environmental stresses, during ripening. EpiBL decreased cold injury in rape plants during winter. It reduced kernel abortion at the tips of corn ears and the abscission of grape fruits. EpiBL promoted growth of the root system in tobacco plants, thus increasing tolerance against water stress. The results also show that EpiBL is useful for the improvement of yield and quality of the watermelon, grape and summer orange. Spraying EpiBL on leaves also increased the contents of sugar in beet plants. Vegetables applied with EpiBL grew better than the control. EpiBL reduced the abscission of cotton bolls and fruits of grapes. In general, the effects of epibrassinolide varied between diverse areas and also during different years, due to changing environmental conditions. New brassinosteroid (TS303) has been synthesized with enduring effects. Soaking of seeds of rape and barley, with TS303 solution, the yields was increased, but in rice the growth of seedling, at early stage, was promoted without an effect on grain yield. Foliar application of TS303 also promoted the growth of vegetables.

*S.Hayat and A.Ahmad (eds.), Brassinosteroids*, 159-170.

## INTRODUCTION

Brassinolide is regarded as a new plant growth regulator. It has attracted attention of many biologists, since its discovery in the pollen of *Brassica napus* (Grove *et al.*, 1979). Studies on the physiological action of BR have been reviewed earlier (Yopp *et al.*, 1981; Zhao and Wang, 1986; Mandava, 1988; Adam *et al.*, 1994; Sasse, 1997; 1999; Yokota, 1999). This survey is focused on the results regarding the physiological responses to the application of brassinolide. It should be noted that the mechanism of action of BR, is still under study. For example, it remains unclear whether BRs can affect stress resistance and, if so, what the mechanism of action is? Some of the observations, related to the stress resistance to BRs are also covered in this survey.

In addition to the information concerning the physiological action of BR, this survey will also present some results obtained by the exogenous application of 24-Epibrassinolide (EpiBL), which has already been synthesized. Ikekawa (1987) succeeded in identifying the presence of EpiBL in the pollen of broad bean, where it was isolated. It has also shown strong physiological activity in many cases. We conducted some trials to confirm the physiological potential of EpiBL. This survey will explain the results of some of these researches in details.

## RELATIONSHIP BETWEEN STRUCTURE AND BIOLOGICAL ACTIVITY

The clarification of the structure–activity relationship of brassinosteroids is theoretically important as it contributes not only to the chemical synthesis and the improvement of the potential of related compounds, but also to illuminate the mechanism of their biosynthesis. Among the synthetic analogues of brassinolide, EpiBL is considered ideal for practical application, in agriculture. In China, hyodeoxycholic acid was used as starting material to synthesize brassinosteroids because it is rich in resources. Zhou's group from hyodeoxycholic acid synthesized nineteen brassinosteroids. The activity of these compounds was compared, using rice lamina inclination and intact radish elongation tests. Among them, 26, 27-bisnorbrassinolide, 23-phenylbrassinosteroid and 23-phenylbrassinolide showed higher activities, compared with epibrassinolide. The biological activity of these compounds increased with an increase in the concentration from 0.0001 to 1 $mgl^{-1}$ in rice lamina inclination test and from 0.01 to 1 $mgl^{-1}$, in intact radish test. The order of activity with respect to B-ring oxygen functional group was lactones >ketones. In lactone and ketone types of brassinosteroids, brassinolide and homobrassinolide as well as 26,27-bisnorbrassinolide showed strongest biological activity. EpiBL, 23-phenyl of the lactone and 23-phenyl of the ketone types also showed stronger activity, while 23-carboxyl compounds were less active. The compound, which lost hydroxyl group at 2-carbon position, remained more active in rice test, while compound with modified side chain at 22-carbon position significantly promoted elongation of the cotyledon petiole and hypocotyl in intact radish. Similar effects of brassinosteroids were observed in different varieties of rice (Wang and Zhao, 1989, Wang *et al.,* 1994).

## EFFECT OF EPIBL ON THE ELONGATION

Excised coleoptile segments of etiolated seedlings were used in this experiment. Shen *et al.* (1988) found that EpiBL was more effective in stimulating the elongation of coleoptile than IAA, when applied at low concentrations ($<10^{-2}$ mg / l). It was shown that the treatment with EpiBL and IAA resulted in synergistic stimulation of segment elongation and $H^+$ secretion in coleoptile segments. This indicated that the action of EpiBL was related to acid induced growth. The experiment also revealed that both EpiBL and IAA induced elongation together with ethylene production. However, the lag phase of ethylene production by IAA was shorter, compared to EpiBL. On the other hand, EpiBL had antagonistic action on the inhibitory effect of ABA (Shen *et al.*, 1988).

## EFFECT OF EPIBL ON ETHYLENE PRODUCTION

Effect of EpiBL on the synthesis of ethylene was investigated. Etiolated mung bean hypocotyl segments were used to conduct this investigation. EpiBL was applied at the concentrations of 0.001~10 mg/l. It was observed that the synthesis of ACC and the capability of the tissue to conjugate ACC to form MACC were stimulated by this treatment. In the EpiBL treated tissues, MACC was formed, from ACC in greater quantities. However, there was no effect of EpiBL on ethylene forming enzymes (Wu *et al.*, 1987).

## EFFECT OF EPIBL ON ACTIVITY OF PEROXIDASE AND IAA OXIDASE

Stimulating effect of EpiBL on the elongation in cucumber seedlings was studied. It was noted that the steroid played an important role in the elongation of hypocotyl of light-grown seedlings with a lag period of more than 10 hours but was shorter in those incubated with IAA. It was, therefore, suggested that growth promotive effect in cucumber hypocotyl segments, by EpiBL and IAA, might be regulated by different modes. EpiBL remarkably inhibited peroxidase activity at 0.1-1 mg/l and this shift was comparable with that of IAA oxidase activity. However, the seedlings, incubated with IAA, did not show any change in the activity of both the enzymes (Xu and Zhao, 1989).

## EFFECT OF EPIBL ON ENDOGENOUS $GA_3$ AND ABA

The effects of EpiBL on the endogenous gibberellin, ABA and starch contents, in light-grown cucumber hypocotyl, were investigated. Results showed that the stimulatory effect of EpiBL on the elongation growth was greater than that of IAA or Gibberellin. However, the lag period of the effect of EpiBL was longer than that of IAA, but similar to that of gibberellin.

The contents of endogenous $GA_3$ and ABA were higher in the hypocotyl segments, treated with EpiBL. After 24 h of the treatment, the ratio of $GA_3$/ABA in EpiBL treated hypocotyl was twice, as much as compared with the control. EpiBL treated hypocotyl possessed lower level of starch. Electron microscopic examination of elongating region of hypocotyl segments indicated that both EpiBL and $GA_3$ reduced the content of starch grain, in chloroplast. It was also noted that the effect of epibrassinolide and $GA_3$ was additive in enhancing hypocotyl elongation and starch hydrolysis. These results suggest that the action of EpiBL was somewhat similar to that of gibberellin. It is known that the growth of rice seedling is insensitive to brassinolide, but was strongly favoured with gibberellin. The observed similarity in the effect of EpiBL and $GA_3$ could be related to the maintenance of a lower osmotic potential in hypocotyl of cucumber seedlings (Xu *et al.*, 1990).

## EFFECT OF EPIBL ON THE ACTIVITY OF PLASMA MEMBRANE ATPASE

The membrane of wheat roots, on being incubated with EpiBL, did not show significant increase in the activity of ATPase. On the contrary, the activity decreased in EpiBL, at a concentration of $2 \times 10^{-6}$ mol/l. However, in the reaction mixture of EpiBL with IAA ($5 \times 10^{-5}$ mol/l), the activity of ATPase was higher than that treated with IAA alone. These results demonstrate that there was possibly synergistic action between EpiBL and IAA on ATPase, which is known to regulate the linear growth (Xu *et al.,* 1995).

## TRANSPORT AND DISTRIBUTION OF EPIBL

Studies with $^{14}C$-labeled epibrassinolide have confirmed its movement from root to shoot. Brassinosterone was identified as the precursor, in the synthesis of EpiBL. Brassinosterone exhibited biological activity similar to that of EpiBL, in rice lamina inclination test. The radiograph revealed that $^{125}I$-brassinosterone was distributed mainly in the elongation zone of the epicotyls of mung bean seedlings whereas, in cucumber seedlings, it was mostly confined to the apex and cotyledon bases (Xu *et al.*, 1994a; 1995).

## BRASSINOSTERONE BINDING SITE

Different fractions of mung bean epicotyl extract were tested for brassinosterone binding. The fraction 1000-x g showed stronger brassinosterone binding than others and gets saturated after an incubation period of 1 h. The binding was temperature-dependent and achieved maximum response at 4 $^0$C. The quantity of brassinosterone bound was also dependent on pH, with the optimal pH being 7. 5. Trypsin treatment weakened the brassinosterone binding. This result indicated that protein might be involved at the brassinosterone-binding site. Several analogues of brassinosterone could also, to various extents, bind to the site that brassinosterone bound to. (Xu *et al.*, 1994b).

## EFFECT OF EPIBL ON NUCLEIC ACID

Effect of EpiBL on the metabolism of nucleic acid, in epicotyls of mung bean seedlings, was investigated. The results showed that actinomycin D (0.5 ~ 5 mg /l) and cycloheximide (0.01 ~ 0. 1 mg /l) inhibited growth of mung bean seedlings, induced by

EpiBL. The content of DNA and RNA was markedly enhanced by EpiBL and the activity of RNA polymerase, in mung bean epicotyl tissue, was promoted by EpiBL. On the other hand, the activities of DNase and RNase were slightly inhibited by EpiBL. TIBA partly inhibited the action of EpiBL. The observed accumulation of DNA and RNA inthe tissues, treated with EpiBL could be the expression of the promotion of RNA polymerase and the slight inhibition of DNase and RNase. (Wu and Zhao, 1991; 1993).

## APPLICATION OF EPIBL IN TISSUE CULTURE

EpiBL showed stronger activity in stimulating the growth of explants, from tobacco seedlings, compared with other known plant hormones. Rootlet number in explants, incubated on MS medium containing 0.01-0.05 mg/l EpiBL apparently increased (Chen *et al.*, 1990). EpiBL (0.02 mg/l) also induced bud differentiation and the generation of callus, incubated on MS medium. Chen and Chi (1986) have also drawn similar conclusions in hypocotyl explants of *Astragalus adsurgens.*

*Arabidopsis thaliana* was used to investigate the role of EpiBL in cell differentiation and regeneration under *in vitro* conditions. It was observed that the callus cultured on MS medium supplemented with 0.05, 0.5 or 5.0 mg/l EpiBL, formed much greener buds and shoots. Moreover, supplementing the culture medium with EpiBL and 0.1mg/l Kinetin improved the greening of callus and bud formation. However Kinetin alone failed to induce differentiation. The electron microscopic examination of green callus, cultured on the medium added with EpiBL only possessed normal chloroplasts (Chen et al., 1996; Zhao., 1995).

## EFFECT OF EPIBL ON SENESCENCE

Using detached cucumber cotyledons, we found that EpiBL accelerated the destruction of their chloroplasts (Zhao *et al.*, 1990; Ding and Zhao, 1995). Moreover, EpiBL also promoted senescence in mung bean seedling, accompanied with enhanced activity of peroxidase and malondialdehyde and decreased the activity of superoxide dismutase and catalase (He *et al.*, 1996). In contrast, Ershova and Khripach (1996) reported a decrease in the level of malondialdehyde in pea seedling, treated with EpiBL. Zhao *et al.* (1987) noted that EpiBL apparently inhibited the accumulation of anthocyanin in the hypocotyl segments of mung bean seedling. The comparison of ultrastructure of cells, in elongated region, of treated hypocotyl segments with that of control indicated that EpiBL exhibited retarding action on the deterioration of various cell organelles (Zhao *et al.*, 1987; Chen *et al.*, 2001). The morphological examination of the cells demonstrated that the promotive effect of EpiBL on stem growth was mainly due to the stimulation of cell elongation.

## EFFECT OF EPIBL ON STRESS RESISTANCE

Takematsu (1989) called brassinosteroid as a stress-elevating hormone, however, there is no evidence to support this hypothesis. Using $^{3}H$ labeled $H_2O$, the promotive effect of EpiBL on water absorption was found. EpiBL has also been shown to affect the fatty acid composition of membrane lipid, in mung bean hypocotyls tissue. Under stress condition, proline content in EpiBL treated mung bean hypocotyl segments was enhanced, remarkably (Zhao *et al.*, unpublished data).

## EPIBL AND DE-ETIOLATION

Brassinosteroids are considered as negative regulators of the de-etiolation, in dark-grown *Arabidopsis* seedlings. Luo *et al.* (1998) researched the inhibition of hypocotyl elongation of det2, in darkness, because of the absence of BR-dependent elongation. It was independent of gene expression for photomorphogenesis. BRs resembled cytokinins, as positive regulators of de-etiolation. The inhibition of hypocotyl elongation and the development of leaves and epicotyls, in de-etiolation, were found to be independent processes.

## SELECTION OF BRASSINOLIDE-INSENSITIVE MUTANT OF ARABIDOPSIS

EpiBL -insensitive *Arabidopsis* mutant was isolated by screening the growth of seedlings on a medium containing 0.05 mg/l EpiBL. The mutant expressed phenotypic differences in the morphology of roots and leaves. Genetic analysis of these mutants indicated that the EpiBL insensitivity was due to recessive mutation, which was designated as Br (Zheng *et al.*, 1996).

## APPLICATION OF EPIBL AND TS303

In China, large-scale field trials, spread over a period of 10-years, have demonstrated significant effect of EpiBL on the production of crops. Partially, it is assigned to the improved tolerance against environmental stress. In Henan Province, treatment with EpiBL, during booting or flowering stages, resulted in an increase in seed yield of wheat. It was primarily the result of promoted flower development and reduced abortion rate of grains. The weight of spike and per-thousand seeds increased, as compared with the control. Spraying on the foliage also increased resistance against leaf wilt, one of the most common harmful diseases induced by environmental stresses, during ripening. EpiBL, dramatically alleviated cold injuries in rape plants, during winter. It also reduced kernel abortion, at the tips of corn ears and the abscission of grape fruits. Moreover, when sprayed on the leaves, EpiBL promoted the growth of roots in tobacco plants, thus increasing tolerance against water deficiency. The results also showed that EpiBL might be used to improve the yield and quality of watermelon, grape and summer orange. Spraying EpiBL on leaves enhanced sugar content in beet. Vegetables applied with EpiBL grew better than the control. EpiBL prevented abscission of cotton bolls and fruits of grapes. In general, the response of plants, to EpiBL changed with the area of cultivation because of variation in climatic conditions.

Yokoda (1990) specified the area, to be more than 2664, ha where wheat and corn crops were grown with the application of EpiBL. This area was rapidly expanding but not with required pace because of various reasons, including the relevant patents

(Yokoda, 1999). China, registered (22s, 23s)-28-homobrassinolide (trade name: BR-120) as a plant growth regulator for tobacco, sugar cane, rapeseed, tea, and some fruits (Yokoda 1999). In fact, EpiBL was first registered in China by Nippon Kayaku company for wheat and corn (trade name: Nong Le Li) in 1990. Then, the mixture of 24- EpiBL and 3- EpiBL was also registered by Jiang Men pesticide factory (trade name: Tian Feng Su). 24- EpiBL is the main component of Tian Feng Su. Epi-homobrassinolide was also registered in China (Trade name: Yun Da–120).

Since the discovery of brassinolide, much effort has been focused to its practical applicabilityin agriculture. Some preliminary trials with brassinolide by USDA (Beltsville group) suggested that EpiBL could be used for improving crop production (Maugh 1981). Takematsu's group (Takematsu and Takeuchi, 1989) carried out extensive studies on its application, in agriculture in Japan. The research on synthesis, biological activity and application of EpiBL in China has been undertaken under the supervision of Ikekawa, since 1984. EpiBL samples, for collaborative research between Japan and China on basic biological research and practical use were kindly supplied by Ikekawa of Tokyo Institute of Technology. Nippon Kayaku Company also sponsored extensive large-scale field trials in China. Large-scale tests on various cereals and vegetables have been carried out at more than 100 stations. Typically, when EpiBL was sprayed on wheat at blossoming or seed filling stage, the yield increased by about 10% (Jin *et al.*,1988 ; Ikekawa and Zhao, 1991). In the case of corn, the kernel abortion of ear tips was reduced (Zhang *et al.*, 1987; Ikekawa and Zhao, 1991).

Han of Henan Agricultural University has reported that spraying the EpiBL solution on tobacco leaves after transplanting increased the root weight, leaf area and nicotine content (Ikekawa and Zhao, 1991). Collaborative work on the usage of EpiBL in improving crop production has greatly progressed. The data on field trials conducted in China has been reviewed by Ikekawa and Zhao (1991). We would like to summarize some results of the experiment conducted in China.

## EFFECT OF EPIBL ON WHEAT PRODUCTION

Large-scale field trials, over a 10-year period, have demonstrated significant effect of EpiBL on the production of wheat. Demonstrative trials were conducted in Henan Province, under the joint supervision of Nippon Kayaku Company and Wheat Research Institute of Henan Agricultural Academy. The large-scale field trials were also conducted in the same Province under the supervision of Shanghai Institute of Plant Physiology and Wheat Institute of Henan Agricultural Academy. It was noted that the percentage of grain setting, especially the setting of less viable flowers in the upper part of the ear, improved. The number of caryopsis per ear and weight of per 1,000 grains also increased. In contrast, the number of infertile spikelets per ear reduced significantly (Jin *et al.*, 1988). Luo of Huazhong Agricultural University has suggested that brassinolide promotes photosynthetic rate in leaves of wheat plants and direct the translocation of photosynthates to the ears (unpublished data) predominantly to the upper part of the ears.This was the possible reason for the observed increase in the number of grains per spikelet, most significantly in the upper parts of the ears (Jin *et al.*,

1988).The other viable reason to explain the improvement in grain production could be improved tolerance, generated by EpiBL, against environmental stress. In Henan Province, supplying excessive N fertilizer (or disproportionate, N/P ratio) to wheat plants results in leaf wilt when V type temperature change occurred in late seed filling period. Leaf wilt in wheat plants is one of the serious physiological diseases affecting wheat production. In general, it makes 1,000-kernel weight to decrease by about 3-7 g . However, it should be emphasized that V type temperature is the most important factor in inducing this disease. In Henan Province, temperature frequently exceeds 30°C in the first ten-day period of May, before torrential rain decreases the temperature. When it is cleaning up, the temperature rises again. This is called V type temperature. In this process, leaf wilt in wheat plants could occur, suddenly. Although this disease happens to be usually in the filling period, significant differences can, however, be detected in the middle filling period. These plants exhibit higher levels of N, lower tolerance to environmental stress and longer duration of physiological greenness. Moreover, the intermediate products of N-metabolism (such as amino acids, $NH_3$ and putrescine) accumulate excessively in late filing period. High putrescine, which is very poisonous to plants, can lead to the physiological leaf wilt, in wheat plants. 1,000 kernel weight was negatively correlated with the putrescine content in the flag leaf (Jin *et al.*1990; Zhao and Ikekawa, 1993).

## EFFECT OF EPIBL ON TOBACCO PLANTS

Han conducted field trials, on tobacco plants at Henan Agricultural University, during 1984-1990. It was noted that percent seed germination was significantly enhanced by soaking them in 0.01~ 0.05 mg/l solutions of EpiBL. The growth of leaves, roots and leaf nicotine content increased by foliar application of EpiBL (Han *et al.*, 1987; Ikekawa and Zhao, 1991). It may be emphasized here that the application of EpiBL, after transplantation, resulted in increased tolerance against environmental stress.

## EFFECT OF EPIBL ON CORN PRODUCTION

Danyu-13 is a high-yielding variety of corn, grown in Henan Province, but abortion of its ear tips seriously affects, the yield. Zhang (1987) noted that the spray of EpiBL on the foliage, prior to the emergence of the tassel, significantly decreased the kernel abortion of its ear tips and increased crop yield. This improved grain yield of corn seemed to be an expression of an increase in the weight of 1,000 grains and number of grains per ear. The statistical data for corn, obtained from all the stations, in China reported from 1986-1990, indicated at the most, 10 % increase in yield by spraying 0.01 $mgl^{-1}$ of EpiBL. Moreover, EpiBL treatment also promoted photosynthetic rate and stimulated the elongation of pollen tube (Zhang *et al.*, 1987; 1989).

## EFFECTS OF EPIBL ON RAPE PLANTS

Cold injury is one of the important factors that used to affect the growth of rape plants, in Jiangsu Province. In general, at least 15~30% of rape seedlings die due to cold injury, during winter. Jiangsu Yanchen Agricultural Institute, on the basis of their trials, suggested that EpiBL could be used to partially overcome cold injury in rape seedlings during winter. EpiBl may be useful for increasing the tolerance of rape plants to cold injury and improve the seed yield of rape. Based on the experimental findings, since 1986, it was resolved that application of EpiBL, three times prior to winter, significantly improved not only the tolerance to cold injury but also enhanced seed yield by about 10 % (unpublished data).

## EFFECT OF EPIBL ON GRAPE AND ORANGE

Kyoho is becoming an important and extensively planted grape cultivar in China. However, there is very serious fruit abscission, resulting in the reduction of yield. Moreover, under unfavorable climatic conditions, fruit abscission is one of the key factors affecting yield in the grape cultivation in Shanghai area. The spray of 0.01 mg/l of EpiBL, at flowering stage, remarkably checked premature abscission of flowers and young fruits. The treatment decreased the activity of cellulase in the abscission zone and the fruit attained maturity early (Xu *et al.*, 1994). Similarly, the group of scientist at Citrus Research Institute, Zhejiang Academy of Sciences, prevented the abscission of young orange fruits by the application of EpiBL and noted very low cellulase activity in abscission zone, like grapes (Hu *et al.*, 1990). In Sichuan province, long-term cultivation of summer oranges, which were introduced from America, resulted in decrease of yield and quality but was partially restored by the spray of EpiBL (unpublished data).

## EFFECT OF EPIBL ON BEET

Effect of EpiBL on seed germination, growth of seedling, root yield and sugar content was investigated in sugarbeet. The seed soaking in the solution of 0.04 mg/l EpiBL exhibited maximum germination. Spraying foliage with 0.04 mg/l EpiBL improved the growth of seedlings and the root, which also possessed larger quantities of sugar (unpublished data).

## EFFECT OF EPIBL ON VEGETABLE AND WATERMELON

Epibrassinolide had significant, growth-promoting effect on vegetables, such as celery, spring onion, cabbage and lettuce (Wang *et al.*, 1988; Huang and Li, 1998)). Large-scale field trials were also conducted on watermelon at all the stations, in China to implicate the EpiBL in improving the production of watermelon and to increase resistance to environmental stress.

A new brassinosteroid (TNZ303), has recently been synthesized as a representative of brassinosteroids with greater endurance (Kamuro and Takatsuto, 1999).

Cotton seeds were soaked in TNZ303 solution, overnight, before sowing. The resulting plants produced more cotton (unpublished data) because of its inhibitory effect on abscission of the balls. It may be suggested that by soaking the seeds of rape and barley in TNZ303 solution, one could also achieve an increase in seed yield. However, the grain yield in rice did not increase significantly, even though the growth of rice seedling at early stage was promoted by TNZ303. Foliage application of TNZ303 also promoted the growth of tomato and eggplant seedlings, especially under stress condition (Zhao, 2000).

However, since TNZ303 is a formulation composed of two chemicals, it is difficult to compare the effect of TNZ303 with that of EpiBL because of the synergistic effects between TS303 and PDJ (jasmonate analogue). Recently, TNZ303 has also been registered in China (trade name: Bao Min Feng). We hope that TNZ303 will exhibit its effect by increasing yield and tolerance, against environmental stresses.

## REFERENCES


Adam, G., Marquardt, V., Vorbrodt, H.M. (1991). Aspects of synthesis and bioactivity of brassinosteroids, In Brassinosteroids: Chemistry, Bioactivity and Applications, pp. 74-85. Eds H G Cutler, T Yokota and G Adam, American Chemical Society, Washington.

Chen, J.C., Wang, L.F., Zhao, Y.J. (1990). Effects of 24-epibrassinolide on growth of tobacco root explants. Acta Agriculturale Shanghai 6 (4): 89-90.

Chen, J.C., Xu, M.D., Zhao, Y.J. (1996). Effect of epibrassinolide on cell differentiation in *Arabidopsis thaliana*. Acta Phytophysiologica Sin 22 (4): 399-403.

Chen, L.P., Wang, B.L., Chen, J.C., Zhao, Y.J. (2001). Effect of epibrassinolide on the ultrastructure of epidermous cell in hypocotyl segments of mung bean seedlings. Journal of Zhejiang University (Agric &Life Sci.) 27 (4): 451-453.

Ding, W.M., Zhao, Y.J. (1995). Effect of Epi-BR on activity of peroxidase and soluble protein content of cucumber cotyledon. Acta Phytophysiologica Sin 21(3): 259-264.

Ershova, A.N., Khripach, V.A. (1996). Effect of epibrassinolide on lipid peroxidation in *Pisum sativum* at normal aeration and oxygen deficiency. Russian Journal of Plant Physiology 43:750-752.

Grove, M.D., Spencer, G.F., Rohwedder, W.K., Mandava, N.B., Worley,J.F., Warthen, J.D.Jr., Steffens,G.,L., .Flippen-Anderson,J.L., Carter Cook,J.,Jr. (1979). Brassinolide, a plant growth-promoting steroid isolated from *Brassica napus* pollen. Nature 281:216-217.

Han, J.F., Zhang, X.M., Q-Q,G., Zhao, Y.J.(1987).Studies on effects of 24-epibrassinolide on the properties and chemical component of flue-cured tobacco. Zhong Guo Yan Cao 2 :4-6.

Hu, A.S., Jiang, B.F., Guan, Y.L., Mou, H. (1990). Effects of epibrassinolide on abscission of young fruit explants and cellulase activity in abscission zone of citrus. Plant Physiology Communications 5: 24-26.

He, Y.J., Xu, R.J., Zhao, Y.J. (1995). Effect of epibrassinolide on the growth and the content of soluble saccharide and proteins of *Brassica campestris* seedlings. Plant Physiology Communications 31(1): 37-38.

He, Y.J., Xu, R.J., Zhao, Y.J. (1996). Enhancement of senescence by epibrassinolide in leaves of mung bean seedling. Acta Phytophysiologica Sin 22: 58-62.

Huang, B.C., Li, Y.C. (1998). Improvement of crop production by Tian Feng Su.I6th International Conference on Plant Growth Substances, Abstracts pp-162.

Ikekawa, N., Zhao, Y.J. (1991). Application of 24-epibrassinolide in agriculture, In Brassinosteroids: Chemistry, Bioactivity and Applications, pp. 230-291. Eds H G Cutler, T Yokota and G Adam, American Chemical Society, Washington.

Kamuro, Y., Takatsuto, S. (1999). Practical application of brassinosteroids in agricultural fields, In Brassinosteroids: Steroidal Plant Hormones, pp 223-241 Eds A Sakurai, T Yokota and S D Clouse, Springer-Verlag, Tokyo.

Luo, J., Chen, J.C., Zhao, Y.J. (1998). Brassinolide induced de-etiolated of *Arabidopsis thaliana* seedlings resembles the long-term effects of cytokinins. Australian Journal of Plant Physiology 25: 719-728.
Jin, X.C., Xu, W.S., Liu, A.F., Zhao, Y.J. (1988). The effect of epibrassinolide upon the physiological properties and yield of wheat. Acta Agriculturae Boreali Sin 3(2): 18-22.
Mandava, N.B. (1988). Plant growth-promoting brassinosteroids. Annual Review of Plant Physiology and plant Molecular Biology 39: 23-52.
Maugh II, T.H. (1981). New chemicals promise larger crops. Science 212: 33-34.
Sasse, J.M. (1997). Recent progress in brassinosteroid research. Physiologia Plantarum 100: 696-701.
Sasse, J.M. (1999). Physiological action of brassinosteroids In Brassinosteroids: Steroidal Plant Hormones, pp-137-161, Eds A Sakurai, T Yokota and S D Clouse, Springer-Verlag, Tokyo.
Shen, Z.D., Zhao, Y.J., Ding, J. (1988). Promotion effect of epibrassinolide on the elongation of wheat coleoptiles. Acta Phytophysiologica Sin 14(3): 233-237.
Takematsu, T., Takeuchi, Y. (1989). Effects of brassinosteroids on growth and yields of crops. Proceeding of Japan Academy 65(Series B): 149-152.
Wang, Y.Q., Luo, W.H., Zhao, Y.J.,Ikekawa, N. (1988). Effects of epibrassinolide on growth of celery. Plant Physiology Communications 1: 29-31.
Wang,Y.Q., Luo,W.H., Xu,R.J., Zhao,Y.J.(1994a).Effect of epibrassinolide on growth and fruit quality of watermelon. Plant Physiology Communications 30(6): 423-425.
Wang, Y.Q., Luo, W.H., Xu, R.J., Zhao, Y.J., Zhou, W.S., Huang, I.F., Shen, J.M.(1994b).Biological activity of brassinolides and relationship of structure to plant growth promoting effects. Chinese Science Bulletin, 39:1573-1577.
Wu, Y.M., Bao, Y.W., Liu, Y. (1987). Effect of epibrassinolide on formation of 1-(malonyl amino)-cyclopropane –1-carboxylic acid, 1-amino cyclopropane-1-carboxylic acid and ethylene in etiolated mung bean hypocotyl segments. Acta Phytophysiologica Sin 13 (1): 107-111.
Wu, D.R., Zhao, Y.J. (1991). Effects of epibrassinolide on endogenous IAA and its oxidase in epicotyls of mung bean seedlings. Acta Phytophysiologica Sin 17(4): 327-332.
Wu, D.R., Zhao, Y.J. (1993). Effects of epibrassinolide on the metabolism of nucleic acid in epicotyls of mung bean seedlings. Acta Phytophysiologica Sin 19(1): 49-52.
Xu, R.J., Zhao, Y.J. (1989). Effects of epibrassinolide on the activity of peroxidase and IAA oxidase in hypocotyl of cucumber seedlings. Acta Phytophysiologica Sin 15(3): 263-267.
Xu, R.J., Guo, Y.S., Zhao, Y.J. (1990). Epibrassinolide induced changes in the elongation, endogenous $GA_3$, ABA and starch content of cucumber hypocotyls. Acta Phytophysiologica Sin 16(2): 125-130.
Xu, R.J., He, Y.J., Wang, Y.Q., Zhao, Y.J (1994). Preliminary study of brassinosterone binding sites from mung bean epicotyls. Acta Phytophysiologica Sin 20(3): 298-302.
Xu, R.J, He, Y.J., Wu, D.R., Zhao, Y.J. (1995). Effect of epibrassinolide on the distribution of cAMP and activity of plasma membrane ATPase in plant tissue. Acta Phytophysiologica Sin 21(2): 143-148.
Xu, R.J., Wang, Y.Q., Wu, D.R., He, Y.J., Zhao, Y.J. (1994). Preparation of $^{125}$I-brassinolide and its biological activity. Acta Phytophysiologica Sin 20(2): 121-127.
Xu, R.J., Li, X.D., He, Y.J., Wang,Y.Q., Zhao, Y.J.(1994). Effects of treatments with epibrassinolide and chololic lactone on the fruit-set and ripening in some grape cultivation. Journal of Shanghai Agricultural college 12 (2): 90-95.
Yokoda, T., Takahashi, N. (1985). Chemistry, physiology and agricultural application of brassinolide and related steroids, In Plant Growth Substances pp 129-138. Ed. M Bopp, Springer –Verlag, Berlin.
Yokota, T. (1999). The history of brassinosteroids: discovery to isolation of biosynthesis and signal transduction mutants, In Brassinosteroids: Steroidal Plant Hormones, pp 1-20. Eds A Sakurai, T Yokota and S D Clouse, Springer-Verlag, Tokyo.
Yopp, J.H., Mandava, N.B., Sasse, J.M. (1981). Brassinolide, a growth–promoting steroidal lactone.1.Activity in selected auxin bioassays. Physiologia Plantarum 53: 445-452.
Zhao, Y.J., Luo, W.H., Wang, Y.Q., Xu, R.J.(1987). Retarding effects of epibrassinolide on maturation and senescence of hypocotyl segments of mung bean seedlings. Acta Phytophysiologica Sin 13(2): 129-135.
Zhao, Y.J., Wang, Y.Q. (1986). Physiological effects of epibrassinolide and its application in agriculture. Exploration of Nature 5(3): 133-136.
Zhao, Y.J., Xu, R.J., Luo, W.H. (1990). Inhibitory effects of abscissic acid on epibrassinolide induced senescence of detached cotyledons in cucumber seedlings. Chinese Science Bulletin 35 : 928-931.

Zhao, Y.J., Ikekawa, N. (1993). Application of 24-epibrassinolide in agriculture in China. XV International Botanical Congress, Yokohama, Japan pp112.

Zhao, Y.J., Chen, J.C., Zheng, H.Q. (1995). Hormonal regulation of cell differentiation and growth in *Arabidopsis thaliana*. 15th International Conference on Plant Growth Substances, Minnesota USA, Abstract 211.

Zhao, Y.J. (2000). Application of 24-epibrassinolide and TNZ303 in Agriculture in China. Plant Biology 2000 program. July 2000. The Annual Meeting of the American Society of Plant Physiologists, San Diego USA, pp141.

Zhang, X.M., Ren, H.P., Chen, Z.K., Zhao, Y.J. (1987). A preliminary study on the effect of spraying 24-epibrassinolide on kernel abortion of corn ear tip. Acta Agriculturae Universitatis Henanensts 21(1): 56-64.

Zhang, X.M., Ren, H.P., Chen, Z.K., Zhao, Y.J. (1989). Effects of epibrassinolide on the development of corn ear. Plant Physiology Communications 5 : 42-43.

Zheng, H.Q., Chen, J.C., Zhao, Y.J., Xu, Z.H. (1996). Selection of brassinolide insensitive mutant of *Arabidopsis thaliana*. Acta Phytophysiologica Sin 23(3): 293-298.

CHAPTER 8

MARCO ANTÓNIO TEIXEIRA ZULLO AND MARIANGELA DE BURGOS MARTINS DE AZEVEDO

# BRASSINOSTEROIDS AND BRASSINOSTEROID ANALOGUES INCLUSION COMPLEXES IN CYCLODEXTRINS

A new approach for enhancing the biological activity of a brassinosteroid, involving its administration as a guest in an inclusion complex of a plant growth inactive compound, is described. The method of choice of the host molecule, the preparation of brassinosteroid or brassinosteroid analogue inclusion complexes in cyclodextrins, their physical characterisation and an example of their biological activity are presented. The biological activity elicited by 24-epibrassinolide/ β-cyclodextrin inclusion complex is at least equivalent to that by 24-epibrassinolide itself.

## INTRODUCTION

The brassinosteroids, the new class of plant hormones discovered after the isolation of brassinolide (**1**, Figure 1; Grove *et al.*, 1979), potentially useful compounds in agricultural practices, accelerate maturation of crops, increase crop yield, or diminish abiotic stress effects (Khripach *et al.*, 1999, 2000). Similar compounds, the brassinosteroid analogues, prepared either chemically or biochemically, may also have the same effects as natural brassinosteroids (Adam *et al.*, 1991; Núñez Vázquez and Robaina Rodríguez, 2000). Several biological assays, developed for testing the activity of other classes of plant hormones, have been adapted for assaying brassinolide or brassinosteroid activity (Mandava *et al.*, 1981; Yopp *et al.*, 1981; Takatsuto, 1994). In many instances the observed brassinosteroid activity, in a given bioassay was not observed in greenhouse or field experiments, due to differences in physiological state of development of the individual plant at the time of application, what led to the development of brassinosteroid analogues with long-lasting activity (Kamuro and Takatsuto, 1999). These compounds are characterized by the possibility of being converted by the plant, during a relatively long lapse of time, to the active compound. Another possibility for improving the bioactivity of a brassinosteroid is its co-application with another plant growth regulator, usually an auxin (Takeno and Pharis, 1982). An entirely new approach for enhancing the bioactivity of a brassinosteroid, but not exactly for extending the lapse of this activity, is described here, by which the

*S.Hayat and A. Ahmad (eds.), Brassinosteroids*, 171-188.

brassinosteroid is introduced to the plant carried into an inclusion complex of a plant growth inactive compound, like the cyclodextrins.

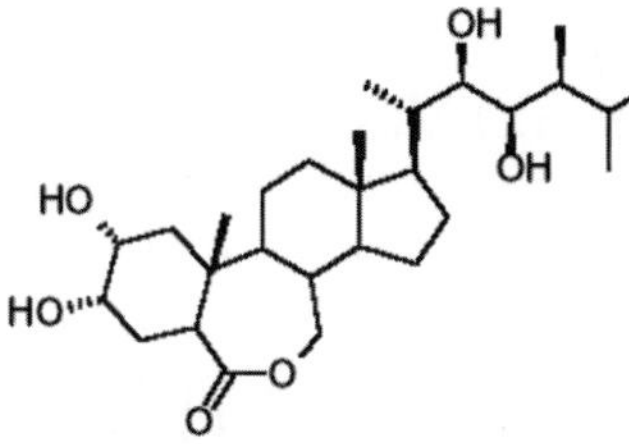

1

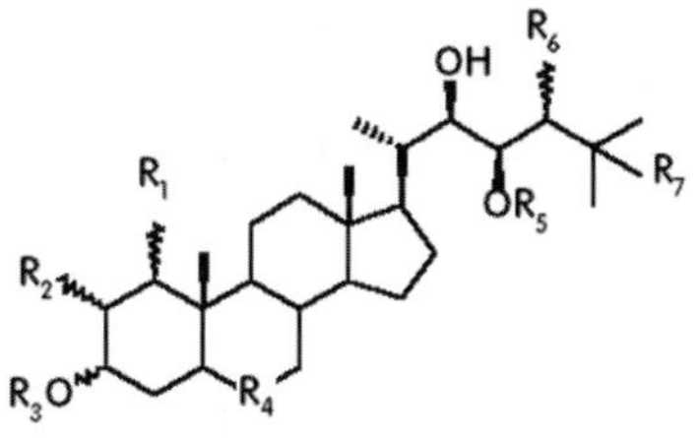

2 $R_1$ = H, α-OH or β-OH
$R_2$ = H, α-OH or β-OH
$R_3$ = H, n-$CH_3(CH_2)_{10}CO$, n-$CH_3(CH_2)_{12}CO$ or β-D-glucopyranosyl
$R_4$ = $CH_2$, CHOH, CO or $CO_2$
$R_5$ = H or β-D-glucopyranosyl
$R_6$ = H, α-Me, β-Me, α-Et, $CH_2$ or (E)-$CHCH_3$
$R_7$ = H or Me

*Figure 1 – Brassinolide (1) and general formula for natural brassinosteroids (2).*

## THEORETICAL CONSIDERATIONS

The natural brassinosteroids, with general structural formula **2** (Figure 1), are lipophylic compounds, characterized by a 5α-cholestane skeleton oxygenated at least at carbons 3, 22 and 23 (Zullo *et al.*, 2003). The most active brassinosteroids are also oxygenated at carbon 6, as a ketone or a 6-oxo-7-oxa-lactone (Zullo and Adam, 2002). The vicinal hydroxyls in the side chain or, in some brassinosteroids, in the ring A, are plausible points of attachment to an enzyme receptor site or for complexation to a suitable host molecule.

Preliminary calculations using the CHARMM algorithm (Brooks *et al.*, 1983) allowed to know the approximate geometry of a generic set of brassinosteroids, assuming conformations with a fully extended side chain. In such conformations, brassinosteroids (see structure **3** in Figure 2) measure ca. 18 Å from the hydrogens of the ring A hydroxyls to the most distant of the hydrogens at carbons 26 or 27, with a side chain length of ca. 8 Å between the most distant hydrogens at carbons 21 and 26 or 27 and a skeleton length of ca. 11.5 Å measured from the hydrogens of the ring A hydroxyls to the hydrogens at carbon 16. The maximum ring A width is about 4.9 Å, measured from $H_{1\beta}$ to $H_{4\alpha}$, the distance between the carbonyl oxygen to $H_{11\alpha}$ ranges from 6.1 Å for ketones to 6.4 Å for lactones and the distance between $H_{21}$ and $H_{15\alpha}$ is about 6.7 Å. The height of the steroid nucleus is of the order of 4.3 Å, as well as between the most distant of the hydrogens in carbons 26 and 27. Hence the host molecule should have a size able to accommodate at least part of the guest brassinosteroid in an inclusion complex (Figure 2).

The cyclodextrins were thought to be candidates for such host molecules. These compounds are toroidally cyclic oligomers constituted of 6 to 12 $\alpha(1\rightarrow4)$ linked D-glucose units and the hexa-, hepta- and octaoligomers are named α-, β- and γ-cyclodextrin, respectively (Saenger, 1980). These oligosaccharides have a slight conical form whose cavity is surrounded by the glucose units in chair conformation, with the narrow opening containing the C-6 hydroxyls and the wide opening the C-2 and C-3 hydroxyl groups. Due to the hydrophilic character of their external surface and the lipophylic characteristics of their cavities, they are able to act as hosts in inclusion complexes for allowing the solubilisation of many organic compounds in aqueous solutions. The height of the cyclodextrins is ca. 8 Å, and the internal diameter of their cavities range from ca. 5 Å for α-cyclodextrin to ca. 8 Å for γ-cyclodextrin, while the external diameter ranges from ca. 14.6 Å for α -cyclodextrin to 17.5 Å for γ-cyclodextrin (see structure **4** in Figure 2).

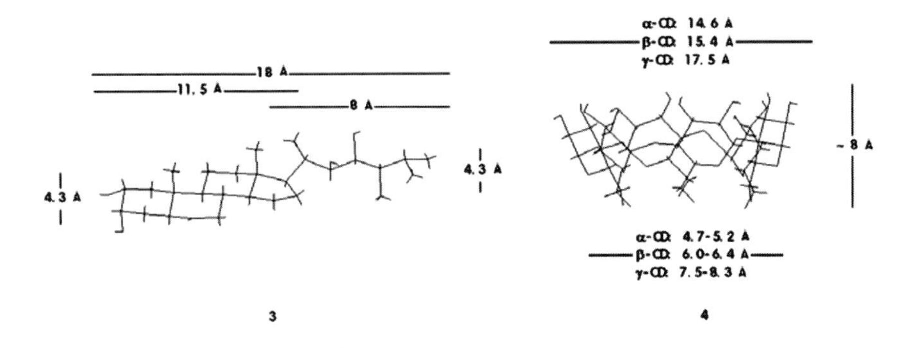


*Figure 2 – Dimensions of a typical brassinosteroid (3) and of cyclodextrins (4).*

The cyclodextrins are well known complexing agents widely used in the food and pharmaceutical industry (Stella and Rajewski, 1997), and have been used for the complexation of, among others, steroid hormones (Alberts and Muller, 1992; Bednarek *et al.*, 2002; Cavalli *et al.*, 1999; Marzona *et al.*, 1992) and plant growth regulators (Szejtli *et al.*, 1989; Brutti *et al.*, 2000). They can form true inclusion complexes with another compound, when the later is enclosed inside the cyclodextrin cavity, or only aggregates, when the complexation occurs at the outer surface of the cyclodextrin molecule. In the case of the true inclusion complexes the law of constant proportions do apply, and their formation can be represented by the equation (1):

$$CD + G \rightarrow CD/G \quad (1)$$

where CD is the cyclodextrin, G the guest compound, and CD/G is the inclusion complex, and the energy change $\Delta E_{CD/G}$ associated with the complex formation is given by the relation (2):

$$\Delta E_{CD/G} = E\ (CD/G) - [E\ (CD) + E\ (G)] \quad (2)$$

where E(CD/G), E(G) and E(CD) are the total energies of the complex, of the isolated guest (G) and of the isolated host (CD), respectively.

For testing the viability of forming the inclusion complexes between cyclodextrins (hosts) and brassinosteroids (guests), a molecular mechanics study of the complex structures between β-cyclodextrin (**5**) and 24-epibrassinolide (**6**) or 28-homocastasterone (**7**) was performed using the MM+ force field and the HyperChem software. The brassinosteroid/ β-cyclodextrin interactions were evaluated by taking into account the energy involved in this process (Zhou *et al.*, 2000). No cut-off was used and geometry optimisation was carried out until the energy gradient was less than 0.01 kcal $mol^{-1}$ $Å^{-1}$, for both isolated and complex structures. Among the several possible orientations for the 1:1 complexes of 24-epibrassinolide (**6**) or 28-homocastasterone (**7**) and β-cyclodextrin (**5**), those involving insertion of the A/B rings (*rAB* conformations), the side chain (*SC* conformations), the C/D rings (*rCD* conformations) and ring D (*rD* conformations) into the cyclodextrin cavity were considered. The energies of the possible complexes, resulting from the insertion of one brassinosteroid molecule into two cyclodextrin cavities (*rAB+SC* conformations) were also calculated. In this case equations (1) and (2) were adjusted, respectively, to equations (3) and (4):

$$2CD + G \rightarrow CD/G \quad (3)$$

$$\Delta E_{CD/G} = E\ (CD/G) - [2E(CD) + E\ (G)] \quad (4)$$

The structures of compounds **5**, **6** and **7** and the optimised structures of the inclusion complexes **8** and **9** are given in Figure 3 together with the notation used to designate the different conformations of 1:1 inclusion complexes, and the computed energy changes associated with the inclusion phenomenon are presented in Table 1.

A negative energy change ($\Delta E_{CD/G}$) indicates that the inclusion phenomenon might result in a new species more stable than the isolated host and guest(s). By this thermodynamic criterion, in all cases studied the complex formation between the brassinosteroid and β-cyclodextrin is favoured, as judged by the data shown in Table 1. This stabilisation may be due to the establishment of hydrogen bonds between carbonyl or hydroxyl functions of the guest with hydroxyl or other functions of the host(s) molecules. For 1:1 complexes, insertion of rings C or D (conformations *rCD* or *rD*, respectively) into the β-cyclodextrin cavity causes an energy change in the range of 21-24 kcal.mol$^{-1}$ for both complexes of 24-epibrassinolide (**6**) or 28-homocastasterone (**7**).

*Table 1. Total energies for the isolated and inclusion complexes between β-cyclodextrin (5) and 24-epibrassinolide (6) or 28-homocastasterone (7) and energy changes associated with the inclusion phenomena.*

| *Species* | $E_{total}$ *(kcal mol$^{-1}$)* | $\Delta E_{CD/G}$ *(kcal mol$^{-1}$)* |
|---|---|---|
| β-Cyclodextrin (5) | 83.94 [a] | |
| 24-Epibrassinolide (6) | 99.13 [a] | |
| 28-Homocastasterone (7) | 70.95 [a] | |
| 1:1 Complexes between β-cyclodextrin (5) and 24-epibrassinolide (6) | | |
| *rAB* conformation | 160.95 [a] | -22.12 |
| *rCD* conformation | 159.84 [a] | -23.23 |
| *rD* conformation | 158.89 [a] | -24.18 |
| *SC* conformation | 154.92 [a] | -28.15 |
| 1:1 Complexes between β-cyclodextrin (5) and 28-homocastasterone (7) | | |
| *rAB* conformation | 145.97 | -8.92 |
| *rCD* conformation | 133.75 | -21.14 |
| *rD* conformation | 130.27 | -24.62 |
| *SC* conformation | 136.31 | -18.58 |
| *rAB+SC* complexes between β-cyclodextrin (5) and brassinosteroids | | |
| 24-Epibrassinolide (6) | 218.65 [a] | -48.36 |
| 28-Homocastasterone (7) | 192.99 | -45.84 |

[a] data taken from De Azevedo *et al.*, 2002

For inclusion of the ring A into the β-cyclodextrin cavity there is a striking difference between the predicted energy changes for the complex formation, which is of –22.12 kcal.mol$^{-1}$ for the *rAB* conformation of the 24-epibrassinolide/β-cyclodextrin inclusion complex **8** and of –8.92 kcal.mol$^{-1}$ for the *rAB* conformation of the 28-homocastasterone/β-cyclodextrin inclusion complex **9**. This difference may be accounted for the presence of the more polar 6-oxo-7-oxa lactone function in 24-epibrassinolide (**6**) in contrast to the 6-keto function in 28-homocastasterone (**7**), what would allow the formation of a higher number of hydrogen bonds between 24-epibrassinolide (**6**) and β-cyclodextrin (**5**) than between 28-homocastasterone (**7**) and β-cyclodextrin (**5**). For the formation of complexes of *SC* conformation there is no marked difference between the energy changes when the host is 24-epibrassinolide (**6**) or 28-homocastasterone (**7**). The bulky substituents at C-24 of 28-homocastasterone (**7**), where the distance between the most distant hydrogens in carbons 26 and 29 is ca.

6.3 Å, is of about the same size of the diameter of the β-cyclodextrin cavity making more difficult the complexation of their side chain hydroxyls to β-cyclodextrin (**5**) than those of 24-epibrassinolide (**6**), where the distance between the most distant hydrogens in carbons 26 and 28 is ca. 5.5 Å. The same reasoning can be applied for explaining the relative stabilities of the hypothetical *rAB+SC* conformations of the inclusion complexes of 24-epibrassinolide (**6**) and 28-homocastasterone (**7**). One should note that the energy changes for the formation of these last complexes are not equal to the sum of the energy changes for the formation of complexes of *rAB* and *SC* conformations. Although they are predicted to be more stable than the preceding 1:1 complexes, there is no evidence that they shall be formed. For these reasons it is predicted that, for the 1:1 complexes, in the 24-epibrassinolide (**6**) series the *SC* conformation is the most stable, while in the 28-homocastasterone (**7**) series the most stable is the *rD* conformation.

## PREPARATION OF INCLUSION COMPLEXES OF BRASSINOSTEROIDS IN CYCLODEXTRINS

For the preparation of the inclusion complexes of 24-epibrassinolide (**6**) or 28-homocastasterone (**7**) in β-cyclodextrin (**5**) acetone solutions of the brassinosteroids were mixed to equimolecular aqueous solutions of the oligosaccharide, refluxed and concentrated to dryness (De Azevedo *et al.*, 1999, 2000, 2001a). The resulting powders were submitted to differential scanning calorimetry, X-ray powder analysis, scanning electron micrography and $^1$H nuclear magnetic resonance spectrometry.

## CHARACTERISATION OF BRASSINOSTEROID/β-CYCLODEXTRIN INCLUSION COMPLEXES

### *Differential scanning calorimetry*

The differential scanning calorimetry profile of β-cyclodextrin (**5**) shows a broad endothermic peak, ranging from about 105 °C to 150 °C (Figure 4E), which was attributed to the release of water molecules entrapped inside the cyclodextrin cavity. 24-Epibrassinolide (**6**) showed a broad endothermic peak at 250.6 °C, corresponding to its melting point (Figure 4A), while the 24-epibrassinolide/β-cyclodextrin thermogram (Figure 4B) showed a pronounced reduction in this peak area (De Azevedo et al., 2002). The thermogram of 28-homocastasterone (**7**) presented a narrow endothermic peak at 236.9 °C, attributed to the melting point of the sample (Figure 4C), and the thermogram of its complex with β-cyclodextrin (Figure 4D) showed also a substantial reduction in the peak area corresponding to the brassinosteroid melting point. The narrower, slightly distorted endothermic peaks corresponding to the cyclodextrin in the complexes may be due to the exchange of water molecules with a brassinosteroid molecule as a result of complex formation, what is corroborated by the marked reduction in the peak areas corresponding to the melting points of the brassinosteroids, which implies that the molecular arrangement of the brassinosteroids in the complexes are different from the pure compounds in their own crystals.

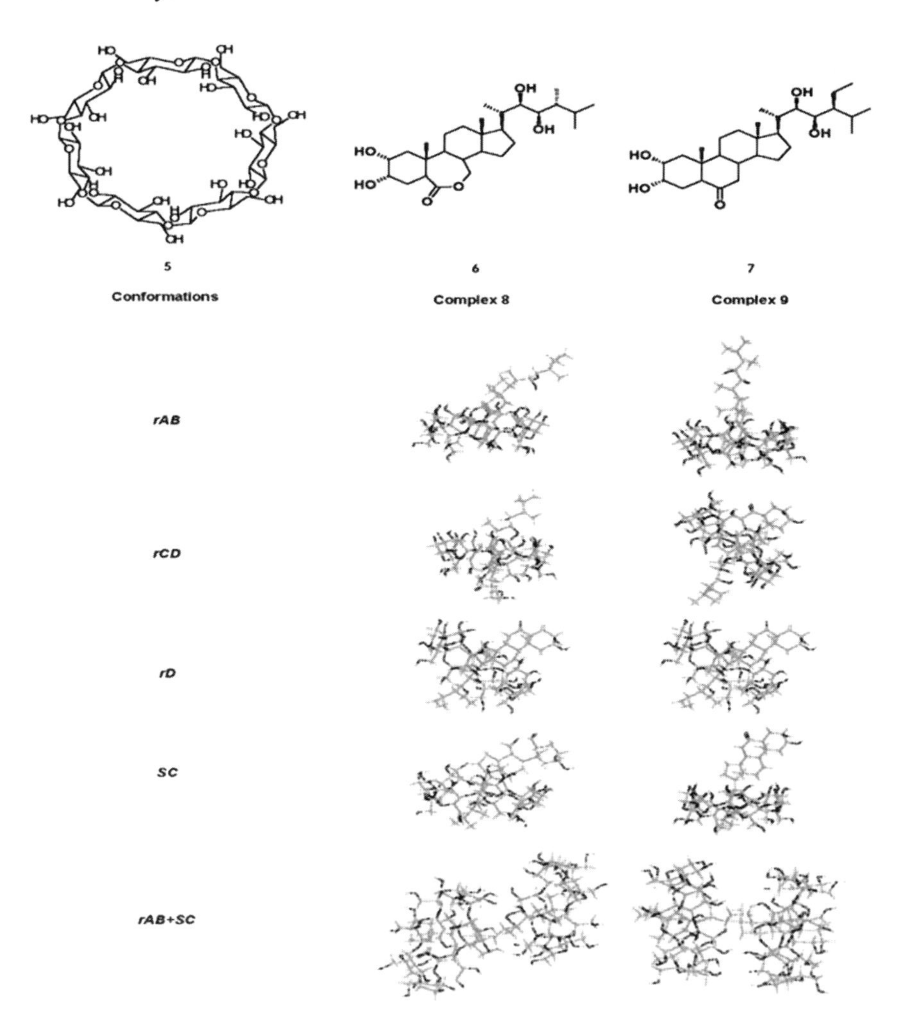


*Figure 3. Structures of β-cyclodextrin (**5**), 24-epibrassinolide (**6**), 28-homocastasterone (7) and of their inclusion complexes **8** and **9**, respectively, in β-cyclodextrin (**5**).*

*X-ray powder diffraction*

The X-ray powder diffraction pattern of the 24- epibrassinolide/β-cyclodextrin complex (**8**, Figure 5B; De Azevedo et al., 2002) is not a simple superposition of the patterns for 24-epibrassinolide (**6**, Figure 5A) and β-cyclodextrin (**5**, Figure 5E) and indicates crystalline properties with two new crystalline peaks at $2\theta = 11.5°$ and 24°. The neat crystalline peak at $2\theta = 8°$ observed in the X-ray powder diffraction pattern of 24-epibrassinolide (**6**, Figure 5A) is clearly absent in that of the complex **8** (Figure 5B). The X-ray powder diffraction pattern of the 28-homocastasterone/β-cyclodextrin complex (**9**, Figure 5D) showed changes in signs in the region $2\theta = 9°$ to 14° that cannot be explained by the superposition of the diffraction patterns shown by 28-homocastasterone (**7**, Figure 5C) and β-cyclodextrin (**5**, Figure 5E). The interferences in the X-ray diffraction patterns of the inclusion complexes were attributed to the insertion of the brassinosteroids **6** or **7** into the β-cyclodextrin (**5**) cavity, by similarity with other β-cyclodextrin inclusion complexes (Rajagopalan *et al.*, 1986).

*Scanning electron microscopy*

The morphological analyses of the compounds were performed by scanning electron micrography. The 24-epibrassinolide (**6**) crystals showed an elongated feature with a dimension ranging from a few micrometers up to 150-200 μm (Figure 6A). The scanning electron micrography of the crystals of the 24-epibrassinolide/β-cyclodextrin inclusion complex (**8**) appeared as platelets (Figure 6B), with no crystals of either 24-epibrassinolide (**6**, Figure 6A) or β-cyclodextrin (**5**, Figure 6F; De Azevedo *et al.*, 2002). The crystals of 28-homocastasterone (**7**) showed a lamellar shape (Figure 6C), while those of its inclusion complex **9** appeared as irregular platelets (Figure 6D), different from either the crystals of 28-homocastasterone (**7**), β-cyclodextrin (**5**) or the mere physical mixture of both (Figure 6E).

*Nuclear magnetic resonance spectroscopy*

The results of differential scanning calorimetry, X-ray powder diffraction and scanning electron microscopy analyses of compounds **8** and **9** are thus indicative that they are inclusion complexes of brassinosteroids in β-cyclodextrin, although they cannot reveal what is or are the conformation(s).

The $^1$H nuclear magnetic resonance spectrum of the 24-epibrassinolide/β-cyclodextrin inclusion complex (**8**, Figure 7B), clearly shows the absence of water molecules inside the β-cyclodextrin (**5**) cavity, is observed as a broad band in the region of δ 4.6-4.2 ppm in the spectrum of the oligosaccharide (**5**, Figure 7C). The upfield shifts of hydrogens linked to C-17, C-20, C-21, C-24, C-25, C-26 and C-27 of the brassinosteroids when the $^1$H nuclear magnetic resonance spectrum of the 24-epibrassinolide/β-cyclodextrin inclusion complex (**8**, Figure 7B) is compared with that of 24-epibrassinolide (**6**, Figure 7A) suggests that the side chain of the brassinosteroid is inserted into the β-cyclodextrin (**5**) cavity, what agrees with the molecular mechanics calculations and the absence of water protons in the spectrum of complex **8** (De Azevedo *et al.*, 2002).

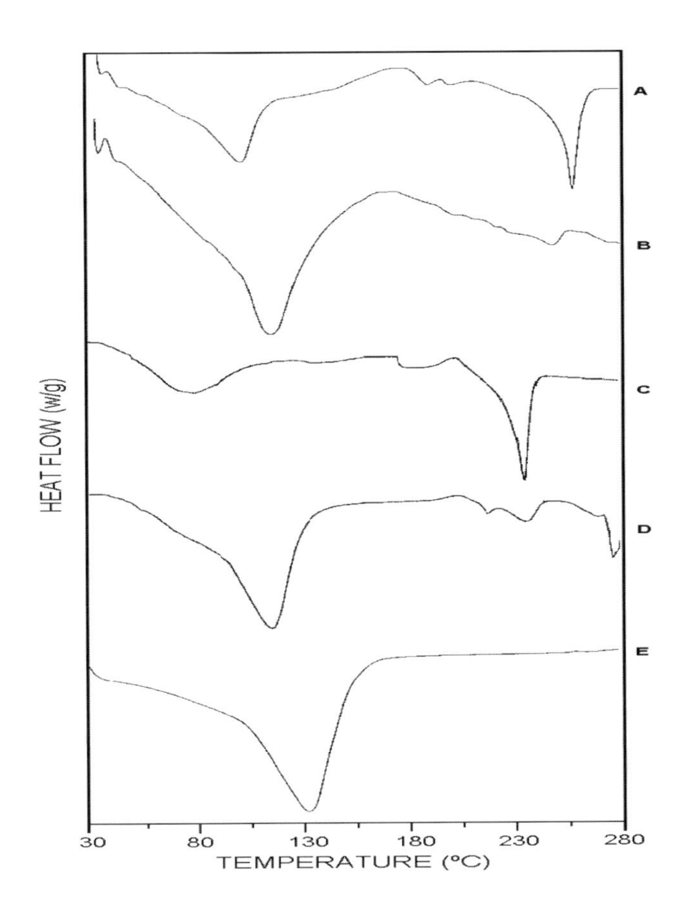


*Figure 4. Differential scanning calorimetry analysis of 24-epibrassinolide (**6**, A), 24-epibrassinolide/β-cyclodextrin inclusion complex (**8**, B), 28-homocastasterone (7, C), 28-homocastasterone/β-cyclodextrin inclusion complex (**9**, D) and β-cyclodextrin (**5**, E).*

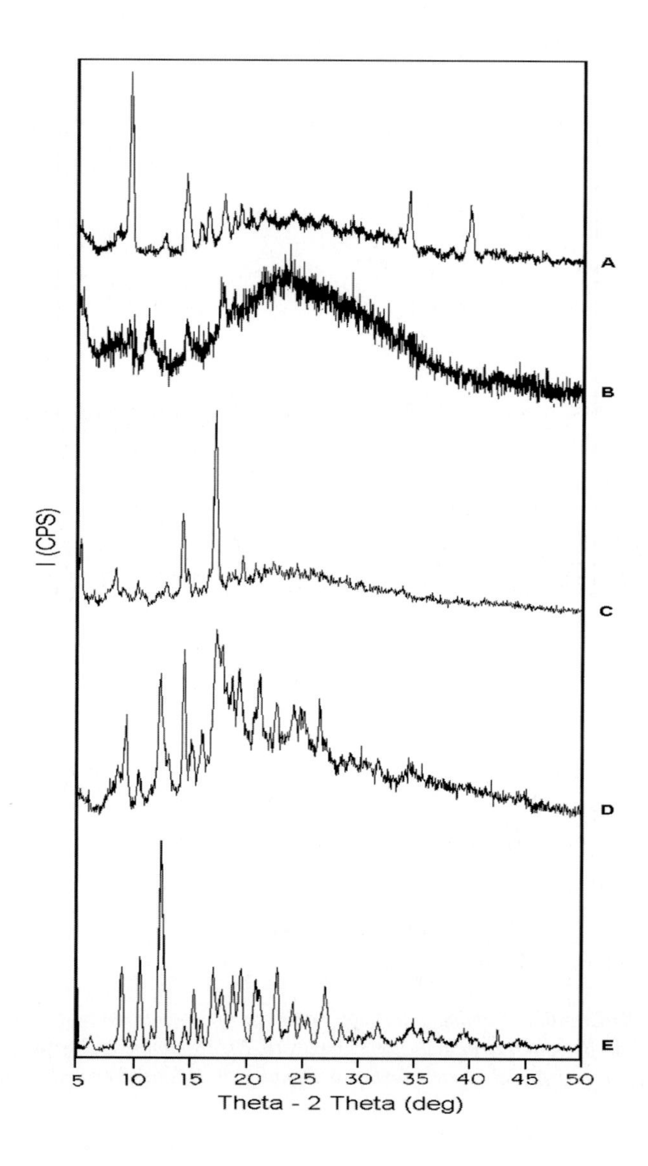


*Figure 5. X-Ray powder diffraction of 24-epibrassinolide (**6**, A), 24-epibrassinolide/β-cyclodextrin inclusion complex (**8**, B), 28-homocastasterone (**7**, C), 28-homocastasterone/β-cyclodextrin inclusion complex (**9**, D) and β-cyclodextrin (**5**, E).*

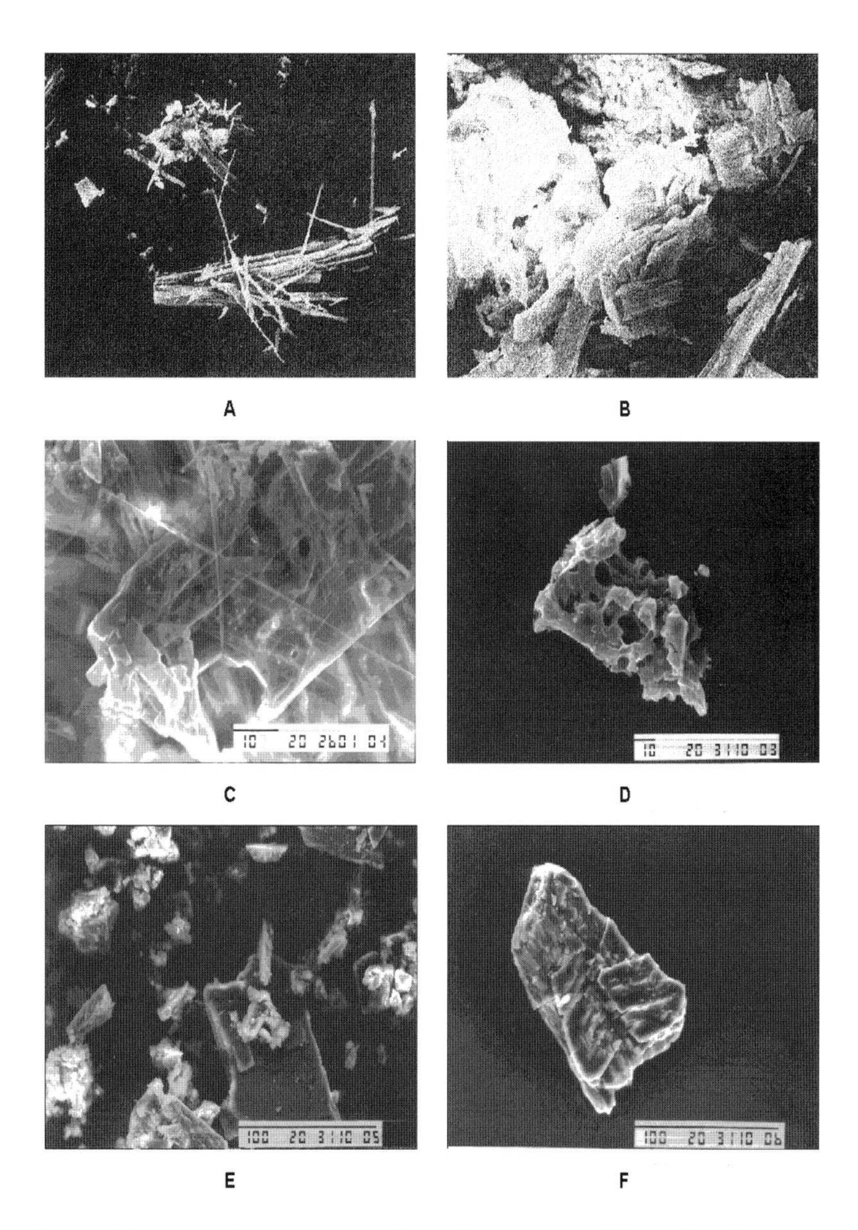


*Figure 6. Scanning electron micrographies of 24-epibrassinolide (**6**, A) and its inclusion complex in β-cyclodextrin (**8**, B), 28-homocastasterone (7, C), its inclusion complex in β-cyclodextrin (**9**, D) and its physical mixture with β-cyclodextrin (E), and β-cyclodextrin(**5**, F).*

## INCLUSION COMPLEXES OF BRASSINOSTEROID ANALOGUES IN CYCLODEXTRINS

As the results of molecular mechanics calculations predict that brassinosteroid A/B rings insertion into the β-cyclodextrin cavity is possible, but the two early models (24-epibrassinolide, **6**, and 28-homocastasterone, **7**) possessed a side chain whose insertion into the saccharide was more favourable than that of the A/B rings, it was tried to verify if a suitable brassinosteroid analogue could complex with β-cyclodextrin in the *rAB* conformation.

One of the analogues chosen for testing this possibility was 2α, 3α-dihydroxy-5α-spirostan-6-one (**10**, Figure 8), the spirostanic analogue of castasterone, which presents 2 unprotected hydroxyl groups at the ring A, what confers some hydrophilic character to its molecule. This structural feature would allow it to be complexed into cyclodextrins. Preliminary molecular model inspection of compound **10** shows that the longest distance between 2 atoms in the ring A ($H_{1\beta}$-$H_{4\alpha}$) would be 5.08 Å, while the smallest cavity in a cyclodextrin (that of α- cyclodextrin) would be of *ca.* 5 Å, and the distances $O_{2\alpha}$-$O_{3\alpha}$ would be 2.48 Å and $O_{3\alpha}$-$O_6$ would be 4.65 Å.

Inclusion of the brassinosteroid analogue **10** into β-cyclodextrin was attempted by the same procedure outlined above (De Azevedo *et al.*, 2001b). The infrared spectrum of product **11** (Figure 8C) revealed that the carbonyl group at ring B is not involved in the complexation, as no shift in the absorption, due to stretching of the carbonyl group, compared to that of compound **10** (Figure 8B) was observed ($\nu_{C=O}$ 1706.9 $cm^{-1}$ for both species). Peaks related to OH vibrations in compound **10** (Figure 8B) are overlapped by those due to the hydroxyls of β-cyclodextrin (**5**, Figure 8A) in product **11** (Figure 8C). The only significant difference in the infrared spectra occurs in the region 2950-2875 $cm^{-1}$, where the peak at 2925.8 $cm^{-1}$ observed for C-H stretching vibrations of β-cyclodextrin (Figure 8A) shifts to 2929.7 $cm^{-1}$ for the product (Figure 8C). Rings E and F of the analogue are also not involved in complexation, as no frequency shift was observed in the fingerprint region of the spectra of product **11**, compared to that of compound **10** (the peaks in this region for product **11** are almost a superposition of peaks due to the cyclodextrin **5** or the brassinosteroid analogue **10**, and are not shown in Figure 8).

As these data appeared insufficient to conclude if complexation has occurred, scanning electron microscopy was used for verification, since this technique proved to be a useful tool in the characterization of cyclodextrin complexes of natural brassinosteroids. Crystals of the spirostanic derivative **10** appear as agglomerates of irregularly shaped blades (Figure 8E). Crystals of product **11** have a quite different appearance, consisting of aggregates of small particles, formed by compact, transparent, glassy crystals, as shown in Figure 8F, differing in appearance, from the β-cyclodextrin crystals (Figure 8D), confirming the suppositions that the complexation of the brassinosteroid analogue **10** with β-cyclodextrin (**5**) has occurred and *rAB* conformations for some brassinosteroids or analogues inclusion complexes in cyclodextrins are possible.

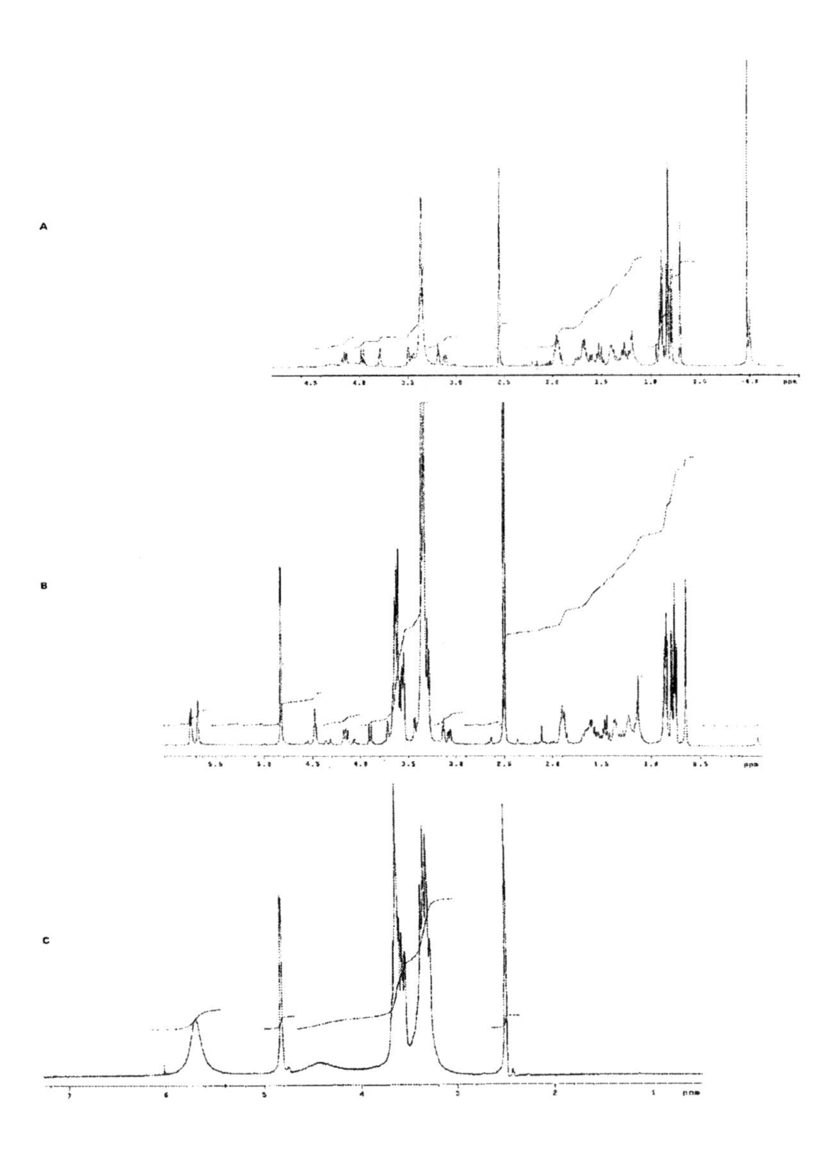


*Figure 7.* *[1]H Nuclear magnetic resonance spectra of 24-epibrassinolide (**6**, A), 24-epibrassinolide/β-cyclodextrin inclusion complex (**8**, B) and β-cyclodextrin (**5**, C), taken in DMSO-$d_6$.*

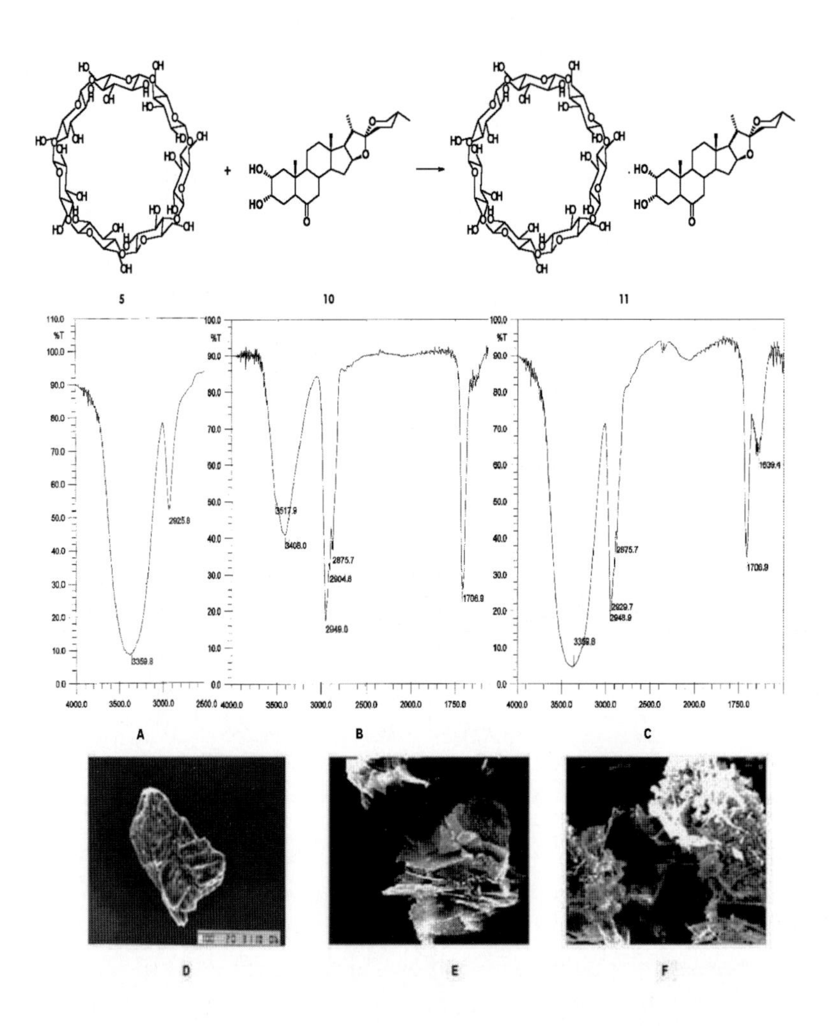


*Figure 8. Structures, partial infrared spectra and scanning electron micrographies of β-cyclodextrin (**5**, A and D), 2α,3α-dihydroxy-5α-spirostane-6-one (**10**, B and E) and its inclusion complex in β-cyclodextrin **11** (C and F).*

## BIOLOGICAL ACTIVITY OF 24-EPIBRASSINOLIDE/β-CYCLODEXTRIN INCLUSION COMPLEX (8)

The biological activities of 24-epibrassinolide (**6**) or that of the 24-epibrassinolide/β-cyclodextrin inclusion complex (**8**) were evaluated by the rice lamina inclination assay, as proposed by Wada *et al.*, 1981 (De Azevedo *et al.*, 1999, 2000, 2001c, 2002), employing seedlings of arbitrarily chosen rice cultivars. The test solutions were prepared from ethanol: water 1:1 (v/v) stock solutions of 24-epibrassinolide (**6**) or 24-epibrassinolide/β-cyclodextrin inclusion complex (**8**), both at a concentration of 500 mg $l^{-1}$.

The analysis of variance of the data shown in Table 2 for each cultivar reveals that only for cultivar DD-91 the effect of treatments on the lamina inclination angle was not significant. For the other six cultivars the lamina inclination angle was a function of 24-epibrassinolide (**6**) or complex **8** concentration, and only for cultivars Koshi-hikari and IAC-103 was a function also on how 24-epibrassinolide (**6**) was administered (neat or complexed). Except for cultivars Arborio and IAC-202, the rice lamina inclination angle was significantly dependent on the interaction between how 24-epibrassinolide (**6**) was administered and the concentration, in the assaying media.

*Table 2. Angles of lamina inclination of rice seedlings treated with 24-epibrassinolide (6) or 24-epibrassinolide/β-cyclodextrin inclusion complex (8).*

| *Concentration* | *Cultivars* | | | | | | |
|---|---|---|---|---|---|---|---|
| *(mg $l^{-1}$)* [a] | *Arborio* | *Cypress* | *DD-91* | *Koshi-hikari* | *IAC-103* | *IAC-104* | *IAC-202* |
| | | | | Control | | | |
| 0 | 56.4 | 37.0 | 2.5 | 2.5 | 109.0 | 26.7 | 12.0 |
| | | | 24-Epibrassinolide (**5**) | | | | |
| $10^{0}$ | 150.4 | 150.2 | 0.0 | 4.8 | 153.7 | 152.0 | 38.8 |
| $10^{-1}$ | 114.9 | 131.5 | 4.3 | 5.0 | 145.8 | 119.0 | 10.5 |
| $10^{-2}$ | 57.1 | 45.8 | 0.5 | 0.0 | 75.5 | 104.2 | 21.3 |
| $10^{-3}$ | n. a. [b] | 68.7 | 2.0 | 11.0 | 114.6 | 63.3 | 14.8 |
| $10^{-4}$ | n. a. | 46.6 | 3.3 | 0.0 | 90.2 | 70.0 | 12.3 |
| | 24-epibrassinolide/β-cyclodextrin inclusion complex (**8**) | | | | | | |
| $10^{0}$ | 145.7 | 136.2 | 0.0 | 17.6 | 165.0 | 173.2 | 49.0 |
| $10^{-1}$ | 130.0 | 99.3 | 3.3 | 4.6 | 151.1 | 157.1 | 14.3 |
| $10^{-2}$ | 70.4 | 75.2 | 3.5 | 10.3 | 106.0 | 50.5 | 16.0 |
| $10^{-3}$ | n. a. | 70.0 | 0.0 | 7.5 | 139.2 | 64.3 | 23.3 |
| $10^{-4}$ | n. a. | 74.2 | 0.0 | 8.5 | 113.8 | 56.7 | 6.0 |
| s | 29.0 | 32.1 | 4.2 | 9.4 | 15.1 | 16.6 | 29.1 |

[a] final concentration of 24-epibrassinolide (**6**); [b] not assayed

An examination of the data shown in table 2, reveals that the rice lamina inclination angles produced by the administration of 24-epibrassinolide/β-cyclodextrin inclusion complex (**8**) are at least statistically equivalent to those produced by 24-epibrassinolide (**6**) at the same concentration, and in many instances the inclination was more pronounced than that produced by administration of 24-epibrassinolide (**6**). At some occasions the inclination observed by the administration of complex **8** was equivalent to that produced by the brassinosteroid **6** at higher concentration. The answers were better for cultivars highly responsive to the rice lamina inclination assay, as expected. For the worst responsive cultivars, such as Koshi-hikari and DD-91, the administration of the complex **8** could discriminate between brassinosteroid responsive and non-responsive materials, since for the brassinosteroid non-responsive cultivar DD-91 the response to administration of either 24-epibrassinolide (**6**) or the complex **8** was arbitrary and inside the experimental error regardless the concentration, while for the responsive cultivar Koshi-hikari the inclination angles were a function of concentration of complex **8**, within the experimental error.

The better answers produced by the complex **8** compared with that of the brassinosteroid **6** can be due either to a facilitated entry of the brassinosteroid into the cell when carried by the cyclodextrin or to a smooth disponibilisation of the brassinosteroid in the cell, through the enzymatic hydrolysis of the oligosaccharide or the displacement by an endogenous compound producing a more stable new inclusion complex.

## ACKNOWLEDGEMENTS

Acknowledgements are due to Dr. Joel Bernabé Alderete Trivinos (Department of Organic Chemistry, Universidad de Concepción, Chile), for the MM+ force field calculations, Dr. Nelson Eduardo Durán Caballero (Laboratory of Biological Chemistry, Universidade Estadual de Campinas, Brazil), for spectral data, Mr. Marcelo M. M. de Azevedo (Laboratory of Biological Chemistry, Universidade Estadual de Campinas, Brazil), for the scanning electron micrographies, Dr. Lúcia Helena Signori Melo de Castro, for supplying rice seeds, and Dr. Nobuo Ikekawa (Niigata College of Pharmacy, Niigata, Japan) and Dr. Gunter Adam (Institute of Plant Biochemistry, Halle, Germany), for samples of 24-epibrassinolide. Support from Fundação de Amparo à Pesquisa do Estado de São Paulo (grants 1999/05119-7 and 1999/07907-2) is gratefully acknowledged.

## REFERENCES

Adam, G., Marquardt, V., Vorbrodt, H.M., Hörhold, C., Andreas, W., Gartz, J. (1991). Aspects of Synthesis and Bioactivity of Brassinosteroids. In Brassinosteroids: Chemistry, Bioactivity and Applications, pp. 74-85. Eds H G Cutler, T Yokota and G Adam. American Chemical Society, Washington.

Alberts, E., Muller, B. W. (1992). Complexation of steroid-hormones with cyclodextrin derivatives. Substituent effects of the guest molecule on solubility and stability in aqueous-solution. Journal of Pharmaceutical Sciences 81:756-761.

Bednarek, E., Bocian, W., Poznanski, J., Sitkowski, J., Sadlej-Sosnowska, N., Kozerski, L. (2002). Complexation of steroid hormones: prednisolone, ethinyloestradiol and estriol with beta-cyclodextrin. Journal of the Chemical Society-Perkin Transactions 2, 999-1004.

Brooks, B. R., Bruccoleri, R. E., Olafson, B. D., States, D. J., Swaminathan, S., Karplus, M. (1983). CHARMM: A program for macromolecular energy, minimization, and dynamics calculations. Journal of Computational Chemistry 4:187-217.

Brutti, C., Apostolo, N. M., Ferrerotti, S. A., Llorente, B. E., Krymkiewicz, N. (2000). Micropropagation of Cynara scolymus L. employing cyclodextrins to promote rhizogenesis. Scientia Horticola 83:1-10.

Cavalli, R., Peira, E., Caputo, O., Gasco, M.R. (1999). Solid lipid nanoparticles as carriers of hydrocortisone and progesterone complexes with beta-cyclodextrins. International Journal of Pharmaceutics 182:59-69.

De Azevedo, M. B. M., Zullo, M. A. T., Durán, N., Salva, T. J. G. (1999). Brassinosteróides, importante classe de hormônios vegetais. Estudo de propriedades físico-químicas do complexo de inclusão de 24-epibrassinolídio com β-ciclodextrina. Annals of III Reunião Latino-americana de Fitoquímica e IX Simpósio Latino-americano de Farmacobotânica, Gramado, Poster 210.

De Azevedo, M. B. M., Alderete, J., Zullo, M. A. T., Salva, T. J. G., Durán, N. (2000). Brassinosteroids: a new class of plant hormones. The biological activity of 24-epibrassinolide and an inclusion complex of 24-epibrassinolide and β-cyclodextrin. Abstracts of the 27th International Symposium on Controlled Release of Bioactive Materials, Paris, pp. 5006-5007.

De Azevedo, M. B. M., Zullo, M. A. T., De Azevedo, M. M. M., Alderete, J., Brosa, C., Durán, N. (2001a). Brassinosteroids: a new class of plant hormones. 28-Homocastasterone and inclusion complex with β-cyclodextrin. Abstracts of the 4th International Congress on Chemistry, 13th Caribbean Conference on Chemistry and Chemical Engineering, Havana, p. 393.

De Azevedo, M. B. M., Zullo, M. A. T., Salva, T. J. G., Robaina, C., Coll, F. (2001b). Cyclodextrins inclusion complexes of spirostane analogues of brassinosteroids. Proceedings of the 28th International Symposium on Controlled Release of Bioactive Materials, San Diego, pp. 4002-4003.

De Azevedo, M. B. M., Zullo, M. A. T., Queiroz, H. M., Salva, T. J. G., De Azevedo, M. M., Duran, N., Alderete, J. B. (2001c). Epibrassinolide/β-cyclodextrin inclusion complex: physico-chemical characterization and rice lamina inclination assay. Abstracts of the 17th International Conference on Plant Growth Substances, Brno, p. 181.

De Azevedo, M. B. M., Zullo, M. A. T., Alderete, J. B., De Azevedo, M. M. M., Salva, T. J. G., Durán, N. (2002). Characterization and properties of the inclusion complex of 24-epibrassinolide with β-cyclodextrin. Plant Growth Regulation 37: 233-240.

Grove, M. D., Spencer, G. F., Rohwededer, W. K., Mandava, N. B., Worley, J. F., Warthen Jr. J. D., Steffens, G. L., Flippen-Anderson, J. L., Cook Jr. J. C. (1979). Brassinolide, a plant growth promoting steroid isolated from Brassica napus pollen. Nature 281:216-217.

Kamuro, Y., Takatsuto, S. (1999). Practical applications of brassinosteroids in agricultural fields. In Brassinosteroids - Steroidal Plant Hormones, pp. 223-241. Eds A Sakurai, T Yokota and S D Clouse. Springer-Verlag, Tokyo.

Khripach, V. A., Zhabinskii, V. N., De Groot, A. E. (1999). Practical applications and toxicology. In Brassinosteroids – A new class of plant hormones, pp. 325-346. Eds. V A Khripach, V N Zhabinskii and A E de Groot. Academic Press, San Diego.

Khripach, V., Zhabinskii, V., De Groot, A. (2000). Twenty years of brassinosteroids: steroidal plant hormones warrant better crops for the XXI Century. Annals of Botany 86:441-447.

Mandava, N. B., Sasse, J. M., Yopp, J. H. (1981). Brassinolide, a growth-promoting steroidal lactone. II. Activity in selected gibberellin and cytokinin bioassays. Physiologia Plantarum 53:453-461.

Marzona, M., Carpignano, R., Quagliotto, P. (1992). Quantitative structure-stability relationships in the inclusion complexes of steroids with cyclodextrins. Annali di Chimica 82:517-537.

Núñez Vázquez, M. C., Robaina Rodríguez, C. M. (2000). Brasinoesteroides – nuevos reguladores del crecimiento vegetal con amplias perspectivas para la agricultura. Campinas: Instituto Agronômico. 83pp.

Saenger, W. (1980). Cyclodextrin inclusion compounds in research and industry. Angewandte Chemie International Edition 19:344-362.

Stella, V. J., Rajewski, R. A. (1997). Cyclodextrins: their future in drug formulation and delivery. Pharmaceutical Research 14:556-567.

Szejtli, J., Szente, L., Harshegyi, J., Daroczi, I., Vorashazy, L., Torok, S. (1989). Inclusion complexes and mixtures of plant growth regulators with cyclodextrins. Hungarian Patent Teljes HU 47961 A2 [apud Chemical Abstracts 112: 50398 (1997)].

Takatsuto, S. (1994). Brassinosteroids: distribution in plants, bioassays and microanalysis by gas-chromatography-mass spectrometry. Journal of Chromatography A 658:3-15.

Takeno, K., Pharis, R. P. (1982). Brassinolide-induced bending of the lamina of dwarf rice seedlings: an auxin mediated phenomenon. Plant and Cell Physiology 23:1275-1281.

Wada, K., Marumo, S., Ikekawa, N., Morisaki, M., Mori, K. (1981). Brassinolide and homobrassinolide promotion of lamina inclination of rice seedlings. Plant Cell Physiology 22:323-326.

Yopp, J. H., Mandava, N. B., Sasse, J. M. (1981). Brassinolide, a growth-promoting steroidal lactone. I. Activity in selected auxin bioassays. Physiologia Plantarum 53:445-452.

Zhou, D., Wu, Y., Xu, Q., Yang, L., Bai, C., Tan, Z. (2000). Molecular mechanics study of the inclusion of trimethylbenzene isomers in α-cyclodextrin. Journal of Inclusion Phenomena and Macrocyclic Chemistry 37:273-279.

Zullo, M. A. T., Adam, G. (2002). Brassinosteroid phytohormones - structure bioactivity and applications. Brazilian Journal of Plant Physiology 14: 143-181.

Zullo, M. A. T., Kohout, L., De Azevedo, M. B. M. (2003). Some notes on the terminology of brassinosteroids. Plant Growth Regulation 39: 1-11.

CHAPTER 9

VLADIMIR A. KHRIPACH, VLADIMIR N. ZHABINSKII AND NATALIYA B. KHRIPACH

# NEW PRACTICAL ASPECTS OF BRASSINOSTEROIDS AND RESULTS OF THEIR TEN-YEAR AGRICULTURAL USE IN RUSSIA AND BELARUS

One of characteristic physiological properties of new plant hormones brassinosteroids (BS), when applied exogenously to vegetating plants, is their ability to stimulate plant growth and development. That is why their possible application in agriculture was considered still in the very beginning of BS investigations. Their progress has been extremely rapid. After discovery of brassinolide, the first member of the series, less than twenty years were necessary to start wide practical use of brassinosteroids in agriculture as crop-yield-increasing and plant-protecting agents. In Russia and Belarus, the first BS-preparation, based on 24-epibrassinolide as active ingredient, has been officially registered in 1992 as potato crop-increasing agrochemical. Later a number of other crops and purposes for its agricultural use have been added. During the developing of practical adaptation of BS many problems connected with commercial-scale production and official status of new agricultural chemical have been solved. Among them: elaboration of economically reasonable synthetic methods for BS preparation, field-scale biological trials with different plant species, toxicological studies, etc. The accumulated experience and results of research programs of different laboratories opened up many new aspects of BS-activity that are very important for further development of this group of phytohormones to a new generation of agricultural chemicals, which are not-conflicting with the environment and explore a new principle in plant protection. These problems and perspectives of agricultural application of brassinosteroids are the subject of the article.

## INTRODUCTION

The discovery of brassinosteroids (BS) (Grove *et al.*, 1979) and their extensive following studies have brought a new vision for the hormonal functions of steroids in living creatures, because to the previously known role as hormones of animals, insects and fungi, their role of plant hormones had been added. The value of the obtained results went largely beyond pure science. One of the characteristics physiological activities of new plant hormones, when applied exogenously to vegetating plants, are their ability to stimulate plant growth and development. That is why their possible application in agriculture was considered in the very beginning of BS investigations.

Two last decades were highly productive in this field. During this period a series of naturally occurring BS of different structure have been identified and synthetically prepared, their physiological properties investigated, and many problems

*S.Hayat and A.Ahmad (eds.), Brassinosteroids*, 189-230

connected with commercial-scale production and official status (state registration) of new agricultural chemicals have been solved. Among them: elaboration of economically reasonable synthetic methods for BS preparation, field-scale trials with different plant species, toxicological studies, etc. For none of the other plant hormonal substances, although being studied for a much longer time, similar development has been seen before. Chemical structures of some representative BS are shown at figure 1 and table 1.

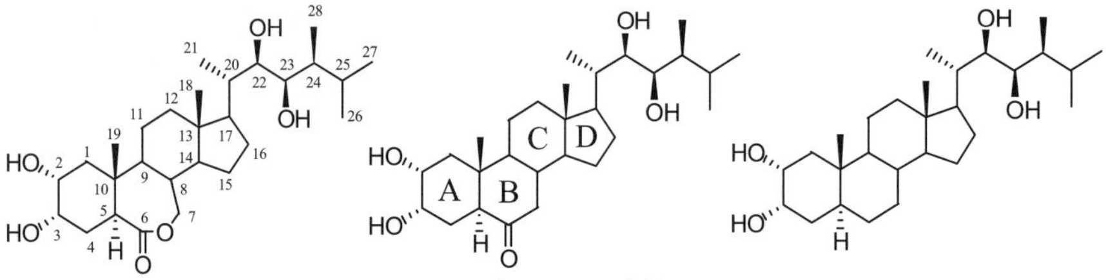

Brassinolide (Bl)* - B-lactone type

Castasterone (Bk) - B-ketone type

6-Deoxocastasterone (Bd)- B-deoxo type

*Figure 1. Three types of cycle B in BS-series [*Here and below, short names of BS are given in accordance with the systematic rules suggested in book (Khripach et al., 1999)]*

Since the discovery of brassinolide, more than a thousand articles on various aspects of BS investigations appeared, which have been mainly summarized in monographs (Khripach *et al.*, 1999; Sakurai *et al.*, 1999). Nowadays, research in BS area continues to be very dynamic that is reflected in many reviews published during the last couple of years (Khripach *et al.*, 2000; Schmidt *et al.*, 2000; Wang and Chory, 2000; Abe *et al.*, 2001; Asami *et al.*, 2001; Bishop and Yokota, 2001; Schnabl *et al.*, 2001; Winter, 2001; Abe and Marumo, 2002; Clouse, 2002; Kim *et al.*, 2002; Rao *et al.*, 2002; Romanov, 2002; Schneider, 2002; Bajguz and Tretyn, 2003; Schaller, 2003; Symons and Reid, 2003). In fact, they represent all the directions of BS research including chemical synthesis of natural BS and their analogs, study on the physiological properties and mode of brassinosteroid action, search for the biosynthetic pathways and new BS in plants. A characteristic feature of the recent period is a tendency for a deeper investigation of molecular-genetic aspects of brassinosteroid effects and wider practical use of their regulatory activities on plant growth and development in agriculture.

*Table 1. Structures of some typical brassinosteroids and their biosynthetic precursors*

| Trivial names of BS | Substituent at | | | | Cycle B |
|---|---|---|---|---|---|
| | C2 | C3 | C24 | C23 | |
| 24-Epibrassinolide | α-OH | α-OH | β-$CH_3$ | α-OH | Bl-type |
| 28-Homobrassinolide | α-OH | α-OH | α-$C_2H_5$ | α-OH | Bl-type |
| 28-Norbrassinolide | α-OH | α-OH | H | α-OH | Bl-type |
| Dolicholide | α-OH | α-OH | $=CH_2$ | α-OH | Bl-type |
| 28-Homodolicholide | α-OH | α-OH | $=CH_2CH_3$ | α-OH | Bl-type |
| 3-Epibrassinolide | α-OH | β-OH | α-$CH_3$ | α-OH | Bl-type |
| 24-Epicastasterone | α-OH | α-OH | β-$CH_3$ | α-OH | Bk-type |
| 28-Homocastasterone | α-OH | α-OH | α-$C_2H_5$ | α-OH | Bk-type |
| Brassinone | α-OH | α-OH | H | α-OH | Bk-type |
| Dolichosterone | α-OH | α-OH | $=CH_2$ | α-OH | Bk-type |
| 28-Homodolichosterone | α-OH | α-OH | $=CH_2CH_3$ | α-OH | Bk-type |
| Typhasterol | H | α-OH | α-$CH_3$ | α-OH | Bk-type |
| Teasterone | H | β-OH | α-$CH_3$ | α-OH | Bk-type |
| 3-Epi-6-deoxocastasterone | α-OH | α-OH | α-$CH_3$ | α-OH | Bd-type |
| 6-Deoxotyphasterol | H | α-OH | α-$CH_3$ | α-OH | Bd-type |
| 6-Deoxoteasterone | H | β-OH | α-$CH_3$ | α-OH | Bd-type |
| 3-Dehydro-6-deoxoteasterone | H | =O | α-$CH_3$ | α-OH | Bd-type |
| 6-Deoxocathasterone (23-dehydroxyteasterone) | H | β-OH | α-$CH_3$ | H | Bd-type |

The early prognosis on practical value of BS (Maugh, 1981) as crop-yield-increasing agents found confirmation in various studies done mainly in Japan, China and the USSR. One of the first brassinosteroid-based agrochemicals was officially registered in 1992 in the USSR, and later its registration was continued in New Independent States, Russia and Belarus. Passed ten-year period gave wide experience and allowed collecting extensive information about possibilities and perspectives of BS use in agriculture, opened up a number of previously unknown effects which are important for crop-protective and crop-quality-increasing applications of BS. In present review we try to summarize major results of this time with special accent on the most recent data. Along with the results taken from Russian literature, which are mostly difficult for Western readers to access, new data from international journals concerning application and physiological properties of BS are also presented.

## PHYSIOLOGICAL EFFECTS OF BRASSINOSTEROIDS: MOLECULAR BIOLOGICAL BASIS FOR THEIR AGRICULTURAL USE

High and specific physiological activity of BS in plants has triggered the interest of researchers to these substances. This chapter makes an attempt to bring into a consideration the new data on physiological effects of BS, which are important for their agricultural use. A strong influence of BS on plant growth and development was considered to be very promising for practice since their discovery. This was the major reason for the initiation of extensive research programs on BS study in different countries that finally has led to the agricultural use of BS. An important feature of this period, which positively influences such a development, is a pronounced social demand for new types of agrochemicals that are not conflicting with the environment. Unfortunately, there are no traditional pesticides acting in this way. Evidently, the search for a solution of this problem is the most promising when natural bio-regulating mechanism and the corresponding bio-regulating compounds are taken as a basis for the elaboration of new environmentally friendly agrochemicals. BS represent a good model of such natural compounds acting in natural doses and in a natural way with aiming the same targets as in wild nature. They are usual constituents of plants, which have been components of animal food chains during their evolution, and this is a kind of guarantee for ecological safety of BS used as agrochemicals. The agricultural use of BS is based on their ability to stimulate physiological processes in plants. As a result, it may become possible to grow not only larger but also better quality crops under normal and unfavorable conditions including stresses and diseases. The data on the ability of BS to change the qualitative and self-protective parameters of plants along with their quantitative characteristics became available not long ago, and till now these properties are not fully recognized as very important ones for practical use, probably, even more than well known crop-yield increasing activity.

### *Effect on Growth and Development*

The earliest found and the most extensively studied effect of BS is their ability to activate growth in a variety of plant systems including whole plants and their separated fragments. This property was used as the most characteristic one for the elaboration of BS-specific bioassays and starting from the beginning of BS investigations were considered to be a key precondition for their agricultural use. During recent decade a number of results indicating an obligatory role of BS in cell growth regulation have been obtained. It was found that plant-growth responses to BS are genetically determined and proceed *via* expression of specific genes under BS action. It looks like BS are involved in all steps of the plant growth including cell expansion, cell division and changes in qualitative and mechanical properties of plant tissues.A recent example bringing a confirmation for such point of view is connected with the isolation and identification of a number of cotton fiber tissue-specific genes, one of which was able to change a degree of expression under treatment with BS such as brassinolide (Bl), 24-epibrassinolide (EBl), 28-homobrassinolide (HBl), and a number of 6-keto- and 6-deoxo-brassinosteroids (Kasukabe *et al.*, 1999).

The most important characteristics reflecting quality of cotton fibers are their length, fineness and mechanical strength, and many attempts have been done to improve these. Among the approaches, an application of different plant hormones was extensively investigated, but no results promising for industrial use have been obtained. In the course of study of BS effects in different plants, data on increase of cotton boll yield were recorded (cited in Khripach *et al.*, 1999), but no influence of BS on quality parameters was found that time. Finding and isolation of the gene which is associated with the formation and elongation of cotton fibers and capable to change its expression under treatment with BS makes possible an improvement in the characteristics and yield of cotton fibers, when it is introduced into cotton plants and expressed in a large quantity. In addition, the introduction of this gene in anti-sense form gives a possibility to suppress the original gene. Application of this new genetic engineering technique to industrial cotton fiber production is expected to be more reliable for fiber quality control than traditional approaches based on crossbreeding and screening.

Although only a few data are reported by now about the genes similar to one discussed above, which are responsible for growth and developmental characteristics of plant and able to change specifically their expression under BS action, their presence in plants looks doubtless. That is why another solution of the same problem on the simultaneous control of quantitative and qualitative parameters of agricultural production can be found *via* search for the conditions (dose, time of treatment, etc.) when these genes are specifically regulated under the direct action of BS on usually cultivating plants. An example of this approach to control yield, length and strength characteristics of flax fibers have been patented (Voskresenskaya *et al.*, 1998).

A major breakthrough in the understanding of mechanism of growth regulation and role of BS in plants occurred when BS-deficient and BS-insensitive mutants were found and used for molecular genetic studies (Altmann, 1999). The isolated BS mutants show severe dwarfism or de-etiolated phenotype in darkness. Dwarfism is a result of mutation in genes involved in the control of normal plant elongation growth, which accumulates the phenomena of cell division and cell elongation. Both phenomena are influenced by exogenous factors, such as light and temperature, and endogenous factors including plant hormones. It means that a number of genes are involved in the dwarfism, especially those related to hormone biosynthesis and hormone receptors. The identification of the affected genes firstly done in the mutants of *Arabidopsis thaliana* triggered further investigations which led to the finding of BS-related defects connected with deviations in the biosynthesis (BS-deficiency) or reception (BS-insensitivity). Important data confirming principal role of BS were obtained in the experiments with specific inhibitors of brassinolide biosynthesis (brassinazole and Brz2001), which showed a dramatic repression of the normal phenotype because of suppression of endogenous brassinolide functions in *Arabidopsis* (Asami *et al.*, 2000; Sekimata *et al.*, 2001), and in the experiments on the restoration of the normal phenotypes of BS-deficient mutants under the treatment with brassinolide (Koka *et al.*, 2000; Catterou *et al.*, 2001; Schultz *et al.*, 2001).

A recent confirmation of important role of BS in growth regulation was obtained in the study of brassinosteroid-related mutants of *Arabidopsis thaliana* (L.) Heynh., det2 and dwf1 which develop small leaves (Nakaya *et al.*, 2002). It revealed that the mutants had fewer cells per leaf blade in comparison with wild-type plants. A restoration of the leaf size by Bl treatment of the det2 mutants could not be explained only by the cell elongation and expansion. The obtained data indicate that defective leaf size in mutant was determined by decreasing both the number and size of leaf cells, confirming the idea about the involvement of BS in cell growth and proliferation.

Till recently, BS-deficient or BS-insensitive mutants have been identified only in dicots but not in monocotyledonous plants, although the latter ones were investigated quite extensively, showed typical growth and developmental reactions under treatment with BS, and even were used as model plants in very sensitive and BS-specific plant test systems (see Khripach *et al.*, 1999 for review). All the obtained results indicated the effects of exogenous brassinosteroids, but not the effects of the endogenous ones. Nevertheless, they suggested that endogenous BS have an important role in growth and development of monocots, particularly in the *Gramineae* family having a unique importance for food production in the world, and initiated special efforts that recently led to the isolation and characterization of both types of mutants, identification of a new gene involved in BS sensitivity, and finding the enzyme and defective step in the biosynthesis of BS causing dwarf phenotype of the biosynthesis mutant. Both types of mutants were isolated from rice and were found to have close similarity with the corresponding mutants of *Arabidopsis thaliana* in molecular-genetic motives of the mutation.

Thus, the first biosynthesis mutant *brd1* (brassinosteroid-dependent 1) was identified (Mori *et al.*, 2002) as a rice (*Oryza sativa* L. cv Nipponbare) dwarf mutant in transgenic lines produced in laboratory conditions. Specific features of the mutant grown under normal conditions are very short leaf sheaths, short and curled leaf blades, weakly developed roots, few tillers and sterility. In darkness, the mutant showed constitutive photomorphogenesis. Histological examination revealed reduced cell length along the longitudinal axis, which explains the short leaf sheath. It also revealed the expanded size of cells, both epidermal and motor cells, along dorsal-ventral axis. The last feature connected with the motor cells might be an explanation of the curly phenotype of leaves, because the motor cells are involved in leaf rolling. The leaf blades of the mutant had more condensed packing of mesophyll cells and smaller intercellular space than the corresponding wild plants. As a result of dorsal-ventral cell expansion, the mutant leaf blades were thicker than the leaf blades of the wild plant.

A key to understanding the biochemical reasons of the mutation was analytical study on endogenous BS and sterol content which showed critical difference in BS biosynthesis of mutant and wild plants (Mori *et al.*, 2002). Thus, the GC-MS analysis revealed no serious difference in the content of major BS precursors, campesterol and campestanol, which implies normally proceeding steroid biosynthesis, at least till the step of sterols, in both types of plants (Table 2). At the later steps of the biosynthesis, the content of BS with the functionalized cycle B (castasterone), which is an obligatory element of chemical structure of BS determinative for their bioactivity, was reduced in the mutant in comparison with the wild-type plants. Just the reverse, a significant increase in the content of BS with the unfunctionalized cycle B of steroid molecule (6-deoxotyphasterol, 6-deoxocastasterone) was found in the mutant.

Taking into account the results of Shimada *et al.*, 2001, which indicated the role of steroid-6-oxidases in transformation of 6-deoxo steroids, such as 6-deoxoteasterone, 3-dehydro-6-deoxoteasterone, 6-deoxotyphasterol, 6-

*Table 2. Content of Endogenous BS in Wild-Type and brd1 Mutant Rice Plants (ng/g fresh weight)*

| BS | *brd1* | Wild |
|---|---|---|
| 6-Deoxocathasterone | 1.09 | 0.89 |
| 6-Deoxoteasterone | 0.18 | 0.12 |
| 3-Dehydro-6-deoxoteasterone | 1.10 | 0.40 |
| 6-Deoxotyphasterol | 20.80 | 6.84 |
| Teasterone | 0.08 | 0.03 |
| 6-Deoxocastasterone | 4.34 | 1.98 |
| Castasterone | 0.04 | 0.45 |

deoxocastasterone, into the corresponding 6-oxo-BS, one could suggest defectiveness of the enzyme that leads to breakage of the C-6 oxidation step of BS biosynthesis. Compensation of the BS deficiency in the mutant by exogenously supplied brassinolide led to the restoration of the normal plant phenotype. Molecular characterization of the gene responsible for C-6 oxidation in rice, *OsBR6ox*, showed its homology with the corresponding genes of *Arabidopsis* and tomato, *AtBR6ox* and *Dwarf*, and showed that it is defective in the mutant. These results and identification of *brd1* by another research group (Hong *et al.*, 2002) clearly indicate that BS are involved in growth and developmental processes, not only in dicots but also in monocots, influencing the elongation of leaves, root differentiation, development of tillers, skotomorphogenesis, and reproductive growth.

Similar conclusion was drawn from the study on brassinosteroid insensitive mutant of rice that was recently identified and compared with the corresponding mutant of *Arabidopsis* in respect to the molecular-genetic motive of the mutation (Yamamuro *et al.*, 2000). This novel dwarf mutant, *d61*, was obtained by mutagenesis with nitrosomethylurea and showed specific shortening of internodes. In growing plant, elongation of the internodes is a result of cell division in the intercalary meristem and further cell elongation in the elongation zone. That is why, dwarfing of the culms could be caused by a defect in one or both processes. It was found that nonelongated internodes in the mutant plants were not able to develop an intercalary meristem and had no an organized arrangement of microtubules.

The dwarf phenotype of the mutant and its erect leaves suggested that the D61 gene product could be involved in either the biosynthesis or reception of BS. Indeed, the mutant showed a weak response to exogenous BS in comparison with the wild plants in all types of experiments including classical lamina joint test. It is known that the level of bending between the leaf blade and leaf sheath in rice is dependent on the activity of exogenously applied BS and their concentrations, and this characteristic feature of rice leaves is used in a specific quantitative bioassay for BS, known as the lamina inclination test (Wada *et al.*, 1981; Takeno and Pharis, 1982). When wild (control) plants were used in the test, the angle between the axis of the leaf sheath and leaf blade was about 90° even in the case of absence of exogenous BS, and it became much higher under application of increasing concentrations of Bl. The maximal angle reached at the concentration of $5 \times 10^{-4}$ μg/ml was about 160°, and it showed no further changes under application of higher Bl concentrations. Although in the mutant plants the degree of bending was also positively influenced by growing concentrations of Bl, the extent of the phenomenon was much smaller than in the wild-type plants under the same conditions. Even at a Bl concentration of $10^{-2}$ μg/ml concentration-angle relationship curve did not plateau reflecting lower sensitivity of the mutant to BS in comparison with the wild plants.

Earlier, accumulation of BS was shown in the experiments with the Bl-insensitive mutants of *Arabidopsis* (Noguchi *et al.*, 1999). To check if there is a similar behavior of the rice mutants, a BS content in the *d61* mutant was quantitatively analyzed by GC-MS (Yamamuro *et al.*, 2000). The concentrations of all measured BS were found to be higher in the mutant than in the wild type plants (Table 3). This can be understood as an attempt of the plants to compensate the decreased sensitivity to BS by activation of their biosynthesis and higher supply of the bioactive substances. An interesting detail was the absence of brassinolide both in the mutant and in the wild plant shoots that might mean together with data (Mori *et al.*, 2002), where Bl was not detected also, a minor role of brassinolide in the BS pool of rice plants. Authors suggested that the *d61* mutation is a result of loss of function of the rice gene homologous to the *BRI1* gene of *Arabidopsis*. The isolation of the *OsBRI1* (*Oryza sativa BRI1*) showed its similarity with the *BRI1* gene of

*Table 3. Content of Endogenous BS in Wild-Type and d61 Mutant Rice Plants (ng/g fresh weight)*

| BS | *d61* | Wild |
|---|---|---|
| 6-Deoxoteasterone | 0.22 | 0.17 |
| 6-Deoxotyphasterol | 1.53 | 1.15 |
| Typhasterol | 1.52 | 0.55 |
| 6-Deoxocastasterone | 0.57 | 0.32 |
| Castasterone | 0.86 | 0.21 |

*Arabidopsis*, which encodes a putative brassinolide receptor kinase. Molecular complementation of the *d61* mutation by introduction of the wild-type *OsBRI1* gene led to the recovery of the normal phenotype. Furthermore, transgenic plants with the antisense strand of the *OsBRI1* transcript showed similar or sometimes even more severe phenotypes than those of the *d61* mutants, and this was a reflection of the suppression of *OsBRI1* function in the transgenic plants. These experiments brought a new indication on the role of BS in physiology of rice plants showing their involvement in various growth and developmental processes, such as (i) internode elongation, which proceeds *via* induction of the formation of the intercalary meristem and the longitudinal elongation of internodal cells; (ii) bending of the lamina joint; and (iii) skotomorphogenesis.

Studies on the intimate details of the mechanism of BS action have not only pure fundamental but also a pronounced practical orientation. Thus, identification of the *OsBRI1* gene, which functions to increase rice brassinosteroid sensitivity and is involved in the elongation of plant internode cells and inclination of leaves, makes possible producing phenotypically transformed plants by controlling this gene. For example, its suppression allows producing dwarf plants, which are resistant to lodging and enabling higher concentration of individual plants per area unit, which could be significant in the production of certain crops. Initiated in this way dwarfism of height or culm length may be useful in production of new ornamental plants with special aesthetic value. An opposite effect can be reached by introducing and expressing the corresponding DNA in plants, and this will result in increasing the overall crop yield that is especially important for animal feed production (Tanaka *et al.*, 2003).

Similar stimuli are involved in the initiation of studies directed to the search of genetic tools for the control of BS biosynthesis in plants. For example, recent identification of BS-deficient *Arabidopsis* mutant (dwf7) and the corresponding gene, which is responsible for an early step in BS biosynthesis, led authors to the idea of its use in the production of transgenic plants with altered morphology (Choe *et al.*, 1999; 2002).

It should be pointed out that nowadays more and more practical applications of genes and their products responsible for BS effects on plant growth and development are seen by researches, and that is why a number of results in this area have been recently patented. Thus, Korean scientists described cloning, characterization and use of pea cytochrome P450 hydroxylase involved in brassinosteroid biosynthesis of plants (Kang *et al.*, 2001; Kang and Park, 2002). This enzyme catalyzes the biosynthetic step of C-2 hydroxylation in the BS biosynthesis that is characteristic for the conversion of typhasterol to castasterone and 6-deoxotyphasterol to 6-deoxocastasterone. Authors identified the gene encoding the cytochrome P450 and the corresponding amino acid sequence. They showed that the cytochrome P450 specifically interacts with the small G protein Pra2. The latter one and cytochrome P450 are localized into endoplasmic reticulum. Transgenic plants with decreased level of Pra2 show dwarfish hypocotyls in the dark, which can be rescued by BR but not by other hormones. Transgenic plants with overexpression of the cytochrome P450 have elongated stem. The obtained results suggest that the Pra2 is a light-regulated molecular switch influencing the hypocotyl growth in etiolated seedlings *via* interaction with the cytochrome P450. The results can find an application in the synthesis of the cytochrome P450 or its biologically active fragments and their use for regulation of the stem growth in transgenic plants. Moreover, they bring a new indication on close cooperation between the molecular mechanisms of light and BS effects in stem elongation.

Although many data showing genetically-determined involvement of BS in growth regulation are collected till now, there is no much information about precise steps of the effect realization and key site in the chain of signaling events where role of BS is critical for the plant development. As shown above, a promising direction is connected with a deeper search for the relationships between BS growth-regulation and light growth-regulation in plants. Starting from the beginning of BS research, a number of results showed their participation in light-regulated plant development, and some of them suggested the role of BS in mediation of the phytochrome regulatory functions (Kamuro and Inada, 1991; Kalituho *et al.*, 1996, 1997a,b; Chory and Li, 1997). Recent studies of *Arabidopsis* mutants gave a new confirmation for these relationships (Neff *et al.*, 1999; Clouse, 2001; Luccioni *et al.*, 2002), but these data did not open the mechanism of light and BS interaction.

A key for understanding of the phenomenon might be found *via* investigation of BS as possible components of light signal transduction system interacting with the photoreceptors. As mentioned above, there is only a few data in the literature on the relationship between the responses mediated by phytochrome and BS action. Recently the first results indicating the existence of their relationship with the responses mediated by blue-light photoreceptor were reported (Tishchenko *et al.*, 2001; Karnachuk *et al.*, 2002) and showed that the mechanism of BS-participation in light signaling could function *via* regulation by BS of the hormonal balance in plants.

The experiments included measurement of hormone content in wild type of *Arabidopsis thaliana* and in photomorphogenetic mutant *hy4*, which is defective in gene encoding synthesis of cryptochrome CRY1 (blue-light photoreceptor), under the action of BS. Ttreatment of the *Landsberg erecta* ecotype (Ler) of *Arabidopsis thaliana* with EBl in the dark led to a shortening of the hypocotyl length, and similar result was obtained under the action of blue light (Figure 2). Although both factors had the same direction, the effect of EBl acting alone was higher than the corresponding effect of blue light. A combination of both retarding factors acting together led to a synergistic effect that indicates a relationship between the responses mediated by cryptochrome and BS action, where BS could play a role of transducing agents in a signaling pathway from the photoreceptor to the hypocotyl. When the mutant *hy4* was treated with EBl, the hypocotyl length was decreased both in the dark and blue light, but there was no retarding effect with blue light acting alone. As a result, the effect of EBl on the hypocotyl growth looks like a compensation of the insensitivity of the mutant to blue light that makes the mutant response similar to those of normal plant. One of the explanations of this phenomenon could imply a role of cryptochrome in triggering the mechanism of BS biosynthesis, certain step of which is blocked when *CRY1* is defective. In this case, application of the exogenous hormone compensates gene deficiency. A parallel study on major phytohormone content in both types of plants led to similar conclusion. For example, the effect produced by brassinosteroid on IAA (Figure 2) and cytokinin content in the mutant can be characterized as the shifting of hormonal spectrum of the mutant towards wild type hormonal balance. Although there are no direct indications on acting principles of CRY-BS interaction, a regulation of calcium flux across membranes by BS, which is an important element of blue-light signal transduction pathway (Jenkins, 1997), looks reasonable hypothesis because of known ion-flow-regulating properties of hydroxylated steroids in mammalian cells and brassinosteroids in plant cells.

An indication on possible cooperation of BS with calcium signaling system was reported by Ilkovets *et al.*, 1999. They investigated the effect of some phytohormones (indole-3-acetic acid, 1-naphtylacetic acid, gibberellic acid, abscisic acid and 24-epibrassinolide) on the cytosolic free $Ca^{2+}$-ion concentration in protoplasts of *Nicotiana plumbaginifolia* in $Ca^{2+}$-deficient and $Ca^{2+}$-enriched media. All the hormones were found to be able to increase $[Ca^{2+}]$cyt but acted mainly as activators of ion transport from intracellular depot, except 24-epibrassinolide, the effect of which was connected with $Ca^{2+}$-transportation from intercellular space. An important feature of EBl action was its efficiency at the concentrations much lower than for other hormones (about $10^3$ times). Taken together, these data suggest the presence of a cross talk between BS- and $Ca^{2+}$-signalling and possible involvement of $Ca^{2+}$-ions in BS-signal transduction pathway.

Earlier, ability of BS to influence ion transportation *via* membranes was investigated by many authors, especially in connection with the activation of membrane-bound proton pump, which is assumed to be one of the initial steps of cell elongation followed by loosening of the cell wall as a result of its acidification and activation of polysaccharide hydrolases. Finding the ability of BS to activate proton pumps of plasmalemma stimulated further interest to study their behavior in respect to the corresponding enzymes of tonoplast, which are responsible for the membrane transport and accretion of sugars, amino acids and other metabolites in vacuole.

The plant vacuolar membrane contains two functionally and physically distinct phosphohydrolases: $H^+$-ATP-ase and $H^+$-pyrophosphatase of the vacuolar-type, which catalyse electrogenic $H^+$-translocation. Study of the effect of epibrassinolide on the activity of the enzymes in the tonoplast isolated from red beet roots (*Beta vulgaris* L.) in wide range of concentrations ($10^{-5}$-$10^{-15}$ M) showed that both hydrolytic and transport function of them can be regulated by the hormone depending on the stage of ontogenesis and potassium content in the media (Ozolina *et al.*, 1999; Pradedova *et al.*, 2002). Red beet is a biennial plant and for this reason three periods corresponding to different stages of ontogenesis were chosen: 1) first year of vegetation characterized by intensive growth and sugar accumulation; 2) dormancy; 3) second year of vegetation (reproductive period). It was found that the hydrolytic activity of $H^+$-pyrophosphatase was more sensitive to the action of EBl than those of $H^+$-ATP-ase and reached its highest level (about 200% of control) during the dormancy period at a hormone concentration of $10^{-11}$ M. Smaller effect (maximal enzyme activation was about 150% of control) took place during the second year of vegetation and corresponded to the

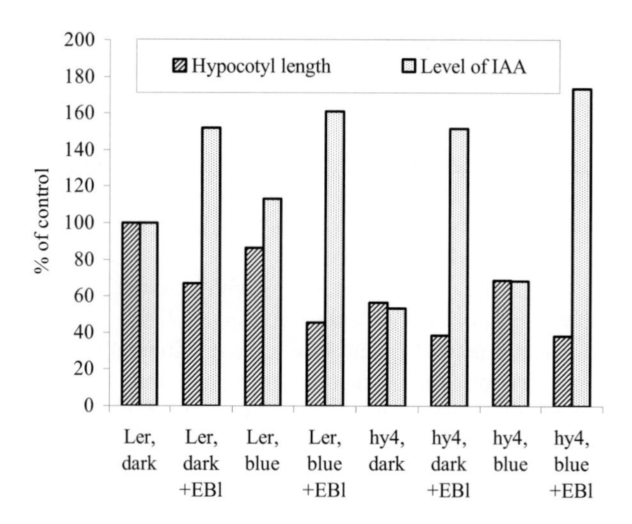


*Figure 2. Effect of EBl ($10^{-6}$ M) on hypocotyl length and level of free IAA in dark-grown and blue-light grown 7-day-old seedlings of Arabidopsis (wild-type and hy4 mutant)*

concentration of $10^{-12}$ M of EBl. During the period of intensive growth and sugar accumulation (first year of vegetation) EBl did not influence the activity of $H^+$-pyrophosphatase, but at this stage maximal responsiveness to EBl was a characteristic of $H^+$-ATP-ase of tonoplast. Its highest stimulation (about 160% of control) took place at EBl concentration of $10^{-13}$ M. This enzyme showed weaker stimulation by EBl during the second year of vegetation and almost no reaction on the hormone during the dormancy stage. A characteristic feature of EBl action on both enzymes was low active concentrations, which were $10^7$ -$10^3$ times lower than the concentrations of other phytohormones needful to initiate similar effect.

The effect of EBl on transport activity of proton pumps of tonoplast was studied at a concentration of $10^{-11}$ M (Figure 3), which was found to be the most efficient in the experiments on hydrolytic activity. Again, $H^+$-pyrophosphatase was more sensitive to EBl, but a degree of stimulation was twice as high as for the case of hydrolytic activity. Similar to it, during the first year of vegetation, EBl stimulated $H^+$-ATP-ase and did not influence $H^+$-pyrophosphatase, and acted in an opposite manner during the dormancy period. Study of EBl action ($10^{-11}$ M) on the accumulation of $H^3$–sucrose in vacuoles showed stimulation (about 140% of control) during both the first and the second year of vegetation, but exclusively *via* activation of $H^+$-ATP-ase and not *via* $H^+$-pyrophosphatase that influenced negatively the sucrose accumulation.

*Adaptation, Quality and Productivity Improvement*

Stimulation or inhibition of certain enzymatic reactions is an important step of BS-signal transduction pathway leading to physiological response, usually *via* a sequence of further biochemical shifts. As mentioned above, many of them, not only those connected with the mechanical growth but also involved in changes of developmental phases, transition of plants from one physiological state to another and adaptation in the broadest sense of the word, are BS-dependent and can be regulated by exogenously applied hormones.

An example of EBl-induced ethylene production in potato tubers that means activation of 1-aminocyclopropane-1carboxylate (ACC) synthase and ACC oxidase,

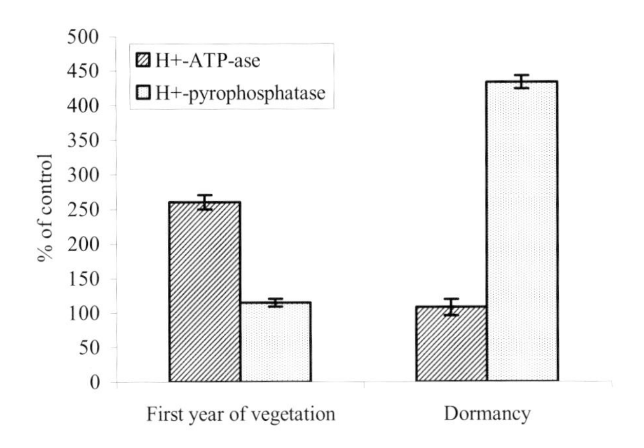


*Figure 3. Effect of EBl ($10^{-11}$ M) on transport activity of phosphohydrolases of tonoplast (% of untreated control)*

the cytosolic enzymes responsible for the last two reactions in ethylene biosynthesis, showed the involvement of BS in regulation of tuber dormancy (Korableva *et al.*, 1998, 2002). Previously, major role of ethylene in induction and maintenance of tuber dormancy was known. Its increased production, usually accompanied by increased abscisic acid (ABA) biosynthesis, is an important condition of induction or prolongation of deep dormancy. Treatment of potato (*Solanum tuberosum* L., cv. Nevskii) tubers with EBl by immersion into solutions of different concentrations ($10^{-2}$ and $10^{-3}$ ppm) resulted in increased production of ethylene and higher content of ABA in both free and conjugated forms. This was manifested in prolongation of deep dormancy and at the most efficient concentration ($10^{-2}$ ppm), applied immediately after tuber harvesting, inhibited sprouting by 36-38 days. The most intensive production of ethylene was observed at the first and the seventh days after treatment (about 300 and 150% of water-treated control, respectively), but at the thirtieth and later days of dormancy EBl-treated tubers had no difference with the control. The increased level (about 80% in average) of both free and conjugated ABA in treated tubers took place during the whole period of storage, but their ratio changed in the course of time in favour of the conjugated form (Table 4). Electron microscopic and morphometric analysis of functionally different zones of potato tuber apices showed that inhibition of cell stretching was one of the EBl-effects at the cell level. The treatment led to enhancement of the number of vacuoles, which had a smaller size, and to diminishing cell volume in tunica and all types of meristems, which was especially significant in the rib meristem. The latter effect is of special importance as the rib meristem plays leading role in the early stages of growth of potato tuber apices. The selective inhibition of the meristem growth correlates with the EBl-influenced changes in the endoplasmatic reticulum (Platonova, 1998), and with the suppression of activity of Golgi apparatus in functionally different zones of tuber apices that confirms a supposition about target cells responding to the action of BS (Platonova and Korableva, 1999a,b). Although these results demonstrate no stimulation but opposite effect on the cell growth, they fit the idea about special role of BS during critical periods of plant development and their involvement in the integration of the new emerging regulation network by triggering shifts in cellular sensitivity to plant hormones (Amzallag, 2002).

*Table 4. Effect of EBl on the content of free and conjugated ABA (μg/g fresh weight) in apical regions of potato tubers during their storage*

| EBl ppm | 30 days | | 60 days | | 120 days | | 150 days | |
|---|---|---|---|---|---|---|---|---|
| | $ABA_{free}$ | $ABA_{con.}$ | $ABA_{free}$ | $ABA_{con.}$ | $ABA_{free}$ | $ABA_{con.}$ | $ABA_{free}$ | $ABA_{con.}$ |
| 0 | 6.1±0.8 | 1.6±0.1 | 5.2±0.5 | 1.7±0.1 | 1.9±0.1 | 0.8±0.1 | 1.1±0.1 | 0.8±0.1 |
| 0.01 | 12.5±1.0 | 2.7±0.3 | 9.8±1.1 | 2.6±0.2 | 7.6±0.9 | 2.5±0.3 | 4.2±0.8 | 2.4±0.3 |

Many recent data show that growth-regulating effects of BS are connected not only with their action on the mechanisms directly responsible for growth, such as activation of specific enzymes involved in cell wall loosening and cell expansion, but can be realized *via* changing balance of some physiologically active substances in indirect way. The best-documented reactions of this type include mediation of BS-action *via* hormonal shifts that cause the resulting effect. However in some cases, a hormone-like action of BS proceeds *via* a hormone-independent route and does not involve the corresponding hormone initiation.

An interesting example of this mechanism is presented by the study of EBl-effects on the levels of ABA and wheat germ agglutinin (WGA) in the roots of four-day-old seedlings of *Triticum aestivum* L. in normal and salinity-stress conditions (Shakirova and Bezrukova, 1998). WGA is a well-studied and readily available cereal lectin, which physiological functions are still discussed. It is known that intensive synthesis and accumulation of lectin take place during formation and ripening of wheat germs (the process controlled by ABA) and also in vegetating plants as a response to stresses with previous ABA accumulation. Under salinity-stress conditions a rapid ABA accumulation followed by the enhancement of the lectin content was found. EBl-treatment almost doubled the accumulation of lectin, but produced no effect on the

ABA-content. A combination of salt- and EBl-treatment diminished the salinity-induced accumulation of ABA and WGA in roots and led to partial growth recovery in comparison with the action of salt alone. The finding of ABA-independent WGA accumulation stimulated further search for possible role of EBl in controlling WGA-gene expression (Shakirova *et al.*, 2002). Soaking the seeds in EBl solutions showed maximal stimulation of wheat-seedling root growth at two concentrations differing by three orders of magnitude: 0.4 and 400 nM. At these concentrations the root length was about 120% of the untreated control, while the intermediate values produced smaller or no effect. For this reason mentioned concentrations were applied to the four-day-old seedlings in further experiments on the WGA and ABA quantification in their roots that was done using ELISA technique. A transcriptional activity of the WGA genes was estimated by dot-blot analysis. The experiments showed that both active concentrations of EBl-stimulated WGA accumulation in roots, which evidently resulted from its *de novo* synthesis, since accumulation of the protein was registered only after 3 h of root treatment, i.e. during the period of active expression of its gene. At the same time, no changes in ABA content were found in the investigated samples. Both concentrations produced similar shifts in WGA amounts: the WGA level reached its maximum (about 200% of control) in 7 h of EBl action, when the gene expression was the highest, and then declined. A comparison of the data on RNA analysis illustrating the transcription of WGA gene and protein accumulation showed a good correlation, which is an indication on the EBl involvement into the WGA-content regulation at the transcriptional level.

Revealing close relations between BS action and WGA biosynthesis strongly suggest their probable co-operation in growth regulation and developmental changes. The interaction of WGA and EBl has been investigated recently in a model of cell division in wheat roots (Bezrukova *et al.*, 2002). Seeds of wheat were sterilized and germinated for 3 days. Then the endosperm was excised and seedlings were kept for 24 hours in nutritive solution (2% sucrose) or in nutritive solution with the additive of a protein (1 mg/l of WGA or 1 mg/l of bovine serum albumin (BSA), as a control non-lectin protein). For some experiments EBl-pretreated seeds (soaking in 0.4 and 400 nM solutions of EBl for 3 hours) were used. Mitotic index for cells of apical meristem of seedling roots and cell areas in stretching zone were measured. It was found that treatment of seedlings by exogenous WGA significantly activated growth of root cells. Thus, in WGA-treated plants the mitotic index was on average more than 160%, and cell area about 130% of the untreated control. At the same time, BSA did not affect the growth parameters. Study of alterations in mitotic activity and endogenous WGA content in EBl-treated plants showed synchronous changes of both characteristics at both concentrations used for seed pretreatment and similar levels of mitosis stimulation (about 160% of control). An important difference between the concentrations was that in the case of 0.4 nM the highest mitotic activity and WGA accumulation were observed on the third day of growing, while 400 nM produced maximal effect on the fourth day. Combined treatment with both factors (EBl and exogenous WGA) stimulated cell division (about 175 % of control) only in four-day-old seedlings pretreated with 0.4 nM of EBl, which was totally inefficient when applied alone at this phase of plant development (Table 5). An explanation could be seen in low level of endogenous lectin during this period and compensation of the deficiency by the exogenous one. This was not the case in combined treatment of four-day-old seedlings with 400 nM of EBl, because at this concentration plants reached the highest level of endogenous WGA namely at the fourth day and additional (exogenous) WGA did not stimulate cell division. These facts suggest again possible co-operative action of BS and lectin in growth regulation.

Although a physiological function of WGA in plants is not fully disclosed at present time, there is a clear understanding of an important role of carbohydrate-binding proteins of such type in numerous recognition processes. Their interactions with carbohydrates existing on cell surface are crucial steps of many physiological reactions such as cell differentiation, fertilization, immune response, etc. The finding of interaction between BS and lectins could indicate a new branch of BS-signaling cascade overlapping the other system involved in developmental and adaptive processes in plants. Among the latter ones, a very promising area for further theoretical and applied studies of BS is connected with the investigation of BS-facilitated adaptation of plants to different stresses (abiotic and biotic). Although many data obtained from the beginning of BS research show their high efficiency as stress-protecting agents (reviewed in Khripach *et al.*, 1999, 2000), little is known about the molecular mechanism of the protective action. The data on modulation of gene expression and properties of membranes indicate that probably both genetic and non-genetic routes are involved, and a modification of regulation networks is a result of BS action leading to adaptability increase. BS show protective properties in different conditions of environmental stresses such as thermal, salt, herbicidal, drought and others.

Recent investigations made broader the circle of studied plants and brought better insight into the mechanistic aspect of the phenomenon. Thus, study on the effect of EBl on *Brassica napus* seedlings under thermal stress (Dhaubhadel *et al.*, 1999, 2002) showed that the treatment led to a significant increase of the seedling thermotolerance and accumulation of specific heat-shock proteins. This resulted from the activated biosynthesis and was connected with the suppressed degradation of some of the elements of the translational apparatus during stress period followed by their higher expression during recovery. The final result was manifested as a quicker resumption of cellular protein biosynthesis and higher survival rate of EBl-treated plants compared with untreated ones.

*Table 5. Effect of seed pretreatment on mitotic activity of apical meristem cells in roots of four-day-old wheat seedlings in the presence or absence of proteins (+ S.E.).*

| Treatment | Control | 0.4 nM EBl | 0.4 nM EBl + WGA | 0.4 nM EBl + BSA | 400 nM EBl | 400 nM EBl + WGA | 400 nM EBl + BSA |
|---|---|---|---|---|---|---|---|
| Mitotic Index, % | 4.03 ±0.22 | 3.23 ±0.20 | 7.07 ±0.34 | 4.60 ±0.28 | 6.35 ±0.31 | 5.98 ±0.29 | 6.60 ±0.32 |

The ability of BS to increase plant resistance to drought stress and prevent the loss of crop yield has a high value for production of different crops, especially cereals, in drought climate conditions. It was shown by studies on the EBl application to different varieties of spring wheat (Prusakova *et al.*, 2000a; Nilovskaya *et al.*, 2001) under normal and stress conditions (drought soil), and clear dependence of the result on the variety and manner of treatment (seeds or vegetating plants and phase) was demonstrated. A characteristic feature of plants treated with EBl (spraying, $10^{-5}$ ppm solution) in the beginning of booting stage or flowering, was higher water content in leaves of the upper tear (Prusakova *et al.*, 2000b). The water content increase had the highest value (about 10%) at drought conditions in the least drought-resistant variety, but it also took place in lower scale in all wheat genotypes at normal soil humidity. Independent data on osmotic determination of protoplasmic drought tolerance of leaf cells of resurrection grass *Sporobolus stapfianus* correlate with the previous results and give a new example of drought tolerance induction under BS action (Ghasempour *et al.*, 2001). During recent period other data on protective properties of BS under stress conditions were obtained, and some of them confirmed previously known alleviating effects of BS on plants under salinity stress (Anuradha and Rao, 2001) and hypoxia (Ramonell *et al.*, 2001).

As mentioned above, a combination of stimulative and protective BS activities in plants is often realized in plant productivity and product quality improvement. An example opening some details of the gross effect in lupine plants treated with EBl was reported recently (Zabolotnyi *et al.*, 2003). The amount of fruit elements on leguminous plant permanently decreases after ovule fertilization during development of ovaries. The amount of beans preserved till ripening depends strongly on plant ability to supply them with necessary nutrition components. This ability can be increased either by improvement of soil composition and feeding *via* roots or by stimulation of the uptake of nutrients in other way. It was shown that application of EBl facilitates setting and viability of beans. It goes together with growth increase of the whole plant and with increase of amino acid content. When plants were sprayed with a solution of EBl ($10^{-7}$-$10^{-9}$ M) twice, at the end of flowering and seven days later, the amount of beans on treated plants at the beginning of fruit formation was 31-36% higher than in control. The weight of juvenile beans exceeded that of control by 51%, whereas other parts of a plant were not more than 24% heavier. During ripening the difference preserved and developed into 29% seed yield enhancement. EBl treatment increased significantly the amount of free amino acids in plants.

Table 6 shows the content of amino acids in different parts of a plant at the phase of fructification. Especially high difference with the control values was observed in inflorescence axes, whereas amino acid content in stems of treated plants was even slightly reduced. These changes are supposed to be connected with the activation of metabolic processes in plants, but not with nitrogen fixation in the tubercles, because the latter one was not influenced by EBl treatment. The discussed effects produced by BS in plants create natural basis of their agricultural use. Although the effects do not cover the whole spectrum of BS activities, they reflect an essential part of it. Further details connected with practical applications of BS are presented in the next part of the article.

*Table 6. Content of free amino acids in lupine plants treated with EBl*

| Amino acid | Leaves | | Stems | | Inflorescence axes | |
|---|---|---|---|---|---|---|
| | mg/100g fresh weight | % of control | mg/100g fresh weight | % of control | mg/100g fresh weight | % of control |
| Alanine | 11.36 | 179 | 1.91 | 105 | 8.21 | 239 |
| Aspartic acid | 39.03 | 153 | 10.10 | 106 | 396 | 230 |
| Glutamic acid | 4.51 | 92 | 1.52 | 83 | 3.87 | 103 |
| Proline | 1.24 | 113 | 0.32 | 114 | 4.50 | 206 |
| Threonine | 2.75 | 180 | 0.82 | 87 | 9.03 | 205 |
| Valine | 1.05 | 109 | 0.21 | 65 | 4.55 | 198 |
| Sum total | 95.26 | 141 | 22.84 | 92 | 459.82 | 220 |

## SOME RESULTS OF AGRICULTURAL APPLICATION OF BRASSINOSTEROIDS IN RUSSIA AND BELARUS

Although the whole potential of agricultural use of BS is still not realized, the experience of 20-year-long investigations and trials has changed the question from “Is it really possible?” to “How to do it better?” Getting the answer on the last question implied solution of several problems. First, a complicated synthesis of BS could not immediately be realized in industrial scale. It took about a decade to elaborate appropriate synthetic methods. Second, exogenously applied BS act in a specific manner differing them from traditional agrochemicals. They influence plant itself and lead to the effect *via* activation of internal potentials of plants. Even in the cases of protective action against pathogens they produce no toxic effect, but act by stimulation of plant immunity. Their action in plants could be compared with the action of vitamins in human, and to achieve the desirable effect they have to be applied namely at the time, when a plant needs it and by the method which allows a plant to get them. Many failures of the attempts to improve crops by BS application could be a consequence of unsuitable time, dose, and mode of treatment. Nevertheless, there is a huge amount of evidence of positive BS application from different parts of the world. Preparations with BS as active ingredients are produced nowadays in China, Japan, India (Rao *et al.*, 2002). In Russia and Belarus, plant growth regulator Epin with EBl as active substance is used in agriculture started from 1992. Since that time many reports from practical plant-breeders and agricultural scientists have confirmed the advantage of BS application in plant cultivation. Further discussion concerns mainly the results on practical application of BS obtained during the last 10 years in Russia and Belarus.

In general, the properties that helped BS to move from the laboratories to agricultural fields can be shortly formulated as follows:

(a) BS stimulates physiological processes in plants leading to improvement of crop yield and quality.
(b) BS increase plant resistance against phytopathogens and can be used as partial substitutes for some traditional pesticides, diminishing in this way the unfavorable influence of pesticides on the environment.
(c) BS are natural products, widely spread in plant kingdom. They are included in the food chains of men and mammals, with which their metabolic pathways were interconnected during a long common evolution.
(d) Positive effect may be achieved with very small active doses of exogenous BS, which are comparable with their natural content in plants.

Table 7 shows crops, effects, dose and phase of development recommended for treatment of plants by Epin (0.25 g/l solution of EBl, containing a detergent) (Anon, 2000, 2002b). As follows from this table, it is approved for treatment of crops for the enhancement of germination rate, growth stimulation, increasing resistance to diseases and drought or cold conditions, stimulation of fruit setting and ripening. The effects result in crop yield and quality increase for many plant species, including cereals, potato and vegetables, fruit trees, ornamentals, and mushrooms. Further, more detailed discussion on the benefits of BS application for different crops is presented.

*Cereals*

BS application results in crop yield increase by 5-20% depending on cultivar, climatic conditions, type of soil and level of applied fertilizers, mode and time of treatment (Anon. 2002a).Besides increasing a germination rate, especially for old seeds, promoting growth and development in normal and stress conditions, BS treatment contributes to overcoming several specific for cereals problems. One of them is stability to lodgening, which is very important for the normal development of plants at the final stage. Treatment with BS enhanced the stem strength of barley plants and their resistance to lodgeability. Spraying with HBl at flowering or at booting stage gave approximately the same result, EBl spraying at flowering proved to be more effective than at booting (Prusakova *et al.*, 1995).Stem strength, as well as strength and elasticity of cell walls are dependent on content of calcium, which is a component of the cell walls. It was shown that treatment of spring barley seeds resulted in the enhancement of Ca content in roots and in aerial parts of the seedlings (Ageeva *et al.*, 2001). A distribution of Ca between different parts of a plant was also influenced: at the end of the earing stage Ca content was higher in the stems and lower in the leaves in comparison with the control. An opposite situation took place with potassium content. BS treatment favored the accumulation of K in leaves. Diameters of the first and the second internodes that can serve as an index of stem stability were increased by 4-9%. In the year with wet weather conditions application of EBl and HBl slightly inhibited the stem growth (5-16%). The productivity enhanced under the action of BS by 13-21%.

*Table 7. Application of Epin for different crops*

| Dose | Mode of treatment | Results |
|---|---|---|
| **Potato** | | |
| 20 ml/t | Treatment of tubers before planting | Crop yield and quality enhancement, immunity stimulation, resistance to phytophtora increase, decrease of nitrate, heavy metal and radionuclide accumulation |
| 80 ml/ha | Spraying at budding | |
| 1ml/l of nutrition medium | Addition to standard nutrition Murashige-Skoog medium when seeds material are grown by microcloning | Shortening growth period, increase of internode number, stimulation of root development, increase of number of plants suitable for planting |
| **Tomato** | | |
| 0,5 ml/kg | Soaking seeds for 2 h | Germination rate increase. Increase of resistance to unfavorable environment. |
| 100 ml/ha | Spraying at budding - beginning flowering of the first cluster | Increase of fruit setting, prevention of fruitlet abscission, acceleration of fruit ripening and enhancement of their quality. Enhancement of resistance to diseases, fungi protective action. Decrease of nitrate, heavy metal and radionuclide accumulation |
| **Cucumber** | | |
| 0.25 ml/kg | Soaking seeds for 2 h | Germination rate increase. Increase of resistance to unfavorable environment. |
| 100 ml/ha | Spraying twice - at the phase of 2 - 3 true leaves and at budding | Acceleration of fruit setting, prevention of fruitlet abscission, increase of early and whole crop, immune system stimulation, increase of resistance to peronosporosis, decrease of nitrate and heavy metal accumulation |
| **Pepper in greenhouses** | | |
| 0.1 ml/kg | Soaking seeds for 2 h | Germination rate increase. Increase of resistance to unfavorable environment. |
| 50 ml/ha | Spraying twice - at the beginning of budding and at flowering | Acceleration of fruit setting, prevention of fruitlet abscission, increase of early and whole crop, immune system stimulation, decrease of heavy metal and radionuclide accumulation |
| **Apple tree** | | |
| 200 ml/ha | Spraying twice - at the beginning of | Acceleration of fruit setting, prevention of fruitlet abscission, cold and drought |

| Dose | Mode of treatment | Results |
|---|---|---|
| | budding and after flowering with the 20-day interval | resistance increase, immune system stimulation, increase of resistance to scab, decrease of heavy metal and radionuclide accumulation |
| **Barley, spring and winter wheat** | | |
| 200 ml/t | Treatment of seeds | Germination rate increase. Crop yield increase, immune system stimulation, increase of resistance to leaf blight and root rot, decrease of heavy metal and radionuclide accumulation |
| 50 ml/ha | Spraying at tillering | |
| **Sugar beet** | | |
| 12 ml/t | Treatment of seeds before sowing | Crop yield increase, sugar content increase, decrease of nitrate, heavy metal and radionuclide accumulation |
| 100 ml/ha | Spraying at 2–3 true leaves | |
| **Buckwheat** | | |
| 12 ml/ha | Spraying at budding | Ripening acceleration, crop yield increase, improvement of consumer properties |
| **Gladioli** | | |
| 0.5 ml/kg | Soaking bulbs for 6 h | Acceleration of germination and pedicle emergence, increase of yield and quality of bulbs and bulblets, increase of resistance to bacteriosis and fusariose |
| **Roses** | | |
| 0.25 ml/500 pieces | Soaking green cuttings for 12-14 h | Acceleration of rooting the cuttings and increase of their quality |
| **Tulips** | | |
| 1 ml/kg | Soaking bulbs for 24 h | Acceleration of root formation, formation of uniform pedicles, increased resistance to diseases. Improvement of decorative properties of flowers, increase of bulb and bulblet yield. |
| 60 ml/ha | Spraying at budding | |
| **Mushrooms *Agaricus bisporus* and *Pleurotus ostreatus*** | | |
| 0.002 ml/1, 2 kg | Mycelium treatment | Crop yield increase, stimulation of fruit body formation |

| Dose | Mode of treatment | Results |
|---|---|---|
| 0.005 ml/m$^2$ | Spraying at fruit body formation. Addition to watering at the phase of fruit body formation, 3–4 times. | |
| **Cardiac motherwort, *Leonurus cardiaca* Moldavian dragonhead, *Dracocephalum moldavice*** | | |
| 50 ml/ha | Spraying twice - at 2–3 true leaves and in 6 days after the first treatment | Growth stimulation, crop yield increase |

In cereals, synchronous development of the main stem and lateral shoots is an important condition for a good crop. It is highly influenced by the weather conditions and can be improved in the unfavorable years by BS application. Synchrony of the ear formation and crop yield of four varieties of spring barley treated by Epin was estimated in the experimental conditions for 4 years: 1995-1998 (Vlasova *et al.*, 2000, 2002). Seedlings of four barley varieties differed in the synchrony of earing, e.g. time between appearance of the first ear on the main shoot and the last ear on the lateral shoots, were sprayed at the phase of two true leaves. Remarkable bushing increase under Epin action was marked in 1997 when the temperature conditions were very unfavorable (the temperatures fall down 2 to $5^0$C at tillering stage). In these conditions the amount of shoots was 80-100% higher in treated plants in comparison to the control. All treated plants had maximum amount of leaves characteristic for the variety. Although number of shoots increased, this did not initiate prolongation of plant development. Crop increase was achieved by stimulation of the productive bushing and enhancing number of grains in the ear and weight of one ear (Table 8). When two modes of treatment were compared, spraying of seedlings was found to be more efficient than seed soaking in the influence on synchrony of shoot development, shoot-forming capacity and productivity.

Application of BS stimulates plant-growth-promoting uptake of nutrients from soil. The increase of growth rate of barley treated with Epin on the background of various levels of mineral fertilizers has been documented (Vildflush *et al.*, 2001). Measurements were done in field trials on sward-podzolic light-loamy soils in Belarus in 1997–1999 with different temperature and rainfall conditions. Plants were treated with EBl at tillering stage. An accumulation of dry matter was monitored during plant life till maturity. An average content of dry matter within 3 years is shown in table 9.

Treatment of barley plants with Epin facilitated use of nutrients from soil during vegetation and increased grain yield by 360 kg/ha on the background of $N_{60}P_{40}K_{60}$ ($N_{60}P_{40}K_{60}$ means application of 60 kg of nitrogen, 40 kg of phosphorous, and 60 kg of potassium per hectare). A structure of a crop is presented in table 10. The

*Table 8. Effect of Epin on crop structure of different varieties of spring barley*

| Treatment | Variety | No. of productive shoots | Period[b] | No. of grains in one ear | Weight of grains(g) | |
|---|---|---|---|---|---|---|
| | | | | | One ear | One plant |
| Water | Vizit[a] | 4 | 65 | 20.2 | 0.84 | 3.35 |
| EBl | | 7 | 62 | 20.3 | 1.04 | 7.26 |
| Water | Zazerskiy[a] | 4 | 66 | 28.9 | 1.04 | 5.41 |
| EBl | | 7.2 | 64 | 29.2 | 1.35 | 7.89 |
| Water | Vezha | 9.3 | 47 | 19.8 | 0.81 | 7.48 |
| EBl | | 9.5 | 46 | 23.1 | 1.20 | 11.41 |
| Water | Lipen | 7.1 | 42 | 23.5 | 0.88 | 6.31 |
| EBl | | 9.9 | 41 | 26.1 | 1.07 | 10.53 |

[a] results of the unfavorable year
[b] days from sowing till emergence of the last ear

*Table 9. Accumulation of dry matter in barley plants at different stage of development grown on soil with different level of fertilizers and with or without Epin treatment (dry weight of 100 plants, g)*

| Treatment | Phase of plant development | | | | | |
|---|---|---|---|---|---|---|
| | Seedlings | Tillering | Booting | Earing | Milk ripeness | Wax ripeness |
| Without fertilizer | 8.4 | 30.2 | 86.8 | 122.6 | 153.2 | 170.0 |
| $N_{60}P_{40}K_{60}$ | 9.7 | 38.1 | 131.2 | 211.4 | 261.6 | 289.4 |
| $N_{60}P_{40}K_{60}$ + Epin | 10.2 | 38.4 | 142.2 | 237.4 | 284.9 | 313.4 |
| $N_{90}P_{40}K_{60}$ + Epin | 9.9 | 38.3 | 136.8 | 228.3 | 274.0 | 301.4 |
| $N_{90}P_{50}K_{90}$ | 10.2 | 39.2 | 140.5 | 234.0 | 281.4 | 309.4 |

data in table 9 and 10 show that application of Epin on the background of $N_{60}P_{40}K_{60}$ gives the same result as application of much higher dose of fertilizer - $N_{90}P_{50}K_{90}$. This allowed reducing amount of mineral fertilizers by about 30%, increasing the efficiency of crop production, and diminishing contamination of the environment. Use of Epin with higher than $N_{60}P_{40}K_{60}$ level of mineral fertilizers did not bring any improvement. The total protein content in the crop was not influenced by BS in this case.

Although Belarus and the middle Russia are characterized by a temperate climate with average temperature and rainfall conditions in summer suitable for cereals, very often brief frost at late spring or summer beginning takes a substantial part of crops. Draught is also a problem, especially in the last years. For this reason, stress-protective properties of BS continue to attract attention (Popova and Zotova,

*Table 10. Effect of mineral fertilizers and Epin on barley crop*

| Treatment | No of grains in the ear | Weight of grains in one ear | Weight of 1000 grains | Protein content, % | Crop yield, % |
|---|---|---|---|---|---|
| Without fertilizer | 14.1 | 0.64 | 45.2 | 9.5 | 100 |
| $N_{60}P_{40}K_{60}$ | 15.8 | 0.74 | 46.9 | 11.4 | 117.1 |
| $N_{60}P_{40}K_{60}$ + Epin | 17.3 | 0.81 | 47.3 | 11.6 | 120.6 |
| $N_{90}P_{40}K_{60}$ + Epin | 16.9 | 0.75 | 44.5 | 12.0 | 119.7 |
| $N_{90}P_{50}K_{90}$ | 16.7 | 0.77 | 46.6 | 12.5 | 119.9 |

2001; Chizhova *et al.*, 2001; Vedeneev and Deeva, 2001; Deeva *et al.*, 2001; Sanko, 2001). The reduction of plant damage caused by chilling stress was confirmed for spring barley. When young seedlings from treated seeds were kept at -5$^0$C for 12 h, their root length and growth were accordingly 22% and 31% higher in comparison with untreated plants (Tzareva, 2001). Combined application of EBl and fungicide allowed reducing unfavorable consequences of fungicide application to barley plants (Sudnic and Deeva, 2001).

Stimulation of plant internal potentials by exogenously applied BS results not only in better survival in stress conditions, but also in diminishing disease damage. For example, spraying barley plants at tillering phase with EBl decreased an extent of leaf diseases induced by mixed fungi infections (Pshenichnaya *et al.*, 1997; Khripach *et al.*, 1997c; Volynets *et al.*, 1997b; Manzhelesova, 1997). With a very small dose of 5 mg/ha, a percentage of damaged plants estimated at heading was diminished from 50% in control to 40% in the experiment. With the dose of 15 mg of EBl per hectare only 35% of experimental plants were damaged, the effect was similar to one induced by the fungicide Bayleton applied at the dose of 0.5 kg/ha. EBl-treated plants had better growth parameters and gave higher crop yield. The barley crop was 40% higher than those in control. Since BS do not possess fungicide activity, the effect can be completely attributed to activation of internal mechanisms of plant resistance. An ability of BS to regulate cell membrane permeability and transport of ions found an agricultural application in the areas polluted with heavy metals and radioactive debris. It was shown that treatment with EBl reduced significantly the absorption of heavy metals by barley, sugar beet, tomato, and radish (Khripach *et al.*, 1996a). For example, when barley plants were sprayed at the booting stage with 10 mg/ha of EBl, the absorption of lead, zinc, and copper was correspondingly 48, 60, and 59% of EBl-untreated control.

The absorption of radionuclides was studied for barley grown on the model soil enriched with cesium and strontium ions (Khripach *et al.*, 1997b). Table 11 shows Cs and Sr content in young plants and in different parts of mature plants. Three kinds of conditions were compared: 1–control plants grown on normal soil, 2 - plants grown on soil polluted with Cs (27 mg/kg) and Sr (8.3 mg/kg) salts, and 3 - soil the same as in no. 2, but plants were treated with 0.01 ppm of EBl solution at the booting stage. In all cases, the content of Cs and Sr in treated plants was diminished in comparison with the control grown on the polluted soil. In young plants, Cs content in treated plants was close to that grown on the clean soil. In mature plants, EBl treatment had higher efficiency in diminishing the pollutant content in grain than in straw. Cs and Sr content in grain was correspondingly 60 and 36% lower than those in the polluted control. For straw, diminution of contamination was only 10%. A mechanistic reason of the phenomenon of alteration in ion uptake from the environment to plants is the influence of EBl on active ion transport *via* membranes that favours the "biometal-competitor" ions in pairs $K^+$-$Cs^+$ and $Ca^{2+}$-$Sr^{2+}$.

*Table 11. Cesium and strontium content in barley plants*

| Treatment | Two weeks after treatment, mg/g dry wt. | | Mature plants, mg/g dry weight | | | |
|---|---|---|---|---|---|---|
| | | | Grain | | Straw | |
| | Cs | Sr | Cs | Sr | Cs | Sr |
| 1.Normal soil | 0.063 | 0.025 | 0.016 | 0.003 | 0.069 | 0.062 |
| 2.Polluted soil; control | 0.095 | 0.036 | 0.038 | 0.008 | 0.181 | 0.082 |
| 3. Polluted soil; Epin | 0.068 | 0.030 | 0.027 | 0.006 | 0.174 | 0.075 |

Application of EBl to wheat increased crop yield by 10–17% (Tzyganov *et al.*, 1999; Anon. 2002a). Treatment of wheat seeds with Epin promoted their germination. The results varied for different varieties from 5% enhancement for Enita to 30% for Saratovskaya 29. A protective action of EBl on cereals in drought soil conditions was demonstrated on spring wheat of drought-resistant variety Saratovskaya 29 and Enita adapted to sod-podzol soil of Russia (Table 12) (Prusakova *et al.*, 1999b, 2000b). In addition, treatment with Epin increased protein content (15-30%) in the crop and decreased starch content (6-19%). When two mode of BS treatment were compared, soaking seeds and spraying plants, the first one was found to be preferable (Nilovskaya *et al.*, 2001). Plants of spring wheat, Enita and Lada, obtained from EBl treated seeds, showed better growth parameters, than plants treated with EBl at the beginning of dry period. The crop yield increased mainly due to the enhanced grain number per ear (Table 13).

Buckwheat is a traditional for Russia cereal, which gained popularity because of excellent consumer properties including a large content of digestible proteins. In spite of these properties, its cultivation is not as abundant as other cereals because of a low crop yield, which is mainly a consequence of the heterogeneous development of fruiting elements. Treatment of buckwheat with EBl at the beginning of flowering helped to synchronize partially growth of shoots, setting, and ripening which resulted in the enhancement of plant productivity (Prusakova *et al.*, 1999a) (Table 14). The

*Table 12. Effect of Epin on productivity of spring wheat in drought conditions*

| | Saratovskaya 29 | | Enita | |
|---|---|---|---|---|
| | 12%[a] | 50%[a] | 12%[a] | 50%[a] |
| Number of lateral shoots | 1.67 (1.50)[b] | 2.30 (2.05) | 1.63 (1.28) | 2.65 (1.48) |
| Ear length, cm | 6.20 (6.10) | 9.10 (6.98) | 9.50 (8.93) | 9.30 (8.20) |
| Number of grains in the ear | 24.1 (21.9) | 42.5 (24.2) | 36.9 (35.1) | 41.0 (34.8) |
| Weight of grains from one plant, g | 1.96 (1.61) | 3.49 (1.91) | 2.51 (2.23) | 3.63 (2.13) |

[a] Percentage of optimal water supply
[b] In parentheses, values for control experiments are given

*Table 13. Increase of wheat grain weight (g per plant) in normal and dry conditions depending on the mode of EBl treatment*

| Variety | Opt. water, soaking seeds | Dry conditions, soaking seeds | Opt. water, spraying | Dry conditions, spraying |
|---|---|---|---|---|
| Enita | 0.59 (0.29)[a] | 0.47 (0.26) | 0.43 (0.30) | 0.35 (0.26) |
| Lada | 0.38 (0.28) | 0.33 (0.25) | 0.32 (0.28) | 0.27 (0.24) |

[a] In parentheses, values for control experiments are given.

content of starch and proteins in the crop was not influenced substantially by EBl treatment. In contrast, seed treatment and/or spraying at budding increased content of chlorophyll and carotinoids in the leaves by 20-50% depending on the *cv*. The best result was observed with two treatments (Kolotovkina *et al.*, 2001).

*Table 14. Effect of EBl on the growth, development and productivity of buckwheat*

| Treatment | Height of plants, cm | Length of the first order shoots | No. of inflo-rescences per plant | No. of seeds per plant | Weight of 1000 seeds, g |
|---|---|---|---|---|---|
| Control | 79.4 ± 2.2 | 16.3 ± 1.5 | 9.1 ± 0.7 | 34.8 ± 1.5 | 22.3 ± 0.05 |
| EBl | 94.3 ± 2.1 | 42.7± 1.7 | 16.3 ± 0.8 | 51.1 ± 2.1 | 22.7 ± 0.1 |

*Potato*

It was the first culture officially registered in Russia for treatment with BS for crop improvement. BS effect on potato is connected with various aspects of plant development. Spraying potato plants with BS enhanced crop yield due to increasing size of tubers. Crop had better quality with respect to diminishing nitrate content and the enhancement of starch and vitamin C content (Khripach *et al.*, 1996b). Alteration of potato food value resulted from the treatment of plants at budding with BS is presented in table 15. From these data it can be concluded that dose of 10–20 mg/ha is an optimal choice.BS enhanced resistance of potato to diseases. EBl spraying at budding or flowering stage diminished development of fungi infection on the plants. With respect to phytophthora suppression, the efficient method consisted in spraying plants at the beginning of budding phase with 10-20 mg/ha of BS (Bl, EBl, or HBl) (Khripach *et al.*, 1996c, 1997a). An extent of diminishing of disease development varied for different cultivars and was especially high for the *cv.* Rosinka that was the most fungi-sensitive in control (Table 16).

Unlike the standard fungicides, BS does not have direct fungistatic activity and act *via* stimulation of natural resistance mechanism. A comparison of potato plants treated twice with 2 kg/ha of fungicide arcerid (composition of ridomil and polycarbacine) and potato plants sprayed at the beginning of budding with 20 mg/ha of HBl revealed very close fungi protective effect for both cases and higher crop for HBl-treated plants. BS increases an economic expediency of an application of fertilizers and microelements (Anon. 2002a). In field experiments, on the background with optimal NPK content in soil, treatment of potato plants at budding beginning of flowering with microelements resulted in 17% yield enhancement.

*Table 15. Effect of BS on crop yield and food value of potato*

| Treatment | Crop increase, % | Dry matter, % | Starch, % | Nitrates, mg/kg | Vitamin C, mg/kg |
|---|---|---|---|---|---|
| Control | | 22.8 | 15.7 | 113.6 | 19.0 |
| Bl, 5 mg/ha | 8.5 | 23.6 | 15.9 | 98.8 | |
| Bl, 10 mg/ha | 18.3 | 24.2 | 16.9 | 85.3 | |
| EBl, 5 mg/ha | 6.0 | 22.8 | 15.8 | 98.5 | |
| EBL, 10 mg/ha | 14.5 | 23.2 | 16.4 | 96.7 | |
| EBl, 20 mg/ha | 16.7 | 23.7 | 16.7 | 89.3 | 25.0 |

*Table 16. Effect of BS on the development of phytophthora infection in potato*

| Treatment | Tubers damaged by phytophthora, % of total amount (% of control) | | |
|---|---|---|---|
| | *cv.* Orbita | *cv.* Sante | *cv.* Rosinka |
| Control | 1.9 (100) | 2.5 (100) | 2.9 (100) |
| EBl | 1.7 (89) | 1.9 (76) | 1.9 (66) |
| HBl | 1.6 (84) | | |

When it was combined with EBl treatment, crop yield increase was 25% (Goncharic, 2001). The result of BS application to potato by treatment of potato tubers is not as unequivocal as one obtained when young plants are sprayed. When potato tubers were treated with EBl directly before planting, this resulted in diminishing the phytopthora infection of plants during the vegetation period (Filipas and Ul'yanenko, 2001). The percentage of damaged tubers of new crop was diminished also: 11% for treated plants and 21% for control. Treatment immediately after harvesting increased a production of ethylene and stimulated biosynthesis of protective phenolic and terpenoid compounds that prolonged a period of deep dormancy, promoted storage of tubers, and enhanced resistance to phytopthora infection and to other diseases (Korableva *et al.*, 1992, 2002). However, it was shown that EBl and HBl treatment at certain conditions promoted mycelia (*Phytophtora infestans*) growth on potato tubers (Vasyukova *et al.*, 1994). A protective action of BS was demonstrated by the stimulation of potato resistance to virus infection (Rodkin *et al.*, 1997; Bobric *et al.*, 1998). In the production of potato planting material from meristem *in vitro,* the introduction of BS in the nutrition medium proved to have beneficial effect on all growth and developmental parameters of cuttings, increased the efficiency of the reproduction, and had long-term effect on the productivity and resistance to virus infection of plants grown from the tubers of the first generation. BS stimulated growth of roots and aerial parts, cuttings achieved planting stage in 8 days (14 in control) and contained more internodes (7-9 versus 3-4 in control). Plants obtained from treated cuttings were not damaged by mosaic virus. In the next generation, the obtained seed tubers gave higher yield of crop, which was less damaged by virus infection.

*Tomato*

Spraying tomato plants grown on soil with optimal NPK mineral content at the beginning of flowering with EBl solution increased fruit weight by 45% (Likchacheva *et al.*, 2001). In the conditions of short and cold summer of Karelia, growing of tomatoes in greenhouses implies treatment with a complex of growth regulators. Due to lack of sunlight, for young seedlings a retardant is usually used for preventing excessive elongation. After planting seedlings on the permanent place, treatment with EBl neutralized retardant action, promoted growth, fruiting, and increased crop yield (Budykina *et al.*, 2001). Study of the influence of Epin on growth and productivity of tomatoes in greenhouses revealed for treated plants better growth parameters including height of plants, number and total square of leaves, diameter of stem, weight of aerial part of plants and roots, which are presented in table 17 (Matevosian *et al.*, 2001). Two treatments were used: first–soaking seeds, and second–spraying at the beginning of budding of the first cluster. Detailed analysis of Epin effect on flowering and fruit setting of tomato plants in greenhouses is presented in table 18. Acceleration of flowering, fruit setting and ripening on the first four clusters allowed obtaining the first crop 3-5 days earlier. The whole crop from the plants treated with Epin was 24% higher than those in the control. Tomatoes had higher content of ascorbic acid and carbohydrates.

A quality of tomato fruits grown on soil with enhanced content of heavy metals can be improved by soaking seeds in EBl solution (Table 19). Crop of plants grown from treated seeds took smaller amount of metal (Cd, Zn) from a soil in comparison with control (Khripach *et al.*, 1996a). Unexpectedly, a combination of two treatments, soaking seeds and spraying plants at budding stage, was less effective than soaking seeds only.

*Table 17. Effect of Epin on growth parameters of 60-day-Old tomato plants*

| Treatment | Height, cm | Leaves no. | Leaves sq., $cm^2$ | Stem diameter, mm | Fresh weight, g | |
|---|---|---|---|---|---|---|
| | | | | | Aerial part | Roots |
| Control | 40.7 | 7.9 | 755 | 42.5 | 42.5 | 7.1 |
| Epin | 41.3 | 8.2 | 797 | 47.3 | 47.3 | 8.3 |

*Table 18. Effect of Epin on flowering and fruit setting of tomato*

| Treatment | Clusters | | | | | | | | | Total on 4 clusters | | |
|---|---|---|---|---|---|---|---|---|---|---|---|---|
| | 1 | | | 2 | | | 3 | | | | | |
| | A[a] | B | C | A | B | C | A | B | C | A | B | C |
| Water | 10,3 | 6,7 | 65,0 | 11,3 | 6,9 | 61,1 | 10,7 | 5.8 | 54,2 | 43,9 | 25,3 | 57,6 |
| Epin | 12,0 | 8,3 | 69,2 | 14,2 | 9,9 | 69,7 | 12,3 | 6,9 | 56,1 | 50,5 | 31,3 | 62,0 |

[a] A - number of flowers, B - number of fruits, C - percent of fruit setting.

*Table 19. Effect of EBl on content of heavy metals in tomatoes*

| Treatment | Cd, mg/kg | Zn, mg/kg |
|---|---|---|
| Polluted soil; control | 0.05 | 5.0 |
| Polluted soil; EBl-treated seeds | 0.02 | 3.3 |
| Polluted soil; EBl-treated seeds; EBl-sprayed plants | 0.04 | 3.6 |

*Cucumber*

Soaking seeds for 20 h in EBl solution resulted in 15% improvement of germination (Elagina and V'iygina, 2001). In 13-day-old seedling dry weight was 10%, Leaf Square 8%, length of the main root 16%, chlorophyll content 11%, photosynthesis intensity 10%, and transpiration intensity 7% higher than in the control. The alteration of the physiological processes under Epin action resulted in the enhancement of early crop yield (2.18 kg/$m^2$ versus 1.71 kg/$m^2$ in the control). Treated plants were more resistant to cold (Popova and Zolotar, 2001) and drought soil (Pustovoitova *et al.*, 2001) conditions. Several treatments of cucumber: soaking seeds in EBl solution, spraying plants at 3-4 true leaves and at the beginning of flowering increased early crop by 24% and the whole crop by 16% (Timeiko *et al.*, 2001). EBl treatment resulted not only in crop yield increase, but also in improvement of consumer properties of cucumbers. Content of dry matter, vitamin C and potassium was increased by 14, 15, and 5%, respectively.

*Carrot*

Application of BS for crop yield increase in this case is not obvious at the first glance because of the contradictory results concerning root response to BS (reviewed in Khripach *et al.,* 1999). Root vegetables give crops, which result from the differential development of root tissue, and there are several indications that BS treatment facilitates the formation of these crops. In carrot, first aerial parts of plants were influenced, that resulted in better-developed and stronger plants, which gave higher crop (Budai, 2000). Soaking carrot (*Daucus carota* L.) seeds in EBl solution accelerated germination, at the 5-th day from its beginning a number of seedlings from treated plants was two-fold higher than in the control (Table 20). Later the difference became less substantial. At the stage of 7-8 true leaves, when the formation of fruiting root body started, the difference in the quantitative parameters of plants from treated and untreated seeds was still remarkable, being more substantial for leaves than for roots. As fruiting root body developed, experimental plants showed better growth of roots, and in 70-day-old plants their diameter and fresh weight were 40% higher than for untreated ones.

*Table 20. Effect of EBl on the development of carrot plants*

| Treatment | Germination rate (%) | | | Phase of 7-8 true leaves | | | 70-day-old plants | | |
|---|---|---|---|---|---|---|---|---|---|
| | Days from the beginning of germination | | | Fresh weight of a plant (g) | Dry weight (mg) | | Diameter of the carrot root (mm) | Fresh weight (g) | |
| | 2 | 5 | 8-11 | | Leaves | Carrot root | | Leaves | Carrot root |
| Control | 12 | 23 | 42 | 1.0 ± 0.1 | 62.2 ± 21.2 | 7.5 ± 0.6 | 7.8 ± 0.6 | 7.5 ± 0.6 | 7.8 ± 0.6 |
| EBl | 37 | 45 | 54 | 2.5 ± 0.7 | 154.0 ± 48.1 | 9.5 ± 0.6 | 12.0 ± 0.6 | 9.5 ± 0.6 | 12.0 ± 0.6 |

*Beetroot*

Soaking beetroot (*Beta vulgarisl*) seeds in EBl solution enhanced germination rate of old seeds, although for fresh seeds the effect was less expressed. Development of primary root was accelerated that resulted in enhanced synchrony of germination. The EBl treatment facilitated leaf development and accelerated the formation and growth of roots (Budai and Laman, 2001). At the phase of 3 true leaves, dry weight of aerial parts of plants from treated seeds was 38% higher than those in the control (Table 21), the difference for leaf blades was even larger and came to 60%. EBl treatment facilitated an atrophy of cotyledons at the phase of 4–7 true leaves promoting the development of leaves and roots. At the phase of 7 true leaves, when root tissue differentiated to form fruit body, leaves and roots of treated plants showed better growth parameters. The advantage was preserved during the whole life of plants and was realized in a 30% crop enhancement. Crop yield of sugar beet and yield of sugar from one hectare were improved by EBl application by 23% and 14.9%, correspondingly (Anon. 2002a). Crop from sugar beet plants sprayed with 5 mg/ha of EBl at the phase of 3 true leaves, contained 50% lower content of Pb and 8% lower content of Cd than untreated control (Khripach *et al.*, 1996a).

*Herbs*

BS were used for the enhancement of productivity of herbs *Aralia mandzhurica, Atractylodes ovata, Veronica Iogifolia*, and *Eleuteroccocus senticosus* in the Far East area of Russia. There are several problems connected with their cultivation. The first one is very small yield of seeds because of a low percentage of fertilized flowers. Another problem is that seeds have very long period of dormancy, and germination is slow and of a low percent.

*Table 21. Effect of EBl on the development of beetroot plants*

| Treatment | Phase of 3 true leaves, dry wt., mg | | Phase of 7 true leaves | | | |
|---|---|---|---|---|---|---|
| | Total aerial part | Leaf blades | No. of leaf blades | Diameter of the root, mm | Fresh weight, g | |
| | | | | | Leaves | Roots |
| Control | 69.3 | 39.7 | 8.1 | 5.3 | 6.18 | 0.70 |
| EBl | 95.5 | 60.5 | 9.3 | 7.0 | 8.68 | 0.80 |

Spraying the wild plants on inflorescences allowed increasing seed yield. The results for aralia and eleutherococcus for two years are given in table 22. A weight of 1000 seeds increased by 12-17%, and the percentage of unfertilized flowers decreased substantially. As a result, yield of seeds was 2-3 times higher in comparison with the control (Dulin and Stepanova, 2002). An influence of Epin treatment was extended to the next generation of plants. Seeds obtained from sprayed plants had shorter period of dormancy and higher germination rate (Table 23) that was probably connected with changing the endogenous hormonal balance resulted in relieving physiological retardation. Field germination of seeds and productivity of plants were influenced by the treatment of the inflorescences of previous year. Aralia plants grown from the seeds taken from sprayed plants of former generation had better developed roots and higher chlorophyll content. A content of active principles per plant was 37% higher for treated plants in comparison with control.

Soaking seeds of legumes herbs (*Sophora flavescens* Soland.) and licorice (*Glycyrrhisa pallidiflora* Maxim.) in Epin solution did not influence substantially on germination in normal conditions. However, when temperature was lower than the optimal one, treated sophora seeds (which needed higher temperature) germinated better (Dulin *et al.*, 2002). Dry weight of roots of plants from treated seeds was about 20% higher than in the control. This resulted in the enhanced per-plant production of flavonoids, which are a part of the active complex determining the therapeutic value of these herbs.

*Table 22. Effect of Epin on seed yield of aralia and eleutherococcus*

| Treatment | Aralia | | | | Eleutherococcus | | | |
|---|---|---|---|---|---|---|---|---|
| | 1999 | | 2001 | | 1999 | | 2001 | |
| | 1000 seeds Wt (mg) | Unfertil. flowers, % | 1000 seeds Wt (mg) | Unfertil. flowers, % | 1000 seeds Wt (mg) | Unfertil. flowers, % | 1000 seeds Wt (mg) | Unfertil. flowers, % |
| Control | 770 | 50 | 790 | 67 | 6.6 | 65 | 5.7 | 81 |
| Epin | 900 | 32 | 890 | 38 | 7.8 | 40 | 6.5 | 53 |

*Table 23. Effect of Epin applied by inflorescence spraying of aralia and eleutherococcus on germination of seeds of the next generation*

| Treatment | Seeds germinated (%) | | | | | |
|---|---|---|---|---|---|---|
| | Aralia | | | Eleutherococcus | | |
| | 20 days | 30 days | Whole period | 5 month | 6 month | Whole period |
| Control | 0 | 10 | 68 | 0 | 2.0 | 48 |
| Epin | 4.4 | 42.3 | 74 | 5.8 | 44.4 | 58 |

*Legumes*

The bean first- and the bean second-internode bioassays, which were used to monitor BS isolation since the time of their discovery, can be considered as the first investigation of BS effect on legumes. Crop yield increase in the interval of 7%-26% is documented for soybean, bean, pea, lupine, and other legumes (cited in Khripach *et al.*, 1999). New investigations brought additional knowledge of details of BS effects. Thus, treatment of lupine with EBl increased the supply of fruit elements with amino acids promoting their viability and increasing seed productivity by 29% (Zabolotnyi *et al.*, 2003).

Spraying of bean plants with EBl or HBl at budding phase or twice, at budding and at the phase of 6-7 true leaves, stimulated growth and resulted in crop enhancement mainly due to an increase of the weight of 1000 seeds. When bean plants at the phase of true leaf formation were sprayed with EBl solution, their hormone status was changed and the intensity of photosynthesis increased. As a result, at maturity dry weight of treated plants and seeds was 27% higher in comparison with the control (Likchacheva, 2001). Yield of lupine seeds increased by 10% when plants were sprayed with Epin at flowering (Tzyganov *et al.*, 2001; Persikova, 2001).

*Flax*

It is a traditional technical culture in Belarus, which is grown for production of fiber and linseeds. Application of BS to it proved to have favorable effect on yield of straw and seeds, facilitated treatment of straw and improved quality of resulted fibers. Soaking old seeds in EBl solution enhanced their germination capacity by 11%, whereas with fresh seeds the effect was not noticed. Plants from treated seeds gave higher yield of seeds and straw, with the last one being of better quality (Khodiankova and Duktov, 2002). In drought conditions, EBl activated root system, normalized nutrition uptake, and enhanced the yield of seeds and straw (Khodiankov, 2001). Spraying plants at early stage of development with EBl solution had positive effect on their growth and productivity (Kukresh and Khodiankova, 2002). When plants were sprayed with EBl solution in a dose of 10-30 mg/ha, yield of seeds and straw increased (Voskresenskaya, 1998). Visible effect was marked already with 10 mg, but 20-25 mg gave better results (Figure 4). Treated plants had enhanced stability to lodgening, earlier formation of seeds, and ensured production of fibers with improved mechanical properties.

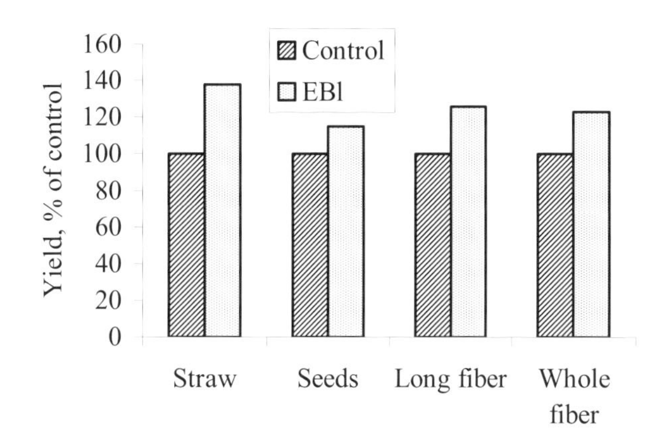


*Figure 4. Influence of EBl on productivity of flax cv. Mogilevskii*

*Miscellaneous*

BS have found an application for ornamentals. Treatment of gladiolus and tulip bulbs and bulbils with EBl solution stimulated their germination and rooting, accelerated sprouting and emergence of floriferous shoots and flowers on 5-7 days, flowers on treated plants were of bigger size. BS, however, exerted the most prominent effect, on the yield and quality of bulbs and bulbils. Their number and weight could be increased up to 60 and 85%, correspondingly, their resistance against fungal infections was enhanced (Runkova, 1995). A development of lily bulbs was also stimulated by BS treatment (Kilchevskii and Frantsuzionok, 1997). BS facilitated reproduction of ornamentals by cloning (Runkova, 2000). Cuttings of chrysanthemum, roses, lilac, some shrubs, and coniferous after treatment with Epin had better rooting capacity: a number and length of roots increased, zone of rooting on the stem was longer. Treated cuttings better survived when planted. Their aerial parts developed faster due to better rooting. An additional spraying with Epin stimulated plant growth. For roses Epin was recommended for enhancement of chilling resistance (Malevannaya and Kositsina-Pinegina, 1996). Application of EBl to grape improved substantially crop yield and diminished a frost damage of buds (Chirilov *et al.*, 1996). Enhancement of berry production and resistance to chill and disease damage was mentioned for black and red currant, and gooseberry (Malevannaya and Bednarskaya, 1995).

BS have found an application not only for increasing productivity of higher plants, but also for growth stimulation of agaricas and algae. Short soaking of inoculum of cultivated mushrooms in Epin solution promotes or inhibits the mycelium growth depending on the solution concentration (Alekseeva *et al.*, 1999). At the concentrations of 0.002–0.004%, which was found to be optimal ones, a diameter of the mycelium of treated culture was 15% bigger at the 5-th day in comparison with the untreated control. The treatment diminished time for emerging the first crop. With the second treatment by spraying at the beginning of fruiting, crop increase was 40 - 50%. Addition of EBl to nutrition medium stimulated growth of different algae (*Spirulina platensis*, *Euglena gracilis*, *Dunaliella salina*, *Chlorella vulgaris*). The effect ranged from 5% to 30% depending on the alga, light condition, and concentration of EBl solution. A growth of biomass of algae resulted, at least partially, from the increase of number of cells in comparison with the control. At concentrations higher than $10^{-7}$ M, EBl inhibited alga growth (Melnikov *et al.*, 1999).

Usual mode of BS application in agriculture consists in soaking seeds in appropriate solution or spraying young plants. A different approach has been elaborated and offered for the use (Pirogovskaya *et al.*, 1996, 1997; Pirogovskaya, 2000). Addition of EBl to fertilizers with prolonged period of action allows increasing crop yield and improving quality of several crops including cereals, potato, vegetables, and technical crops. Owing to a special composition, these slow acting fertilizers release minerals and growth regulators in a soil gradually, providing their long-term consumption and decreasing degradation and washing-out the constituents. The fertilizers can be used on soils with various texture and moisture content. They provided crop yield increase of forage by 11-14 %, oats–by 6-20%, barley–by 8%, potato–by 5-10%, spring wheat–by 17% depending on type of a soil and composition of a fertilizer.

BS application as a component of fertilizers gives results similar to their usage in a usual way. One of the useful properties of BS applied as foliar spray or by soaking seeds, already discussed earlier, is a diminution of heavy metal and radionuclide accumulation by plants. On soils contaminated with radioactive debris, an application of slow-release fertilizers with EBl also diminished a pollution of crops. In table 5, uptake of $^{137}Cs$ and $^{90}Sr$ by potato tubers grown on the same soil, but with different fertilizers, is compared. Application of potassium fertilizers allowed reducing the contamination of tubers, addition of EBl to the same fertilizers gave further diminution of the contamination.

The published data show various aspects of BS application. Accumulated knowledge of BS action on the molecular level clarified in a large extent the effects obtained in fields. Further progress in investigation of mechanism of BS action in plants, on the one hand, and elaboration of economically feasible schemes of synthesis of natural BS and their analogs, on the other hand, will surely make a basis for inclusion of this new class of plant hormones in the regular package of chemicals used for optimization of agricultural production.

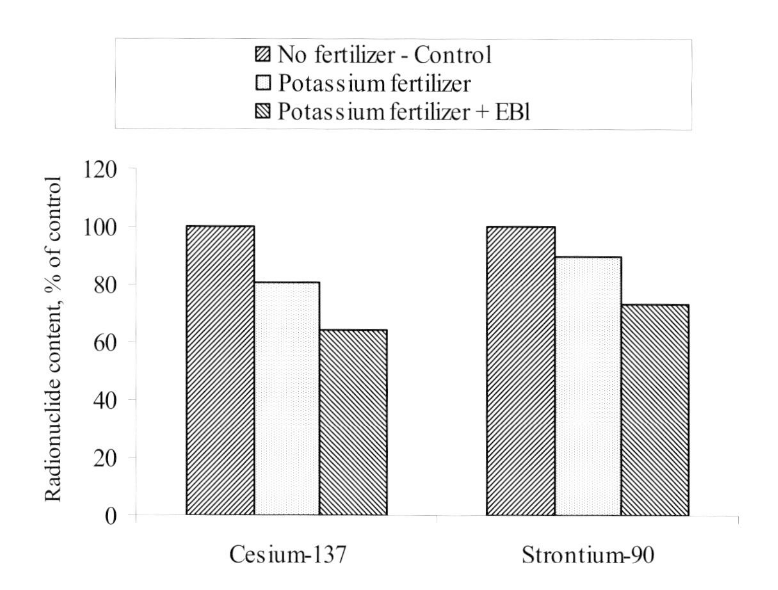


*Figure 5. Effect of EBl addition to potassium slow-release fertilizers on $^{137}Cs$ and $^{90}Sr$ content in potato tubers*

REFERENCES

Abe, H., Marumo, S. (2002). Advance of new plant hormone brassinosteroid. ITE Letters on Batteries, New Technologies and Medicine 3:43-48.

Abe, H., Soeno, K., Koseki, N.N., Natsume, M. (2001). Conjugated and unconjugated brassinosteroids. ACS Symposium Series (Agrochemical Discovery) 774:91-101.

Ageeva, L.F., Prusakova, L.D., Chizhova, S.I. (2001). Influence of brassinosteroids on stem formation and potassium and calcium ions content in spring barley plants. Agrokchimiya 6:49-55.

Alekseeva, K.L., Malevannaya, N.N., Khripach, V.A., Zhabinskii, V.N. (1999). Method of growth stimulation of agarics. Pat. RU 2,160,000.

Altmann, T. (1999). Molecular physiology of brassinosteroids revealed by the analysis of mutants. Planta 208:1-11.

Amzallag, G.N. (2002). Brassinosteroids as metahormones: evidence for their specific influence during the critical period in *Sorghum* development. Plant Biology 4:656-663.

Anon. (2000). List of pesticides permitted for application in Belarus for 2000-2010. Minsk:Uradzhai, pp. 199-201, 287-288.

Anon. (2002a). Application of growth regulators in growing the main agricultural crops. Recommendations for collective farms and farmers. Gorky. 28 pp.

Anon. (2002b). List of pesticides permitted for application in Russian Federation. Suppl. to Zashchita i karantin rastenii, pp. 287-293.

Anuradha, S., Rao, S.S.R. (2001). Effect of brassinosteroids on salinity stress induced inhibition of seed germination and seedling growth of rice (*Oriza sativa* L.). Plant Growth Regulation 33:151-153.

Asami, T., Min, Y.K., Nagata, N., Yamagishi, K., Takatsuto, S., Fujioka, S., Murofushi, N., Yamaguchi, I., Yoshida, S. (2000). Characterization of brassinazole, a triazole-type brassinosteroid biosynthesis inhibitor. Plant Physiology 123:93-100.

Asami, T., Min, Y.K., Sekimata, K., Shimada, Y., Wang, J.M., Fujioka, S., Yoshida, S. (2001). Mode of action of brassinazole: a specific inhibitor of brassinosteroid biosynthesis. ACS Symposium Series (Agrochemical Discovery) 774:269-280.

Bajguz, A., Tretyn, A. (2003). The chemical characteristic and distribution of brassinosteroids in plants. Phytochemistry 62:1027-1046.

Bezrukova, M.V., Aval'baev, A.M., Kildibekova, A.R., Fathutdinova, R.A., Shakirova, F.M. (2002). Interaction of wheat lectin and 24-epibrassinolide in regulation of cell division in wheat roots. Doklady Academii Nauk 387:276-278.

Bishop, G.J., Yokota, T. (2001). Plants steroid hormones, brassinosteroids: current highlights of molecular aspects on their synthesis/metabolism, transport, perception and response. Plant and Cell Physiology 42:114-120.

Bobric, A.O., Khripach, V.A., Zhabinskii, V.N., Zavadskaya, M.I., Litvinovskaya, R.P. (1998). Method of reproduction of improved potato seed material. Pat. Appl. BY 19981189.

Budai, S.I. (2000). Influence of seed treatment with growth regulators on germinating power and morphologic features of upcoming carrot plants (*Daucus carota* L.). Vestsy Natsionalnoi Academii Nauk Belarusi, Seriya biologicheskich nauk 3:38-41.

Budai, S.I., Laman, N.A. (2001). Influence of growth regulators on morphologic features of beetroot plants (*Beta vulgaris* L.V. Esculenta) at the phase of crop formation. Vestsy Natsionalnoi Academii Nauk Belarusi, Seriya biologicheskich nauk 1:15-18.

Budykina, N.P., Drozdov, S.N., Kurets, V.K., Khripach, V.A. (2001). Synthetic growth regulators–means of enhancement of tomato production in greenhouses in Karelia. 2$^{th}$ Conference on Regulation of Plant Growth and Productivity. Minsk, p.26.

Catterou, M., Dubois, F., Schaller, H., Aubanelle, L., Vilcot, B., Sangwan-Norreel, B.S., Sangwan, R.S. (2001). Brassinosteroids, microtubules and cell elongation in Arabidopsis thaliana. I. Molecular, cellular and physiological characterization of the Arabidopsis bull mutant, defective in the $\Delta^7$-sterol-C5-desaturation step leading to brassinosteroid biosynthesis. Planta 212:659-672.

Chirillov, A., Khripach, V., Toma, S., Scurtul, A., Zhabinskii, V., Cozmic, R., Zavadskaya, M., Erlinman, J. (1996). The procedure of cultivation of grape. Pat. MD 701F1.

Chizhova, S.I., Golantseva, E.N., Prusakova, L.D., Tretiakov, N.N., Jakovlev, A.F. (2001). Influence of epibrassinolide and emistim on drought tolerance of wheat of different genotypes. 6$^{th}$ Conference on Regulators of Plant Growth and Development in Biotechnology. Moscow, pp. 131-132.

Choe, S., Feldmann, K.A. (2002). Dwf7 mutants. U.S. Pat. Appl. US 20020068822 A1.

Choe, S., Noguchi, T., Fujioka, S., Takatsuto, S., Tissier, C.P., Gregory, B.D., Ross, A.S., Tanaka, A., Yoshida, S., Tax, F.E., Feldmann, K.A. (1999). The Arabidopsis dwf7/ste1 mutant is defective in the $\Delta^7$ sterol C-5 desaturation step leading to brassinosteroid biosynthesis. Plant Cell 11:207–221.

Chory, J., Li, J. (1997). Gibberellins, brassinosteroids and light-regulated development. Plant Cell Environment 20:801-806.

Clouse, S.D. (2001). Integration of light and brassinosteroid signals in etiolated seedling growth. Trends in Plant Science 6:443-445.

Clouse, S.D. (2002). Brassinosteroids. Plant counterparts to animal steroid hormones? Vitamins and Hormones 65:195-223.

Deeva, V.P., Vedeneev, A.N., Sanko, N.V., Colovei, K.I., Tzareva, E.G. (2001). Mechanisms of regulation of metabolic processes and adaptive properties of different genotypes by growth regulators. 2th Conference on Regulation of Plant Growth and Productivity. Minsk, pp.57-58.

Dhaubhadel, S., Chaudhary, S., Dobinson, K.F., Krishna, P. (1999). Treatment with 24-epibrassinolide, a brassinosteroid, increases the basic thermotolerance of *Brassica napus* and tomato seedlings. Plant Molecular Biology 40:333-342.

Dhaubhadel, S., Browning, K.S., Gallie, D.R., Krishna, P. (2002). Brassinosteroid functions to protect the translational machinery and heat-shock protein synthesis following thermal stress. Plant Journal 29:681-691.

Dulin, A.F., Stepanova, T.A. (2002). Influence of growth regulators on seed germination of aralia and eleutherococcus. Agrokhimiya 4:42-47.

Dulin, A.F., Stepanova, T.A., Matiuschenko, N.V. (2002). Influence of growth regulators on seed quality of some legume herbs. Agrokhimiya 7:56-60.

Elagina, E.M., V'iygina, G.V. (2001). Influence of epibrassinolide on the physiological parameters of cucumber. 6th Conference on Regulators of Plant Growth and Development in Biotechnology. Moscow, p. 28.

Filipas, A.S., Ul'yanenko, L.N. (2001). Influence of growth regulators of potato crop yield and quality. 6th Conference on Regulators of Plant Growth and Development in Biotechnology. Moscow, p. 284.

Ghasempour, H.R., Anderson, E.M., Gaff, D.F. (2001). Effects of growth substances on the protoplasmic drought tolerance of leaf cells of the resurrection grass, *Sporobolus stapfianus*. Australian Journal of Plant Physiology 28 1115-1120.

Goncharic, V.M. (2001). Efficiency of combined application of growth regulators and microelements to potato. 6th Conference on Regulators of Plant Growth and Development in Biotechnology. Moscow, p. 227.

Grove, M.D., Spencer, G.F., Rohwedder, W.K., Mandava, N., Worley, J.F., Warthen, J.D., Steffens, G.L., Flippen-Anderson, J.L., Cook, J.C. (1979). Brassinolide, a plant growth-promoting steroid isolated from *Brassica napus* pollen. Nature 281: 216-217.

Hong, Z., Ueguchi-Tanaka, M., Shimizu-Sato, S., Inukai, Y., Fujioka, S., Shimada, Y., Takatsuto, S., Agetsuma, M., Yoshida, S., Watanabe, Y., Uozu, S., Kitano, H., Ashikari, M., Matsuoka, M. (2002). Loss-of-function of a rice brassinosteroid biosynthetic enzyme, C-6 oxidase, prevents the organized arrangement and polar elongation of cells in the leaves and stem. Plant Journal 32:495-508.

Ilkovets, I.M., Sokolovskii, S.G., Nayt, M.R., Volotovskii, I.D. (1999). Phytohormonal control of $Ca^{2+}$ concentration in plant cell cytoplasm. Vesti NAN Belarusi Seriya Biologicheskich Navuk 58-62.

Jenkins, G.I. (1997). UV and blue light signal transduction in *Arabidopsis*. Plant Cell Environment 20:773-778.

Kalituho, L.N., Chaika, M.T., Mazhul, V.M., Khripach, V.A. (1996). Effect of 24-epibrassinolide on pigment apparatus formation. Proceedings of the Plant Growth Regulation Society of America 23:36-40.

Kalituho, L.N., Chaika, M.T., Kabashnikova, L.F., Makarov, V.N., Khripach, V.A. (1997a). On the phytochrome mediated action of brassinosteroids. Proceedings of the Plant Growth Regulation Society of America 24:140-145.

Kalituho, L.N., Kabashnikova, L.F., Chaika, M.T. (1997b). Action of epibrassinolide on processes of growth and accumulation of photosynthetic pigments in triticale seedlings. Doklady Akademii Nauk Belarusi 41:69-72.

Kamuro, Y., Inada, K. (1991). The effect of brassinolide on the light-induced growth inhibition in mung bean epicotyl. Plant Growth Regulation 10:37-43.

Kang, J.G., Park, C.M. (2002). Cloning, characterization and use of pea cytochrome P 450 hydroxylase involved in brassinosteroid biosynthesis of plants. Eur. Pat. Appl. EP 1209227 A2.

Kang, J.G., Yun, J., Kim, D.H., Chung, K.S., Fujioka, S., Kim, J.I., Dae, H.W., Yoshida, S., Takatsuto, S., Song, P.S., Park, C.M. (2001). Light and brassinosteroid signals are integrated *via* a dark-induced small G protein in etiolated seedling growth. Cell 105:625-636.

Karnachuk, R.A., Golovatskaya, I.F., Efimova, M.V., Khripach, V.A. (2002). Action of epibrassinolide on morphogenesis and hormonal balance in *Arabidopsis* seedlings at green light. Plant Physiology (Rus.) 49:1-5.

Kasukabe, Y., Fujisawa, K., Nishiguchi, S., Maekawa, Y., Allen, R.D. (1999). Cloning, cDNA sequences and expression of cotton fiber tissue-specific genes. Pat. US 5,932,713.

Khodiankov, A.A. (2001). Effect of epibrassinolide and immunotzitofite on drought resistance of flax. 6th Conference on Regulators of Plant Growth and Development in biotechnology. Moscow, p. 289.

Khodiankova, S.F., Duktov, B.P. (2002). Effectiveness of treatment of flax seeds with growth regulators. Ahova Raslin 6:9.

Khripach, V.A., Voronina, L.V., Malevannaya, N.N. (1996a). Preparation for the diminishing of heavy metals accumulation by agricultural plants. Pat. RU 2,119,285.

Khripach, V.A., Zhabinskii, V.N., Litvinovskaya, R.P., Zavadskaya, M.I., Savelieva, E.A., Karas, I.I., Vakulenko, V.V. (1996b). Method of enhancement of food value of potato. Pat. BY 3488.

Khripach, V.A., Zhabinskii, V.N., Litvinovskaya, R.P., Zavadskaya, M.I., Savelieva, E.A., Karas, I.I., Kilchevskii, A.V., Titova, C.H. (1996c). Method of protection of potato from phytophtora infection. Pat. BY 3400.

Khripach, V.A., Zhabinskii, V.N., Malevannaya, N.N. (1997a). Recent advances in brassinosteroids study and application. Proceedings of the Plant Growth Regulation Society of America 24:101-106.

Khripach, V.A., Zhabinskii, V.N., Litvinovskaya, R.P., Zavadskaya, M.I., Deeva, V.P., Vedeneev, A.N. (1997b). Preparation for diminishing of radionuclides accumulation by plants and method of its application. Pat. BY 2806.

Khripach, V.A., Zhabinskii, V.N., Litvinovskaya, R.P., Zavadskaya, M.I., Volynets, A.P., Prochrchick, R.A., Pshenichnaya, L.A., Manzhelesova, N.E., Morozick, G.V. (1997c). Method of protection of barley plants from leave diseases. Pat. BY 5168.

Khripach, V.A., Zhabinskii, V.N., de Groot, A. (1999). Brassinosteroids. A New Class of Plant Hormones. pp. 338, Academic Press, San Diego.

Khripach, V., Zhabinskii, V., de Groot, A. (2000). Twenty years of brassinosteroids: steroidal plant hormones warrant better crops for the XXI century. Annals of Botany 86:441-447.

Kilchevskii, A.V., Frantsuzionok, V.V. (1997). Effect of epibrassinolide on proliferation of lily explants. In "Regulators of plant growth and development", 4$^{th}$, pp. 297-298, Moscow.

Kim, T.H., Kim, B.H., von Arnim, A.G. (2002). Repressors of photomorphogenesis. International Review of Cytology 220:185-223.

Koka, C.V., Cerny, R.E., Gardner, R.G., Noguchi, T., Fujioka, S., Takatsuto, S., Yoshida, S., Clouse, S.D. (2000). A putative role for the tomato genes DUMPY and CURL-3 in brassinosteroid biosynthesis and response. Plant Physiology 122:85-98.

Kolotovkina, Y.B., Prusakova, L.D., Sal'nikov, A.I., Ezhov, M.N. (2001). Influence of growth regulators ecost and epibrassinolide on pigment complex of buckwheat of different genotypes. 6$^{th}$ Conference on Regulators of Plant Growth and Development in Biotechnology. Moscow, p. 100.

Korableva, N.P., Suchova, L.S., Muromtsev, G.S., Koreneva, V.M., Kozakova, V.I., Karsunkina, N.P., Dogonadze, M.Z. (1992). Method of potato treatment for long-term preservation. Pat. SU 1,794,261.

Korableva, N.P., Platonova, T.A., Dogonadze, M.Z. (1998). Effect of brassionalide on the ethylene biosynthesis in potato tuber meristems *Solanum tuberosum* L. Doklady Akademii Nauk (Russia) 361:113-115.

Korableva, N.P., Platonova, T.A., Dogonadze, M.Z., Evsunina, A.S. (2002). Brassinolide effect of growth of apical meristems, ethylene production, and abscisic acid content in potato tubers. Biologia Plantarum 45:39-43.

Kukresh, S.P., Khodiankova, S.F. (2002). Increase of crop yield and quality of flax. Agrarnaya Nauka 7:13-14.

Likchacheva, T.S. (2001). Influence of epibrassinolide treatment on physiological processes in bean plants of "Rubin" variety. 6$^{th}$ Conference on Regulators of Plant Growth and Development in Biotechnology. Moscow, p. 45.

Likchacheva, T.S., Klimachev, D.A., Starikova, V.T. (2001). Influence of level of mineral supply and epibrassinolide treatment on hormone status of tomato generative organs. 6$^{th}$ Conference on Regulators of Plant Growth and Development in Biotechnology. Moscow, p. 44.

Luccioni, L.G., Oliverio, K.A., Yanovsky, M.J., Boccalandro, H.E., Casal, JJ. 2002. Brassinosteroid mutants uncover fine tuning of phytochrome signaling. Plant Physiology 128:173-181.
Malevannaya, N.N., Bednarskaya, I. (1995). Epin. Priusadebnoye Choziaystvo 8-9.
Malevannaya, N.N., Kositsina-Pinegina, E. (1996). Epin–antistress agent. Tsvetovodstvo 7-8.
Manzhelesova, N.E. (1997). Content of phenolic compounds and activity of peroxidase in infected by Helminthosporium teres Sacc. barley treated with epibrassinolide. Vesti AN Belarusi Seriya Biologicheskich Navuk 20-24.
Matevosian, G.L., Kudashov, A.A., Ezhov, A.K., Sotnik, V.G. (2001). Influence of growth regulators on growth, productivity and crop quality of tomato in greenhouses. Agrokhimiya 11:49-58.
Maugh, T.H. (1981). New chemicals promise larger crops. Science 212:33-34.
Melnikov, S.S., Manakina, E.E., Budakova, E.A. (1999). Effect of epibrassinolide on productivity of algae. Vesti AN Belarusi Seriya Biologicheskich Navuk 44-48.
Mori, M., Nomura, T., Ooka, H., Ishizaka, M., Yokota, T., Sugimoto, K., Okabe, K., Kajiwara, H., Satoh, K., Yamamoto, K., Hirochika, H., Kikuchi, S. (2002). Isolation and characterization of a rice dwarf mutant with a defect in brassinosteroid biosynthesis. Plant Physiology 130:1152-1161.
Nakaya, M., Tsukaya, H., Murakami, N., Kato, M. (2002). Brassinosteroids control the proliferation of leaf cells of *Arabidopsis thaliana*. Plant and Cell Physiology 43:239-244.
Neff, M.M., Nguyen, S.M., Malancharuvil, E.J., Fujioka, S., Noguchi, T., Seto, H., Tsubuki, M., Honda, T., Takatsuto, S., Yoshida, S., Chori, J. (1999). *BAS*1: A gene regulating brassinosteroid levels and light responsiveness in *Arabidopsis*. (1999). Proceedings of the National Academy of Sciences of the United States of America 96:15316-15323.
Nilovskaya, N.T., Ostapenko, N.V., Seregina, I.I. (2001). Effect of epibrassinolide on the productivity and drought resistance of spring wheat. 2001. Agrokhimiya 2:46-50.
Noguchi, T., Fujioka, S., Choe, S., Takatsuto, S., Yoshida, S., Yuan, H., Feldmann, K.A., Tax, F.E. (1999). Brassinosteroid-insensitive dwarf mutants of *Arabidopsis* accumulate brassinosteroids. Plant Physiology 121:743–752.
Ozolina, N.V., Pradedova, E.V., Reutskaya, A.M., Salyaev, R.K. (1999). The effects of brassinosteroids on tonoplast proton pumps. Doklady Akademii Nauk 367:829-830.
Persikova, T.F. (2001). Efficiency of growth regulators depending on the feeding conditions of narrow-leaved lupine. Achova raslin 23-25.
Pirogovskaya, G.V. (2000) Slow-Release Fertilizers. Minsk. 287 pp.
Pirogovskaya, G.V., Bogdevitch, I.M., Naumova, G.V., Khripach, V.A., Azizbekyan, S.G., Krul, L.P. (1996). New forms of mineral fertilizers with additives of plant growth regulators. Proceedings of the Plant Growth Regulation Society of America 23:146-151.
Pirogovskaya, G.V., Khripach, V.A., Lapa, V.V., Rusalovich, A.M., Zhabinskii, V.N., Ivanenko, N.N., Bogdevitch, I.M., Krul, L.P. (1997). Fertilizers with biologically active additives of brassinosteroids. Pat. BY 3400.
Platonova, T.A. (1998). Effects of epibrassinolide on the endoplasmic reticulum of apical cells of potato tubers. Prikladnaya Biokhimiya i Mikrobiologiya 34:553-559.
Platonova, T.A., Korableva, N.P. (1999a). Changes of the Golgi apparatus in potato tuber meristems *Solanum tuberosum* L. in the course of transition from rest to growth and under the action of brassinolide. Doklady Akademii Nauk (Russia) 369:557-560.
Platonova, T.A., Korableva, N.P. (1999b). Study on the Golgi apparatus in tuber apices of potato during rest regulation under the action of epibrassinolide. Prikladnaya Biokhimiya i Mikrobiologiya 35:599-603.
Popova, M.P., Zolotar, R.M. (2001). Protection action of brassinosteroids on cucumber at low temperature. $6^{th}$ Conference on Regulators of Plant Growth and Development in Biotechnology. Moscow, pp. 116-117.
Popova, M.P., Zotova, G.S. (2001). Antistress properties of growth regulators. $6^{th}$ Conference on Regulators of Plant Growth and Development in Biotechnology. Moscow, p. 117.
Pradedova, E.V., Ozolina, N.V., Korzun, A.M., Salyaev, R.K. (2002). Effect of epibrassinolide on activities of the tonoplast $H^+$-ATPase and $H^+$-pyrophosphatase under conditions of high and low KCl concentrations. Biologicheskie Membrany (Moscow) 19:216-220.

Prusakova, L.D., Chizhova, S.I., Khripach, V.A. (1995). Stability and productivity of barley and wheat under the action of brassinosteroids. Sel'skochosyaistvennaya Biologiya 93-97.

Prusakova, L.D., Ezhov, M.N., Sal'nikov, A.I. (1999a). Application of emistim, epibrassinolide, and uniconazol for overcoming heterogeneity of buckwheat crop. Agrarnaya Rossiya 1:41-44.

Prusakova, L.D., Chizhova, S.I., Tretiakov, N.N., Ageeva, L.F., Golantseva, E.N., Jakovlev, A.F. (1999b). Antistress properties of ecost and epibrassinolide on spring wheat in the conditions of the Central nonchernozem zone. Agrarnaya Rossiya 1:39-41.

Prusakova, L.D., Chizhova, C.B., Matamoros, K.M.R. (2000a). Spring wheat allocytoplasmic hybrid response to epibrassinolide action in soil drought conditions. Agrokhimiya 52-55.

Prusakova, L.D., Chizhova, S.I., Ageeva, L.F., Golantseva, E.N., Jakovlev, A.F. (2000b). Influence of epibrassinolide and ecost on drought tolerance and productivity of spring wheat. Agrokhimiya 50-54.

Pshenichnaya, L.A., Khripach, V.A., Volynets, A.P., Prokhorchik, R.A., Manzhelesova, N.E., Morozik, G.V. (1997). Brassinosteroids and resistance of barley plants to leave deseases. In Problems of Experimental Botany, pp. 210-217. Byelorussian Science. Minsk.

Pustovoitova, T.N., Zhdanova, N.E., Zholkevich, V.N. (2001). Influence of epibrassinolide on adaptation processes in *Cucumus sativus* L. plants in drought soil conditions. 6[th] Conference on Regulators of Plant Growth and Development in Biotechnology. Moscow, p. 61.

Ramonell, K.M., Kuang, A., Porterfield, D.M., Crispi, M.L., Xiao, Y., McClure, G., Musgrave, M.E. (2001). Influence of atmospheric oxygen on leaf structure and starch deposition in *Arabidopsis thaliana*. Plant Cell Environment 24:419-428.

Rao, S.S.R., Vardhini, B.V., Sujatha,E., Anarudha, S. (2002). Brassinosteroids–a new class of phytohormones. Current Science 82:1239-1244.

Rodkin, A.I., Konovalova, G.I., Bobric, A.O. (1997). Efficiency of application of biologically active substances in primary breeding of potato. 4[th] Conference on Regulators of Plant Growth and Development, Moscow, pp. 317-318.

Romanov, G.A. (2002). The phytohormone receptors. Russian Journal of Plant Physiology (Translation of Fiziologiya Rastenii (Moscow)) 49:552-560.

Runkova, L.V. (1995). Effect of epibrassinolide on flowering of some ornamental plants. In "Brassinosteroids–biorational, ecologically safe regulators of growth and productivity of plants", 4[th]. Minsk, pp. 10-11.

Runkova, L.V. (2000). Effect of Epin on cloning of ornamentals. Tsvetovodstvo: 3.

Sakurai A, Yokota T, Clouse SD (eds). 1999. Brassinosteroids. Steroidal Plant Hormones. Springer-Verlag, Tokyo.

Sanko, N.V. (2001). Resistance of different barley genotypes to water stress. 2[th] Conference on Regulation of Plant Growth and Productivity. Minsk, pp.186-187.

Schaller, H. (2003). The role of sterols in plant growth and development. Progress in Lipid Research 42:163-175.

Schmidt, J., Spengler, B., Voigt, B., Adam, G. (2000). Brassinosteroids - structures, analysis and synthesis. Recent Advances in Phytochemistry (Evolution of Metabolic Pathways) 34:385-407.

Schnabl, H., Roth, U., Friebe, A. (2001). Brassinosteroid-induced stress tolerances of plants. Recent Research Developments in Phytochemistry 5:169-183.

Schneider, B. (2002). Pathways and enzymes of brassinosteroid biosynthesis. Progress in Botany 63:286-306.

Schultz, L., Kerckhoffs, L.H., Klahre, U., Yokota, T., Reid, J.B. (2001). Molecular characterization of the brassinosteroid-deficient lkb mutant in pea. Plant Molecular Biology 47:491-498.

Sekimata, K., Kimura, T., Kaneko, I., Nakano, T., Yoneyama, K., Takeuchi, Y., Yoshida, S., Asami, T. (2001). A specific brassinosteroid biosynthesis inhibitor, Brz2001: evaluation of its effects on Arabidopsis, cress, tobacco, and rice. Planta 213:716-721.

Shakirova, F.M., Bezrukova, M.V. (1998). Effect of 24-epibrassinolide and salinity on the levels of ABA and lectin. Russian Journal of Plant Physiology 45:388-391.

Shakirova, F.M., Bezrukova, M.V., Aval'baev, A.M., Gimalov, F.R. (2002). Stimulation of wheat germ agglutinin gene expression in root seedlings by 24-epibrassinolide. Russian Journal of Plant Physiology 49:253-256.

Shimada, Y., Fujioka, S., Miyauchi, N., Kushiro, M., Takatsuto, S., Nomura, T., Yokota, T., Kamiya, Y., Bishop, G.J., Yoshida, S. (2001). Brassinosteroid-6-oxidases from Arabidopsis and tomato catalyze multiple C-6 oxidations in brassinosteroid biosynthesis. Plant Physiology 126:770–779.

Sudnic, A.F., Deeva, V.P. (2001). Influence of combined application of growth promoters and fungicides on growth, development , and productivity of barley. 2$^{th}$ Conference on Regulation of Plant Growth and Productivity. Minsk, pp.196-197.

Symons, G.M., Reid, J.B. (2003). Hormone levels and response during de-etiolation in pea. Planta 216:422-431.

Takeno, K., Pharis, R.P. (1982). Brassinosteroid-induced bending of the leaf lamina of dwarf rice seedlings: an auxin-mediated phenomenon. Plant and Cell Physiology. 23:1275-1281.

Tanaka, H., Kayano, T., Matsuoka, M. (2003). Gene concerning brassinosteroid-sensitivity of plants and utilization thereof. Eur. Pat. Appl. EP 1275719 A1.

Timeiko, L.V., Khripach, V.A., Talanov, A.V., Drozdov, S.N. (2001). Influence of epibtassinolide on $CO_2$ – exchange, growth, development and productivity of cucumber in greenhouses. 2$^{th}$ Conference on Regulation of Plant Growth and Productivity. Minsk, p.200-201.

Tishchenko, S.Y., Karnachuk, R.A., Khripach, V.A. (2001). Epibrassinolide participation in growth photoregulation and hormonal balance of Arabidopsis under blue light. Vestnik Bashkirskogo Universiteta 166-167.

Tzareva, E.G. (2001). Influence of plant growth regulators at the initial stage of barley ontogenesis on exposure to low temperature. 6$^{th}$ Conference on Regulators of Plant Growth and Development in Biotechnology. Moscow, p. 130.

Tzyganov, A.P., Persikova, T.F., Vildflush, I.P. (1999). Efficiency of nitrogen-fixing microorganisms, foliar growth regulator application on spring wheat depending on the level of mineral fertilizers. Mezhdunarodnyi Agrarnyi Zhurnal 3:20-23.

Tzyganov, A.P., Persikova, T.F., Kakshintsev, A.V. (2001). Influence of physiologically active compounds on productivity of narrow-leaved lupine. 6$^{th}$ Conference on Regulators of Plant Growth and Development in Biotechnology. Moscow, p. 290-291.

Vasyukova, N.I., Chalenko, G.I., Kaneva, I.M., Khripach, V.A., Ozeretskovskaya, O.L. (1994). Brassinosteroids and potato late blight. Prikladnaya Biokchimiya and Microbiologiya 30:464-470.

Vedeneev, A.N., Deeva, V.P. (2001). Pole of a nucleus and cytoplasm in lipid metabolism of separate genotypes under the action of growth regulators in the conditions of water stress. 2$^{th}$ Conference on Regulation of Plant Growth and Productivity. Minsk, pp.31-32.

Vildflush, I.P., Deeva, V.P., Gurban, K.A. (2001). Influence of biologically active compounds on barley plants (*Hordeum vulgare* L.) on sward-podzolic light-loamy soils. Vestsy Natsionalnoi Academii Nauk Belarusi Seriya Biologicheskich Nauk 1:23-26.

Vlasova, N.N., Laman, N.A., Stratilatova, E.V. (2000). Influence of epibrassinolide on earing synchrony and productivity of spring barley (*Hordeum vulgare* L.). Vestsy Natsionalnoi Academii Nauk Belarusi Seriya Biologicheskich Nauk 4:21-24.

Vlasova, N.N., Laman, N.A., Stratilatova, E.V., Trufanova, Y.V. (2002). Influence of kinetin and epibrassinolide on the morphogenetic character of main shoot apex of spring barley (*Hordeum vulgare* L.). Vestsy Natsionalnoi Academii Nauk Belarusi Seriya Biologicheskich Nauk 1:17-19.

Volynets, A.P., Pshenichnaya, L.A., Manzhelesova, N.E., Morozik, G.V., Khripach, V.A. (1997b). The nature of protective action of 24-epibrassinolide on barley plants. Proceedings of the Plant Growth Regulation Society of America 24:133-137.

Voskresenskaya, L.G., Khripach, V.A., Zhabinskii, V.N., Zavadskaya, M.I., Litvinovskaya, R.P. (1998). Method of enhancement of flax productivity and of quality of flax fiber. Pat. BY 5212.

Wada, K., Marumo, S., Ikekawa, N., Morisaki, M., Mori, K. (1981). Brassinolide and homobrassinolide promotion of lamina inclination of rice seedlings. Plant and Cell Physiology 22:323-325.

Wang, Z.Y., Chory, J. (2000). Recent advances in molecular genetic studies of the functions of brassinolide, a steroid hormone in plants. Recent Advances in Phytochemistry (Evolution of Metabolic Pathways) 34:409-431.

Winter, J. (2001). Enzymes involved in the biosynthesis of brassinosteroids. Studies in Natural Products Chemistry (Bioactive Natural Products (Part F)) 25:413-428.

Yamamuro, C., Ihara, Y., Wu, X., Noguchi, T., Fujioka, S., Takatsuto, S., Ashikari, M., Kitano, H., Matsuoka, M. (2000). Loss of function of a rice brassinosteroid insensitive1 homolog prevents internode elongation and bending of the lamina joint. Plant Cell 12:1591-1606.

Zabolotnyi, A.I., L'vov, N.P., Khripach, V.A., Kudryashova, N.N. (2003). Role of trophic and hormonal factors in exogenous regulation of the formation of reproductive organs in yellow lupine (*Lupinus luteus* L.). Prikladnaya Biokhimiya i Mikrobiologiya 39:99-104.

CHAPTER 10

S.HAYAT, A.AHMAD AND Q. FARIDUDDIN

# BRASSINOSTEROIDS: A REGULATOR OF 21$^{ST}$ CENTURY

Brassinosteroids (BRs) were initially assigned a position of a lesser importance than the other recognized plant growth regulators. Over the last 25 years, they have evolved as essential, full flashed regulators of growth and development. Much progress has been achieved in their isolation, characterization and possible mechanism of action. However, their practical applicability has lacked fear behind. The literature available suggests strong potential of steroidal activity in this new generation of growth regulators in improving the biological yield of important plants. Therefore, in this chapter we have explored the importance of BRs in regulating enzyme level, photosynthesis and related aspects determining biological yield under normal and stressed conditions.

## INTRODUCTION

Growth is an organized, well-coordinated complex process where metabolism provides the energy and the building blocks. However, it is the relative hormone level that regulates the pace of growth of each individual part, to produce a form that is recognized as a plant. Earlier, only five groups of hormones (auxins, gibberellins, cytokinins, abscissic acid and ethylene) were designated as regulators of plant growth. However, in the recent past, compelling evidences have been put forward to classify a group of steroidal substances (brassinosteroids), first isolated from rape (*Brassica napus* L.) pollen, as a new class of phytohormones, in addition to named above.

It was in 1970, when Mitchell and co-workers screened the pollens of nearly sixty species, out of which about thirty generated growth in bean seedlings. This growth promoting substance was called "Brassin". The search for its active factor(s) was collectively approached in 1974 by the USDA scientists working at Northern Regional Research Centre (NRRC), Peoria; Eastern Regional Research Centre (ERRC), Philadelphia and Beltsville Agricultural Research Centre (BARC), Maryland. Bee-collected pollens (500 lb) were processed through a pilot plant-size solvents (2-propanol) extraction procedure at ERRC and succeeded in partial purification at BARC. However, it was crystallized at NRRC and was subjected to X-ray analysis to establish its structure. This biologically active plant growth promoter was found to be steroidal lactone ($C_{28}H_{48}O_6$) and was named as "brassinolide" which was renamed as "brassinosteroid". All natural brassinosteroids have a common 5-α choleston skeleton and its structural variants come from the kind and the orientation of functionalities on the

*S.Hayat and A.Ahmad (eds.), Brassinosteroids*, 231-246.

skeleton. Their low level in plants is not uniformly distributed throughout its body but young growing tissues have a larger share than the mature tissues (Yokota and Takahashi, 1986). The richest sources are pollen and immature seeds where its concentration ranges between 1-100 ng per g fresh mass, whereas shoots and leaves have about 0.01-0.1 ng per g fresh mass (Takatsuto, 1994). Till now more than 40 brassinosteroids, structurally and functionally different form each other, have been characterized (Rao *et al.*, 2002). Out of which, three (brassinolide, 24-epibrassinolide and 28-homobrassinolide) are being largely applied to have an impact on plant metabolism, growth and productivity. Steroidal action on plants is being discussed in the following pages.

## BIOASSAYS OF BRASSINOSTEROID

The rice lamina inclination and bean second internode tests are highly specific and sensitive to brassinosteroids. Segments from etiolated seedlings of rice, consisting of the second leaf lamina, lamina joint and sheath are excised and floated on distilled water, containing brassinosteroids. The angle of curvature of lamina joint was measured which is proportional to the concentration of the hormone in the medium. Brassinolide shows activity at a concentration of 0.005 $\mu g\ ml^{-1}$ but at this concentration, indole-3-acetic acid (IAA) shows only weak activity for which it was originally developed by Maeda (1965). This extreme difference in dose response of brassinosteroids and IAA, the test is considered more specific to brassinosteroids. However, the bean second internode test was first developed during the isolation of brassinolide from the pollen of rape (Grove *et al.*, 1979). The treatment of the cuttings of the second internode, from the seedlings of *Phaseolus vulgaris* L. cv. Pinto, with brassinolide, in lanolin paste, showed elongation, curvature, swelling or splitting (Mandava *et al.*, 1983). Brassinolide caused elongation, curvature and swelling at lower concentration (0.01 μg) and splitting at a higher concentration (0-1 μg).

## BRASSINOSTEROIDS AND GROWTH RESPONSE

The desired orientation in plant growth, in response to various phytohormones, is becoming a regular phenomenon for academic and/or economic gains. The involvement of brassinosteroid, in such cases, is very well supported by the observed deviation from normal development in the mutants of *Arabidopsis*, lacking the capability to biosynthesize brassinosteroids but restored by their exogenous application (Altman, 1999; Choe *et al.*, 1999a and b). However, the normal plants of wheat and mustard, applied with brassinosteroids, gained more fresh and dry mass of the leaves and that of the whole shoot (Braun and Wild, 1984) and an increase in length, width and fresh mass of the leaves of tobacco (Diz *et al.*, 1995). Likewise, the application of biobras-6 to tomato (Nunez *et al.*, 1996) and BRs to *Vicia faba* plants (Helmy *et al.*, 1997) favoured plant growth. BRs (brassinolide/epibrassinolide/ homobrassinolide) increased shoot length, dry mass production in *Pinus banksana* (Rajasekaran and Blake, 1998), *Arachis hypogea* (Vardhini and Rao, 1998), *Brassica juncea* (Hayat *et al.*, 2000) and the length of epicotyl of soybean seedlings (Clouse *et al.*, 1992). The growth promoting capacity of the steroids was further corroborated with the expression of multiple growth and developmental defects (Clouse *et al.*, 1996) by the mutant of *Arabidopsis thaliana* as, it was incapable to synthesize BRs because of the lack of genetic expression (Clouse, 1996). Moreover, BRs, alone regulate cell elongation in *Arabidopsis thaliana* (Catterou *et al.*, 2001a) by rearranging the position of the tubules in the cell (Catterou *et al.*, 2001b). This could have been one of the main reasons to explain the BRs induced differentiation of trachary elements, in isolated mesophyll cells of *Zinnia elegans* (Iwasaki and Shibaoka, 1991).

Epibrassinolide/brassinolide alone or in combination with auxin/cytokinin favoured cell division in the protoplasts of cabbage (Nakajima *et al.*, 1996) and that of *Petunia hybrida* (Oh and Clouse, 1998; Oh, 2003) but failed to completely replace either of the hormones. Deviating from these observations (22S, 23S, 24S)-(tri-epi)-brassinolide and 24-epibrassinolide inhibited the growth of callus and suspension cultures of *Agrobacterium tumefaciens*-transformed of *Nicotiana tabacum* (Roth *et al.*, 1989). Similarly, 24-epibrassinolides promoted cell enlargement but no cell division in cultured carrot cells (Bellincampi and Morpurgo, 1988). At the level of the intact plant, the microscopic examination of BR-deficient and BR-insensitive mutants, in *A. thaliana*, the dwarfness in its phenotype was due to reduced cell size and not the cell number (Kauschmann *et al.*, 1996).

Mandava (1988) is of the opinion that BRs-induced cell elongation is dependent on light. Moreover, to be more specific Kamura and Inada (1991), while studying the growth of epicotyl of *Vigna radiata* suggested the involvement of phytochrome in mediating the action of BR. While studying the interaction effect of BRs, with other hormones, it was noted that DA-6/epibrassinolide in association with $GA_3$ increased shoot mass and the length of spinach plants (Liang *et al.*, 1998). Moreover, synthetic brassinosteroid (22, 23-S, homobrassinolide) enhanced the auxin-mediated elongation in *Cucumis sativus* L. but was independent of $GA_3$ (Katsumi, 1985). Yopp *et al.* (1981a) and Sala and Sala (1985) also reached to the same conclusion regarding the BRs and auxins interaction. Such a synergism was also observed in the bending responses of plants (Yopp *et al.*, 1981a and b). BRs mediated elongation is proposed to involve auxins (Takeno and Pharis, 1982; Cohen and Meudt, 1983; Meudt, 1987) by either amplifying the action of auxins or sensitizing the tissue (Katsumi, 1985). However, Sasse (1990) disagreed with the above statement and is of the view that brassinolide do not act through the auxins, in the process of elongation. Moreover, the stem elongation in tomato and soybean, promoted by BRs, most likely did not involve the auxin signal transduction pathways, even though both the hormones are known to influence the cell wall relaxation (Zurek *et al.*, 1994).

Deviating from the above pattern of the site of application of the hormone, BRs were applied through pre-sowing seed treatment. Treating the seeds of *Coffea arabica* with biobras-16 (Soto *et al.*, 1997), *Oryza sativa* with BRs (Wang, 1997), *Cicer arietinum* (Fariduddin *et al.*, 2000) and *Triticum aestivum* (Hayat *et al.*, 2001a) with homobrassinolide resulted in the production of tall plants with higher fresh and dry mass.

The growth of the roots of the seedlings of wheat, mungbean and maize was inhibited by 24-epibrassinolide, supplied at its natural level. However, the response was determined by the species and the age of the root (Roddick and Ikekawa, 1992). Likewise, the response of apical and basal regions of the excised roots of tomato (Roddick *et al.*, 1993) or that of *A. thaliana* (Clouse *et al.*, 1993) to sub-micromolar concentrations of 24-epibrassinolide was negative. The root length and its proliferation in soybean seedlings (Hunter, 2001) and root length (Figure 1) and nodule number in *Lens culinaris* (Hayat and Ahmad, 2003) was reduced by epibrassinolide and 28-homobrassinolide, respectively. In contradiction to the above, the treatment with BRs improved the rooting in the seedlings of *Cucumis sativus* (Ding *et al.*, 1995) and that of rice (Wang and Deng, 1992). Similarly, the cuttings of *Picea abies*, treated with 3, 15 or 60 ppm of (22S, 23S)-homobrassinolide, possessed more adventitious roots than the control (Rönsch *et al.*, 1993).

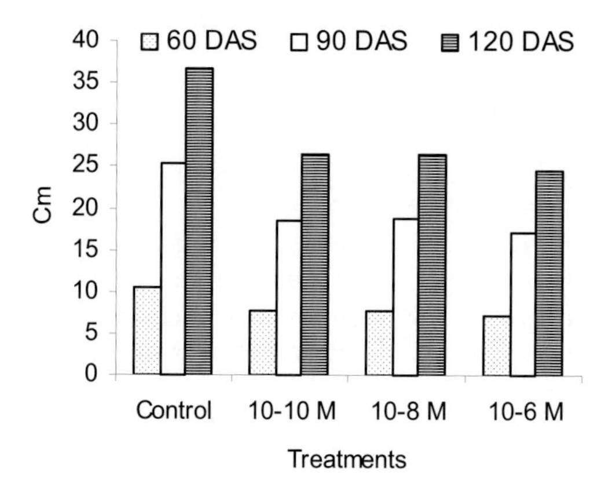


*Figure 1.Root length at three stages of growth in plants of Lens culinaris raised from seeds soaked in three concentarions of 28-homobrassinolide*

## EFFECT OF BRASSINOSTEROIDS ON SEED GERMINATION, FLOWERING AND SENESCENCE

### *Seed Germination*

It is well documented that brassinosteroids promote seed germination, like other hormones. The treatment of the seeds of *Lepidium sativus* (Jones-Held *et al.*, 1996) and *Eucalyptus camaldulensis* (Sasse *et al.*, 1995) with brassinolide improved per cent germination. Similarly brassinosteroids promoted seed germination in case of *Brassica napus* (Chang and Cai, 1998), rice (Dong *et al.*, 1989), wheat (Sairam *et al.*, 1996; Hayat *et al.*, 2003; Figure 2), tomato (Vardhini and Rao, 2000) and tobacco (Leubner-Metzger, 2001). Moreover, brassinolide, 24-epibrassinolide and 28-homobrassinolide promoted seed germination in groundnut (Vardhini and Rao, 1997).

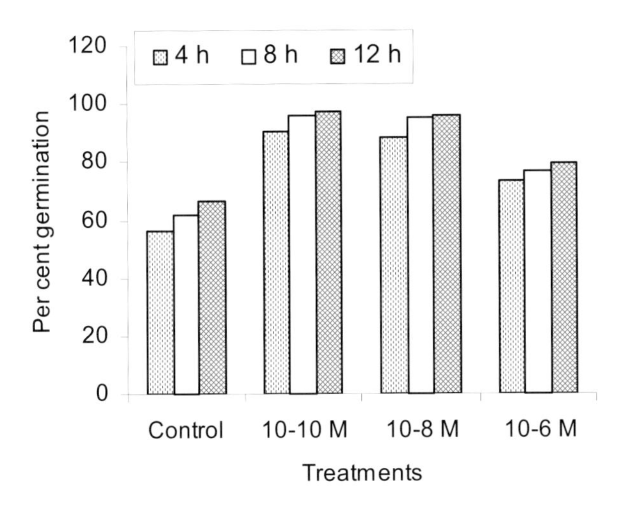


*Figure 2.Effect of different hours of soaking of 28-homobrassinolide on germination in Wheat seeds*

*Flowering*

There has been very limited use of steroids in regulating flowering. The number of flowers in strawberry increased by the application of brassinosteroids at the foliage (Pipattanawong *et al.*, 1996). However, in case of grapes, the application of brassinosteroids in autumn improved the number of flowers but inhibited if the time of application is delayed to late winter (Rao *et al.*, 2002).

*Senescence*

It is the process, which refers to, endogenously regulated deteriorative, changes that become the natural cause of death of cells, tissues, organs or organism (Arteca, 1997). Like other hormones (Rao *et al.*, 2002), brassinosteroids also play a crucial role in regulating the processes leading to senescence. The brassinolide promotes senescence in *Xanthium* and *Rumex* explants (Mandava *et al.*, 1981). In addition to it, brassinosteroids also accelerate senescence in the detached cotyledons of cucumber seedlings (Zhao *et al.*, 1990) and leaves of mung bean seedlings (He *et al.*, 1996). However, brassinosteroid deficit *Arabidopsis* mutants exhibited delayed senescence of chloroplast (Li *et al.*, 1996).

Similarly, the senescence of the leaves of mungbean and mustard was delayed, if supplied with 28-homobrassinolide at early stage of growth (Fariduddin, 2002).

## BRASSINOSTEROIDS AND STRESS TOLERANCE

The resistance to abiotic stresses was increased in the plants, treated with brassinosteroids. The plants of tomato and rice (Kamuro and Takatsuto, 1991), maize (He *et al.*, 1991), cucumber (Katsumi, 1991) and brome grass (Wilen *et al*, 1995) developed resistance to low temperature, on being treated with brassinosteroids. In rice, 24-epibrassinolide improved resistance against chilling stress (1-5°C) and the tolerance was associated with increased ATP and proline levels and SOD activity. This indicates the involvement of brassinosteroid in membrane stability and osmoregulation (Rao *et al.*, 2002).

Brassinosteroids also increased tolerance, to high temperature in wheat leaves (Kulaeva *et al.*, 1991) and brome grass (Wilen *et al.*, 1995) which is associated with an induction of *de novo* polypeptide synthesis (Kulaeva *et al.*, 1991). Application of brassinosteroids, to sugar beet, increased its tolerance to drought stress (Schilling *et al.*, 1991). The ability of 28-homobrassinolide to induce resistance, to moisture stress in wheat was also established (Sairam, 1994).

The ability of brassinosteroids to counteract with the inhibitory effects of salinity on seedling growth of groundnut was attained (Vardhini and Rao, 1997). Similarly, the treatment of the seeds of *Eucalyptus camaldulensis* with 24-epibrassinolide resulted in an increase in their germination, under saline conditions (Sasse *et al.*, 1995). 24-epibrassinolide and 28-homobrassinolide also alleviated the salinity-induced inhibition of germination and seedling growth in rice (Anuradha and Rao, 2001). Seed treatment with very dilute solutions of brassinosteroids considerably improved the growth of rice plants, in saline media (Rao *et al.*, 2002). Likewise Kamuro and Takatsuto (1999), were also impressed by the ability of brassinosteroids to confer resistance in plants, against a wide variety of environmental stresses. They were of the opinion that the role of brassinosteroids in protecting the plants against environmental stresses will be an important research theme and may contribute largely to the usage of brassinosteroids in agricultural production.

## BRASSINOSTEROIDS AND PLANT METABOLISM

It has become a general practice to employ phytohormones to explore the full potential of plants. They are involved in the regulation of photosynthetic processes and partitioning of the assimilates, resulting in the alteration in plant size (Arteca, 1997). However, the response varies with the type of the plant and the hormone.

The aqueous solution of 28-homobrassinolide, applied to the foliage of wheat and mustard (Sairam, 1994; Hayat *et al.*, 2000; 2001a Figure 3) or applied as seed soaking to mungbean (Fariduddin *et al.*, 2003; Figure 4) and dialkylaminoethylalkanoate or epibrassinolide, in association with $GA_3$, to spinach enhanced the photosynthetic rate (Liang *et al.*, 1998). Foliar spray of aqueous solution of BR to wheat and mustard (Braun and Wild, 1984), epibrasisnolide to seedlings of cucumber (Ding *et al*., 1995) and brasisnolide to rice (Fujii *et al*., 1991) increased the rate of $CO_2$ assimilation. However, the epicotyl of cucumber, did not respond to epibrassinolide but the transport of the labeled ($^{14}C$) glucose towards the epicotyl was favoured (Nakajima and Toyama, 1995). Similarly, Hill activity in the foliage of *Vigna radiata* was favourably affected, on being supplemented with aqueous solution of 28-homobrassinolide (Bhatia and Kaur, 1997).

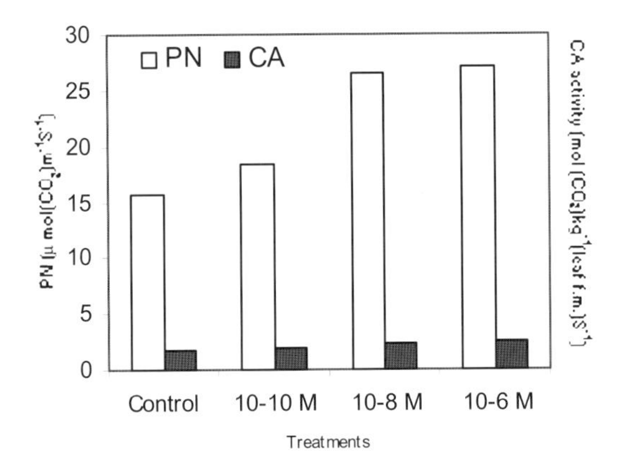


*Figure 3.Effect of foliar application of 28-homobrassinolide on net photosynthetic rate (PN) and carbonic anhydrase (CA) activity in mustard*

The total chlorophyll contents or its fractions increased in the leaves of *Vigna radiata* (Bhatia and Kaur, 1997) and *Brassica juncea* (Hayat *et al*., 2001a) by 28-homobrassinolide, applied directly to their foliage. Similarly, the values for the above parameters increased in the leaves of rice (Wang, 1997), *Cicer arietinum* (Fariduddin *et al*., 2000), *Brassica juncea* (Hayat and Ahmad, 2003b) and *Vigna radiata* (Fariduddin *et al*., 2003) raised from the seeds given presowing treatment with BRs/28-homobrassinolide. Moreover, the water stressed wheat plants treated with 28-homobrassinolide possessed high chlorophyll level (Sairam, 1994).

Carbonic anhydrase (CA, E.C. 4.2.1.1) is the second most abundant soluble protein, other than RuBPcase, in $C_3$-chloroplast (Reed and Graham, 1981; Okabe *et al*., 1984). It is a zinc containing protein with a molecular weight of 180 Kda (Lawlor, 1987) and is ubiquitous enzyme, among living organisms. It catalyzes the reversible inter conversion of bicarbonates ($HCO_3^-$) and $CO_2$ (Sultemeyer *et al*., 1993). The rate of conversion of $HCO_3^-$ to $CO_2$ is normally slow in alkaline conditions. However, CA activates the use of $HCO_3^-$ in the production of $CO_2$ (Lawlor, 1987). In $C_3$ plants, CA has a close association with RuBPCase where it elevates the level of $CO_2$ at its active site (Badger and Price, 1994). An increase in the activity of CA, in the leaves, was attained by the application of 28-homobrassinolide (Hayat *et al*., 2000, 2001a) to the shoot of the *Brassica juncea*. Moreover, the seedlings of wheat and mungbean, raised from the grains treated with 28-homobrassinolide, possessed high CA activity in their leaves (Hayat *et al.*, 2001b; Fariduddin *et al.*, 2003 Figure 5).

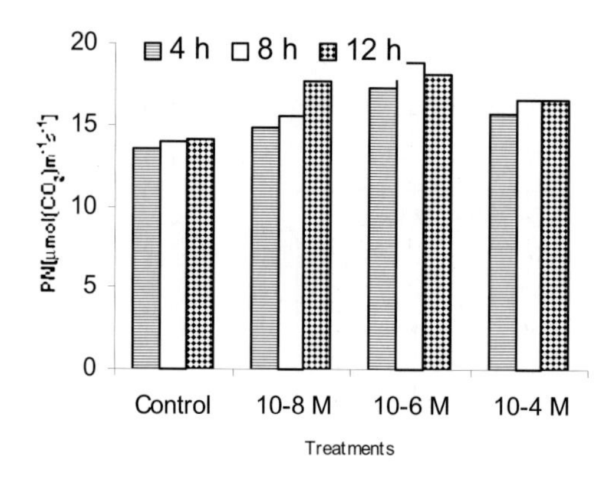


*Figure 4.Effect of different hour of soaking of 28-homobrassinolide on net photosynthetic rate (PN) in mungbean at 30 day stage*

The initiation of the process of reduction of nitrate is catalyzed by the enzyme, nitrate reductase (E.C. 1.6.6.1), the level of which increased in the plants of rice (Mai *et al.*, 1989), maize (Shen *et al.*, 1990), water stressed, wheat (Sairam, 1994) *Lens culinaris* (Hayat and Ahmad, 2003a) and wheat (Hayat *et al.*, 2001b) by the application of BRs.

The activity of two other important enzymes, α-amylase (Hayat *et al.*, 2003) and peroxidase (Churikova and Vladimirovo, 1997; Hayat and Ahmad, unpublished) was also reported to be elevated by the application of 28-homobrassinolide and epibrassinolide, respectively.

With regard to gaseous phytohormone, ethylene, its level is enhanced by brassinosteroids in the etiolated mungbean hypocotyl segments (Arteca *et al.*, 1983) and also that of ACC, an intermediate in ethylene synthesis, supplied to the roots of tomato plants (Schlagnhaufer and Arteca, 1985a). Similarly, brassinosteroids alone or in combination with auxin induced the synthesis of ethylene in rice lamina (Cao and Chen, 1995). A synergism between the ethylene and ACC production in the etiolated mungbean hypocotyls, treated with brassinosteroid and/or IAA (Arteca *et al.*, 1988), which was sensitive to the inhibitors of ethylene synthesis (Schlagnhaufer *et al.*, 1984a; Schlagnhaufer and Arteca, 1985b) was reported. Moreover, the water stressed plants of jackpine, treated with homobrassinolide evolved more ethylene (Rajasekaran and Blake, 1999).

## USE OF BRASSINOSTEROIDS IN AGRICULTURE

Once the presence of brassinosteroids in plants was established the next phase was to explore the possibilities of using these new substances in improving the yield of

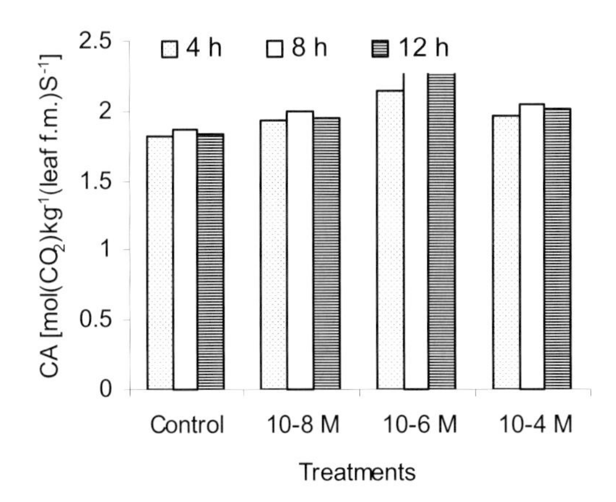


*Figure 5.Effect of different hour of soaking of 28-homobrassinolide on carbonic anhydrase (CA) activity in mungbean at 30 day stage*

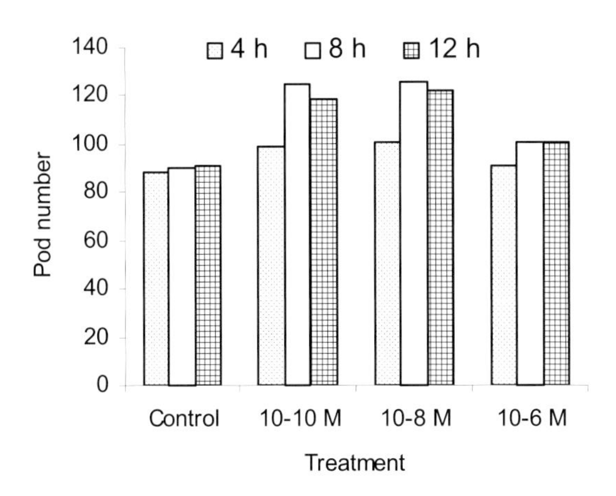


*Figure 6. Pod number in plants of Lens culinaris raised from seeds soaked in three concentrations of 28-homobrassinolide for different times*

economically useful plants. Meudt and his associates (Meudt *et al.*, 1983; 1984) used brassinolide to demonstrate the improvement in the yield of lettuce, radish, bush bean and pepper. Foliar application of dilute aqueous solution of brassinolide improved the yields in wheat and mustard (Braun and Wild, 1984), rice, corn and tobacco (Yokota and Takahashi, 1986). Brassinosteroids were also found to increase the growth and yield of sugarbeet (Schilling *et al.*, 1991), legumes (Kamuro and Takatsuto, 1992) and rape seed (Takematsu and Takeuchi, 1989; Hayat *et al.*, 2000; 2001b). Application of 28-homobrassinolide and 24-epibrassinolide significantly increased yields in potato, mustard, rice and cotton (Ramraj *et al.*, 1997), *Lens culinaris* (Hayat and Ahmad, 2003; Figure 6 & 7), *Vigna radiata* (Fariduddin *et al.*, 2003; Figure 8) and that of corn, tobacco, watermelon, cucumber and grape (Ikekawa and Zhao, 1991) respectively. Foliar application of brassinolide, 24-epibrassinolide (Vardhini and Rao, 1997) and 28-homobrassinolide (Vardhini and Rao, 1998) was highly effective in enhancing the yields of groundnut and tomato (Vardhini and Rao, 2001). Moreover, in China, 28-homobrassinolide has been registered as a plant growth regulator in case of tobacco, sugarcane, rapeseed and tea.

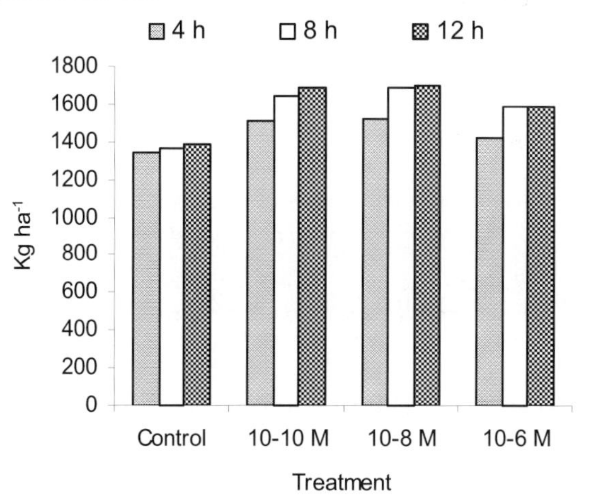


*Figure 7.Seed yield in plants of Lens culinaris raised from the seeds soaked in three concentrations of 28-homobrassinolide for different times*

## FUTURE PROSPECTS OF BRASSINOSTEROIDS

Twenty-five years of research, on brassinosteroids has brought into light several vital functions of this class of phytohormones in the regulation of plant growth, development and the productivity. Hopefully, as the research will progress, many more will be added to the present list. It has been stated earlier that the application of these steroids to plants generates varied physio-morphological changes by involving the genome and also do not initiate co-evolution of pests, enriching our arsenal of plant protection strategies, in the twenty first century. Moreover, the knowledge of the physical and chemical properties of these steroids is tempting us to consider them highly promising, environment friendly protector and promoter of agricultural productivity.

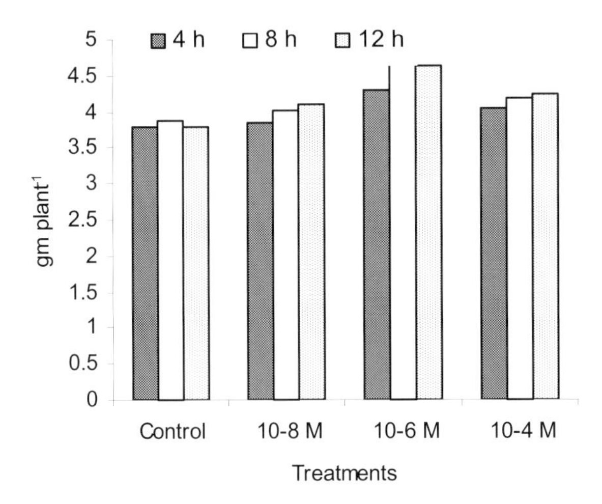


Figure 8.Seed yield in plants of mungbean raised from the seeds scaked in three concentration of 28-homobrassinolide for different times

One of the major constraints, to employ brassinosteroids in large scale in the fields is their high cost. However, recent progress, in the field of chemical synthesis of brassinosteroids and their analogues has led to economically feasible approaches that have brought their practical application in agriculture within reach. Pesticidal companies in China and Japan have started synthesizing brassinosteroids, on a commercial scale. In India, also Godrej Agrovet Ltd., Mumbai, introduced 28-homobrassinolide in the market. We predict a better future for brassinosteroids in realizing crop yields, during the 21st century.

## ACKNOWLEDGEMENT

The senior author (S.Hayat) gratefully acknowledges the financial assistance (No. SR/FTP/LS-A-37/2002) received from the Department of Science & Technology, New Delhi, India.

## REFERENCES

Ahmad, A., Hayat, S. (2003). 28-homobrassinolide improved growth and yield in mustard. 2nd International Congress of Plant Physiology, New Delhi, India, pp. 520.

Altmann, T. (1999). Molecular physiology of brassinosteroids revealed by the analysis of mutants. Planta 208: 1-11.

Anuradha, S., Rao, S.S.R. (2002). Effect of brassinosteroids on salinity stress induced inhibition of seed germination and seedling growth of rice (*Oryza sativa*). Plant Growth Regulation 33 (2): 151-153.

Arteca, R.N. (1997). *Plant growth Substances: Principles and Application.* CBS Publishers & Distributors, New Delhi, India.

Arteca, R.N., Bachman, J.M., Mandava, N.B. (1988). Effect of indole-3-acetic acid and brassinosteroid on ethylene biosynthesis in etiolated mung bean hypocotyl segments. Journal of Plant Physiology 133: 430-435.

Arteca, R.N., Tsai, D.S., Schlagnhaufer, C., Mandava, N.B. (1983). The effect of brassinosteroid on auxin-induced ethylene production by etiolated mung bean segments. Physiologia Plantarum 59: 539-544.

Badger, M.R., Price, G.D. (1994). The role of carbonic anhydrase in photosynthesis. Annual Review of Plant Physiology and Plant Molecular Biology 45: 369-392.

Bellincampi, D., Morpurgo, G. (1988). Stimulation of growth in *Daucus carota* L. cell cultures by brassinosteroids. Plant Science 54: 153-156.

Bhatia, D.S., Kaur, J. (1997). Effect of homobrassinolide and humicil on chlorophyll content, Hill activity and yield components in mungbean (*Vigna radiata* L. Wilczek) Phytomorphology 47: 421-426.

Braun, P., Wild, A. (1984). The influence of brassinosteroid on growth and parameters of photosynthesis of wheat and mustard plants. Journal of Plant Physiology 116: 189-196.

Cao, H., Chen, S. (1995). Brassinosteroid-induced rice lamina joint inclination and its relation to indole-3-acetic acid and ethylene. Plant Growth Regulation 16: 189-196.

Catterou, M., Dubois, F., Schaller, H., Aubanelle, L., Vilcot, B., Sangwan-Norreel, B.S.,Sangwan, R.S. (2001a). Brassinosteroids, microtubules and cell elongation in *Arabidopsis thaliana*. I. Molecular, cellular and physiological characterization of the *Arabidopsis bul1* mutant, defective in the Δ7-sterol-C5-desaturation step leading to brassinosteroid biosynthesis. Planta 212: 659-672.

Catterou, M., Dubois, F., Schaller, H., Aubanelle, L., Vilcot, B., Sangwan-Norreel, B.S.,Sangwan, R.S. (2001b). Brassinosteroids, microtubules and cell elongation in *Arabidopsis thaliana*. II. Effects of brassinosteroids on microtubules and cell elongation in *bul1* mutant. Planta 212: 673-683.

Chang, J.Q., Cai, D.T. (1988). The effects of brassinolide on seed germination and cotyledons tissue culture in *Brassica napus* L. Oil Crops China 18-22.

Choe, S., Dilkes, B.P., Gregory, B.D., Ross, A.S., Yuan, H., Noguchi, T., Fujioka, S., Takatsuto, S., Tanaka, A., Yoshida, S., Tax, F.E., Feldmann, K.A. (1999a). The *Arabidopsis* dwarf1 mutant is defective in the conversion of 24-methylenecholesterol to campesterol in brasisnosteroid biosynthesis. Plant Physiology 119: 897-907.

Choe, S., Noguchi, T., Fujioka, S., Takatsuto, S., Tissier, C.P., Gregory, B.D., Ross, A.S., Tanaka, A., Yoshida, S., Tax, F.E., Feldmann, K.A. 1999b. The *Arabidopsis* dwf7/ste1 mutant is defective in the Δ7-sterol C-5 desaturation step leading to brassinosteroid biosynthesis. Plant Cell 11: 207-221.

Churikova, V.V., Vladimirova, I.N. (1997). Effect of epin on activity of oxidative metaboilsm of cucumber in peronosporous epiphytotia conditions. In "Plant growth and development regulators" $4^{th}$ pp. 78. Moscow.

Clouse, S.D. (1996). Molecular genetic studies confirm role of brassinosteroids in plant growth and development in *Arabidopsis*. Plant Cell 3: 445-459.

Clouse, S.D., Hall, A.F., Langford, M., McMorris, T.C., Baker, M.E. (1993). Physiological and molecular effects of brassinosteroids on *Arabidopsis thaliana*. Journal of Plant Growth Regulation 12: 61-66.

Clouse, S.D., Langford, M., McMorris, T.C. (1996). A brassinosteroid-insensitive mutant in *Arabidopsis thaliana* exhibits multiple defects in growth and development. Plant Physiology 111: 671-678.

Clouse, S.D., Zurek, D.M., McMorris, T.C., Baker, M.E. (1992). Effect of brassinolide on gene expression in elongating soybean epicotyls. Plant Physiology 100: 1377-1383.

Cohen, J.D., Meudt, W.J. (1983). Investigations on the mechanism of the brassinosteroid response. I. Indole-3-acetic acid metabolism and transport. Plant Physiology 72: 691-694.

Ding, J.X., Ma, G.R., Huang, S.Q., Ye, M.Z. (1995). Studies on physiological effects of epibrassinolide on cucumber (*Cucumis sativus* L.). Journal of Zhojiang Agricultural University 21: 615-621.

Diz, G.S., Perez, N., Nunez, M., Torres, W. (1995). Effects of the synthetic brassinosteroid DAA-6 on tobacco (*Nicotiana tabacum* L.). Cultivos Tropicales 16: 53-55.

Dong, J.W., Lou, S.S., Han, B.W., He, Z.P., Li, P.M. (1989). Effects of brassinolide on rice seed germination and seedling growth. Acta Agriculturae University Pekinensis 15: 153-156.

Fariduddin, Q. (2002). The response of *Vigna radiata* and *Brassica juncea* to 28-homobrassinolide and kinetin. Ph.D. thesis, Aligarh Muslim University, Aligarh, India.

Fariduddin, Q., Ahmad, A., Hayat, S. (2003). Photosynthetic response of *Vigna radiata* to presowing seed treatment with 28-homobrassinolide. Photosynthetica 41(2): xxx-xxx.

Fariduddin, Q., Ahmad, A., Hayat, S., Alvi, S. (2000). The response of chickpea, raised from the seeds pretreated with 28-homobrassinolide. In: National seminar on plant physiological paradigm for fostering agro and biotechnology and augmenting environmental productivity in millennium 2000, Lucknow, India pp. 134.

Fujii, S., Hirai, K., Saka, H. (1991). Growth regulating action of brassinolide in rice plants. In: Brassinosteroids: Chemistry, Bioactivity and Applications, pp 306-311. Eds. H G Cutler, T Yokota and G Adam. American Chemical Society, Washington.

Grove, M.D., Spencer, G.F., Rohwedder, W.K., Mandava, N., Worley, J.F., Warthen, J.D., Jr., Steffens, G.L., Flippen-Anderson, J.L., Cook, J.C. Jr. (1979). Brassinolide, a plant growth-promoting steroid isolated from *Brassica napus* pollen. Nature 281: 216-217.

Hayat, S., Ahmad, A. (2003). Soaking seeds of *Lens culinaris* with 28-homobrassinolide increased nitrate reductase activity and grain yield in the field in India. Annals of Applied Biology 143: 121-124.

Hayat, S., Fariduddin, Q., Ahmad, A. (2003). Homobrassinolide affect germination and α-amylase activity in wheat seeds. Seed Technology 25 (1): 45-49.

Hayat, S., Ahmad, A., Mobin, M., Fariduddin, Q., Azam, Z.M. (2001a). Carbonic anhydrase, photosynthesis and seed yield in mustard plants treated with phytohormones. Photosynthetica 39: 27-30.

Hayat, S., Ahmad, A., Hussain, A., Mobin, M. (2001b). Growth of wheat seedlings raised from the grains treated with 28-homobrassinolide. Acta Physiologiae Plantarum 23: 27-30.

Hayat, S., Ahmad, A., Mobin, M., Hussain, A., Fariduddin, Q. (2000). Photosynthetic rate, growth and yield of mustard plants sprayed with 28-homobrassinolide. Photosynthetica 38: 469-471.

He, R.Y., Wang, G.J., Wang, X.S. (1991). Effect of brassinolide on growth and chilling resistance of maize seedlings. In: Brassinosteroids: Chemistry, Bioactivity and Applications, pp 26-35. Eds. H G Cutler, T Yokota and G Adam. American Chemical Society, Washington.

He, Y-J., Xu, R-J, Zhao, Y-J. (1996). Enhancement of senescence by epibrassinolide in leaves of mungbean seedling. Acta Physiologia Sin 22: 58-62.

Helmy, Y.I., Sawan, O.M.M., Abdel-Halim, S.M. (1997). Growth, yield and endogenous hormones of broad bean plants as affected by brassinosteroids. Egyptian Journal of Horticulture 24: 109-115.

Hunter, W.J. (2001). Influence of root applied epibrassinolide and carbenoxolone on the nodulation and growth of soybean (*Glycine max* L.) seedlings. Journal of Agronomy and Crop Science 186: 217-222.

Ikekawa, N., Zhao, Y.J. (1991). Application of 24-epibrassinolide in agriculture. In: Brassinosteroids: Chemistry, Bioactivity and Applications, pp 280-291. Eds. H G Cutler, T Yokota and G Adam. American Chemical Society, Washington.

Iwasaki, T., Shibaoka, H. (1991). Brassinosteroids as regulators of tracheary-element differentiation in isolated *Zinnia* mesophyll cells. Plant Cell Physiology 32: 1007-10014.

Jones-Held, S.,Van Doren,M., Lockwood, T. (1996). Brassinolide application to *Lepidium sativum* seeds and the effects on seedling growth. Journal of Plant Growth Regulation 15: 63-67.

Kamuro, Y., Inada, K. (1991). The effect of brassinolide on the light-induced growth inhibition in mungbean epicotyl. Plant Growth Regulation 10: 37-43.

Kamuro, Y., Takatsuto, S. (1991). Capability for and problems of practical uses of brassinosteroids. In: Brassinosteroids: Chemistry, Bioactivity and Applications, pp 292-297. Eds. H G Cutler, T Yokota and G Adam. American Chemical Society, Washington.

Kamuro, Y., Takatsuto, S. (1999). Potential application of brassinosteroids in agricultural fields. In: Brassinosteroids: Steroidal Plant Hormones, pp 223-241. Eds. A Sakurai, T Yokota and S D Clouse, Springer-Verlag, Tokyo.

Katsumi, M. (1985). Interaction of a brassinosteroid with IAA and GA3 in the elongation of cucumber hypocotyl sections. Plant Cell Physiology 26: 615-625.

Katsumi, M. (1991). Physiological mode of brassinolide action in cucumber hypocotyl growth. In: Brassinosteroids: Chemistry, Bioactivity and Applications, pp 246-254. Eds. H G Cutler, T Yokota and G Adam. American Chemical Society, Washington.

Kauschmann, A., Jessop, A., Koncz, C., Szekeres, M., Willmitzer, L., Altmann, T. (1996). Genetic evidence for an essential role of brassinosteroids in plant development. Plant Journal 9: 701-713.

Khripach, V.A., Zhabinskii, V.N., Malevannaya, N.N. (1997). Recent advances in brassinosteroids study and application. Proceeding of Plant Growth Regulation Society of America 24: 101-106.

Kulaeva, O.N., Burkhanova, E.A., Fedina, A.B., Khokhlova, V.A., Bokebayeva, G.A., Vorbrodt, H.M., Adam, G. (1991). Effect of brassinosteroids on protein synthesis and plant cell ultrastructure under stress conditions.In: Brassinosteroids: Chemistry, Bioactivity and Applications, pp 141-155. Eds. H G Cutler, T Yokota and G Adam. American Chemical Society, Washington.

Lawlor, D.W. (1987). The chemistry of photosynthesis. In: Photosynthesis. Metabolism, Control and Physiology, pp. 127-157. Eds. D W Lawlor Longman Singapore Publishers, Singapore.

Leubner-Metzger, G. (2001). Brassinosteroids and gibberellins promote tobacco seed germination by distinct pathways. Planta 213 (3): 758-763.

Li, J., Nagpal, P., Vitart, V., Chory, J., McMorris, T.C. (1996). A role of brassinosteroids in light-dependent development of *Arabidopsis*. Science 272: 398-401.

Liang, G.J., Li, Y.Y., Shao, L. (1998). Effect of DA-6 and BR+GA3 on growth and photosynthetic rate in spinach. Acta Horticulturae Sinica 25: 356-360.

Maeds, E. (1965). Rate of lamina inclination in excised rice leaves. Physiologia Plantarum 18: 813-827.

Mai, Y., Lin, S., Zeng, X., Ran, R. (1989). Effect of brassinolide on nitrate reductase activity in rice seedlings. Plant Physiology Communication 2: 50-52.

Mandava, N.B. (1988). Plant Growth-Promoting Brassinosteroids. Annual Review of Plant Physiology and Plant Molecular Biology 39: 23-52.

Mandava, N.B., Sasse, J.M., Yopp, J.H. (1981). Brassinolide, a growth promoting steroidal lactone. II. Activity in selected gibberellin and cytokinin bioassays. Physiologia Plantarum 53: 453-461.

Meudt, W.J., (1987). Investigations on the mechanism of the brassinosteroid response. VI. Effect of brassinolide on gravitropism of bean hypocotyls. Plant Physiology 83: 195-198.

Meudt, W.J., Thompson, M.J., Bennet, H.W. (1983). Investigations on the mechanism of the brassinosteroid response.III. Techniques for potential enhancement of crop production. Proceeding of Plant Growth Regulation Society of america 10: 312-318.

Meudt, W.J., Thompson, M.J., Mandava, N.B., Worley, J.F. (1984). Method for promoting plant growth. Canadian Patent No. 1173659. Pp. 11.Assigned to USA.

Nakajima, N., Shida, A., Toyama, S. (1996). Effects of brassinosteroid on cell division and colony formation of chinese cabbage mesophyll protoplasts. Japanese Journal of Crop Science 65: 125-130.

Nakajima, N., Toyama, S. (1995). Study on brassinosteroid-enhanced sugar accumulation in cucumber epicotyls. Japanese Journal of Crop Science 64: 616-621.

Nunez, M., Torres, W., Echevarria, I. (1996). The effect of a brassinosteroid analogue on growth and metabolic activity of young tomato plants. Cultivos Tropicales 17: 26-30.

Oh,Man-OH.(2003). Brassinosteroids accelerate the rate of cell division in isolated petal protoplasts of *Petunia hybrida*. Journal of Plant Biotechnology 5(1) : 63-67.

Oh,Man-OH., Clouse, S.D. (1998). Brassinolide affects the rate of cell division in isolated leaf protoplasts of *Petunia hybrida*. Plant Cell Reports 17: 921-924.

Okabe, K., Yang, S.Y., Tsuzuki, M., Miyachi, S. (1984). Carbonic anhydrase: its content in spinach leaves and its taxonomic diversity studied with anti-spinach leaf carbonic anhydrase antibody. Plant Science Letters 33: 145-153.

Pipattanawong, N., Fujishige, N., Yamane, K., Ogata, R.C. (1996). Effect of brassinosteroid on vegetative and reproductive growth in two day-neutral strawberries. Journal of the Japanese Society for Horticultural Science 65: 651-654.

Rajasekaran, L.R., Blake, T.J. (1998). Early growth invigoration of jack pine seedlings by natural plant growth regulators. Trees 12: 420-423.

Rajasekaran, L.R., Blake,T.J. (1999). New plant growth regulators protect photosynthesis and enhance growth under drought of jack pine seedlings. Journal of Plant Growth Regulation 18: 175-181.

Ramraj, V.M., Vyas, B.N., Godrej, N.B., Mistry, K.B., swmai, B.N., Singh, N. (1997). Effects of 28-homobrassinolide on yields of wheat, rice, groundnut, mustard, potato and cotton. Journal of Agricultural science 128: 405-413.

Rao, S.S.R., Vardhini, B.V., Sujatha, E., Anuradha, S. (2002). Brassinosteroids- A new class of phytohormones. Current Science 82 : 1239-1245.
Reed, M.L., Graham, D. (1981). Carbonic anhydrase in plants distribution, properties and possible physiological roles. Phytochemistry 7: 47-94.
Roddick, J.G., Ikekawa, N. (1992). Modificaiton of root and shoot development in monocotyledon and dicotyledon seedlings by 24-epibrassinolide. Journal of Plant Physiology 140: 70-74.
Roddick, J.G., Rijnenberg, A.L., Ikekawa, N. (1993). Developmental effects of 24-epibrassinolide in excised roots of tomato grown *in vitro*. Physiologia Plantarum 87: 453-458.
Rönsch, H., Adam, G., Matschke, J., Schachler, G. (1993). Influence of (22S, 23S)-homobrassinolide on rooting capacity and survival of adult Norway spruce cuttings. Tree Physiology 12: 71-80.
Roth, P.S., Batch, T.J., Thompson, M.J. (1989). Brassinosteroids: Potential inhibitors of transformed tobacco callus culture. Plant Science 59: 63-70.
Sairam, R.K. (1994). Effects of homobrassinolide application on plant metabolism and grain yield under irrigated and moisture-stress conditions of two wheat varieties. Plant Growth Regulation 14: 173-181.
Sairam, R., Shukla, D., Deshmukh, P. (1996). Effect of homobrassinolide seed treatment on germination, alpha-amylase activity and yield of wheat under moisture stress conditions. Indian Journal of Plant Physiology 1: 141-144.
Sala, C., Sala, F. (1985). Effect of brassinosteroid on cell division and enlargement in cultured carrot (*Daucus carota* L.) cells. Plant Cell Reports 4: 144-147.
Sasse, J.M. (1990). Brassinolide-induced elongation and auxin. Physiologia Plantarum 80: 401-408.
Sasse, J.M., Smith, R., Hudson, I. (1995). Effect of 24-epibrassinolide on germination of seeds of *Eucalyptus camaldulensis* in saline conditions.Proceeding of Plant Growth Regulation Society of America 22: 136-141.
Schilling,G., Schiller, C., Otto, S. (1991). Influence of brassinosteroids on organ relations and enzyme activities of sugar-beet plants. In: Brassinosteroids: Chemistry, Bioactivity and Applications, pp 208-219. Eds. H G Cutler, T Yokota and G Adam. American Chemical Society, Washington.
Schlagnhaufer, C., Arteca, R.N., Yopp, J.H. (1984). Evidence that brassinosteroid stimulates auxin-induced ethylene synthesis in mungbean hypocotyls between S-adenosylmethionine and 1-aminocylopropane-1-carboxylic acid. Physiologia Plantarum 61: 555-558.
Schlagnhaufer, C.D., Arteca, R.N. (1985a). Brassinosteroid induced epinasty in tomato plants. Plant Physiology 78: 300-303.
Schlagnhaufer, C.D., Arteca, R.N. (1985b). Inhibition of brassinosteroid induced epinasty in tomato plants by aminooxyacetic acid (AOA) and cobalt ($Co^{2+}$). Physiologia Plantarum 65: 151-155.
Shen, X.Y., Dai, J.Y., Hu, A.C., Gu, W.L., He, R.Y., Zheng, B. (1990). Studies on physiological effects of brassinolide on drought resistance in maize. Journal of Shenyang Agricultural University 21: 191-195.
Soto, F., Tejeda, T., Nunez, M. (1997). Preliminary study on the use of brassinosteroids in coffee trees. Cultivos Tropicales 18: 52-54.
Sultemeyer, D., Schmidt, C., Fock, H.P. (1993). Carbonic anhydrase in higher plants and aquatic microorganisms. Physiologia Plantarum 88: 179-190.
Takatsuto, S. (1994). Brassinosteroids: distribution in plants, bioassays and microanalysis by gas chromatography-mass spectrometry. Journal of Chromatography 658: 3-15.
Takematsu, T., Takeuchi, Y. (1989). Effects of brassinosteroids on growth and yield of crops. Proceeding of Japan Academy Series B 65: 149-152.
Takeno, K., Pharis, R.P. (1982). Brassinolide-induced bending of lamina of dwarf rice seedlings : an auxin mediated phenomenon. Plant Cell Physiology 23: 1275-1281.
Vardhini, B.V., Rao, S.S.R. (1997). Effect of brassinosteroids on salinity induced growth inhibition of groundnut seedlings. Indian Journal of Plant Physiology 2: 156-157.
Vardhini, B.V., Rao, S.S.R. (1998). Effect of brassinosteroids on growth, metabolite content and yield of *Arachis hypogea*. Phytochemistry 48: 927-930.
Vardhini, B.V., Rao, S.S.R. (2000). Effect of brassinosteroids on the activities of certain oxidizing and hydrolyzing enzymes of groundnut. Indian Journal of Plant Physiology 5: 89-92.
Wang, S.G. (1997). Influence of brassinosteroid on rice seedling growth. International Rice Research Notes 22: 20-21.

Wang, S.G., Deng, R.F. (1992). Effects of brassinosteroid on root metabolism in rice. Journal of Southwest Agricultural University 14: 177-181.

Wilen, R.W., Sacco, M., Gusta, L.V., Krishna, P. (1995). Effects of 24-epibrassinolide on freezing and thermotolerance of brome grass (*Bromus inermis*) cell cultures. Physiologia Plantarum 95: 195-202.

Yokota, T., Takahashi, N. (1986). Chemistry, physiology and agricultural application of brassinolide and related steroids. In: Plant Growth Substances, pp 129-138. Eds. M Bopp, Springer-Verlag, Berlin

Yopp, J.H., Mandava, N.B., Sasse, J.M. (1981a). Brassinolide, a growth promoting steroidal lactone. I. Activity in selected auxin bioassays. Physiologia Plantarum 53: 445-452.

Yopp, J.H., Mandava, N.B., Thompson, M.J., Sasse, J.M. (1981b). Brassinosteroids in selected bioassays. 8th Proceeding of Plant Growth Regulation Society of America pp 110-126.

Zurek, D.M., Rayle, D.L., McMorris, T.C., Clouse, S.D. (1994). Investigation of gene expression, growth kinetics, and wall extensibility during brassinosteroid-regulated stem elongation. Plant Physiology 104: 505-513.